Simulation Foundations, Methods and Applications

More information about this series at http://www.springer.com/series/10128

Dietmar P.F. Möller
With a Chapter Contribution together
with Prof. Dr. Bernard Schroer

Introduction to Transportation Analysis, Modeling and Simulation

Computational Foundations and Multimodal Applications

Springer

Dietmar P.F. Möller
Clausthal University of Technology
Clausthal-Zellerfeld, Germany

ISSN 2195-2817 ISSN 2195-2825 (electronic)
ISBN 978-1-4471-7244-4 ISBN 978-1-4471-5637-6 (eBook)
DOI 10.1007/978-1-4471-5637-6
Springer London Heidelberg New York Dordrecht

Printed on acid-free paper

Springer is part of Springer Science+Business Media (www.springer.com)

Foreword

Modeling and simulation are important to receive research results without the cost and time spent creating working prototypes. Thus, it seems to me a good reason for the NSF Blue Ribbon Panel to clearly constitute the advancement in modeling and simulation as critical for resolving multitude of scientific and technological problems in health, security, and technological competitiveness within the globalized world. Therefore, in our globalized world a variety of Simulation Centers have been established with an economic and scientific focus on modeling and simulation methodology and sophisticated tools and software. One of these centers is the recently founded Simulation Science Center Clausthal-Göttingen in Germany, a common interdisciplinary research facility in simulation science of Clausthal University of Technology and University of Göttingen. The core research areas are advanced methodologies in mathematical modeling and simulation techniques, and their application in real-world problems. The actual research areas the center is dealing with are: simulation and optimization of networks, computational materials science simulation, and distributed simulation, which will be expanded step by step by new hired staff. The available book, written by a member of the Simulation Science Center at Clausthal University of Technoöogy, is a showcase of creative ideas of ongoing research work at the Simulation Science Center with the focus "Introduction into Transportation Analysis, Modeling and Simulation" which belongs to the simulation and optimization of networks group of the center. The book shows how to analyze the complex transportation systems accurately and under varying operation conditions and/or scenarios to predict its behavior for engineering and planning purposes to provide adequate academic answers for today's emerging transportation technology management questions. The chapters are well written showing academic rigor and professionalism of the author. Therefore, the book can be stated as an important reading for new researchers entering this field of transportation research. It offers new perspectives and documents important progress in transportation analysis, modeling and simulation.

Simulation Science Center   Thomas Hanschke
Clausthal-Göttingen, Germany

Preface

The goal of this book is to provide a comprehensive, in-depth, and state-of-the-art summary of the important aspects of transportation analysis and modeling and simulation. The term modeling and simulation refers to computer simulation, with an emphasis on modeling real transportation systems and executing the models. A recent White House report identified computer modeling and simulation as one of the key enabling technologies of the twenty-first century. Its application is universal. For this reason, the book strives to motivate interest in transportation analysis and modeling and simulation as well as to present these topics in a technically correct yet clear manner. This required making some carefully considered choices in selecting the material for this book.

First the fundamentals of modeling are described, as they represent the largest portion of transportation analysis. In addition, the mathematical background describing real transportation systems is introduced on a basic level as well as on a more advanced one; and its correspondence to the respective modeling methodologies is described. Secondly, the most interesting simulation systems are presented at the language and logic level, and their use is described in several case studies. However, a textbook cannot describe all of the available simulation systems in detail. For this reason, the reader is referred to specific supplemental material, such as textbooks, reference guides, user manuals, etc., as well as Internet-based information which addresses several simulation languages. Thirdly, a variety of actual applications are presented which have been conducted during a long period of collaboration with Prof. Bernard Schroer, Ph.D., University of Alabama in Huntsville (UAH), USA.

This book was developed for use by senior and graduate level students in applied mathematics, operations research, computer science, and engineering and business informatics and to serve as the primary text for a course on Transportation Analysis, Modeling and Simulation, held annually at Clausthal University of Technology.

The material in the book can be difficult to comprehend if the reader is new to such an approach. This is also due to the fact that transportation analysis/modeling and simulation is a multidisciplinary domain, founded in computer science, engineering, mathematics, operations research, etc. The material may not be read and comprehended either quickly or easily. Therefore, specific case studies have been embedded with related topics to help the reader master the material.

It is assumed that the reader has some knowledge of basic calculus-based probability and statistics and some experience with computing.

The book can be used as the primary text in a course in various ways. It contains more material than can be covered in detail in a quarter-long (30-h) or semester-long (45-h) course. Instructors may elect to choose their own topics and add their own case studies. The book can also be used for self-study; as a reference for graduate engineers, scientists, and computer scientists for training on the job or in graduate school; and as a reference for transportation analysis and modeling and simulation practitioners and researchers.

For instructors who have adopted the book for use in a course, a variety of teaching support materials are available for download from http://www.springer. com/978-1-4471-5636-9. These include a comprehensive set of Microsoft Power-Point slides to be used for lectures and all video-recorded classes.

The book is divided into six chapters which can be read independently or consecutively.

Chapter 1, Computational Foundation in Transportation and Transportation Systems Modeling, covers the classification of models used for multimodal transportation systems and introduces transportation and transportation systems for the movement of passengers or freight and how to analyze their behavior. A transportation case study planning a seagate harbor expansion by a dry port is introduced.

Chapter 2, Transportation Models, contains a brief overview of the use of models in the transportation sector, the several types of models used in transportation planning, the specific evaluation methods used, queuing theory to predict queuing lengths and waiting times, and the methodological background of congestion, graph theory, and bottlenecks. Finally a ProModel-based case study for a four-arm road intersection is introduced.

Chapter 3, Traffic Assignments to Transportation Networks, introduces traffic assignment to uncongested and congested road networks; the equilibrium assignment, which can be expressed by so-called fixed point models where origin-to-destination demands are fixed, representing systems of nonlinear equations or variational inequalities; and the multiclass assignment based on the assumption that travel demand can be allocated as a number of distinct classes which share behavioral characteristics. The case study involves a diverging diamond interchange (DDI), an interchange in which the two directions of traffic on a nonfreeway road cross to the opposite side on both sides of a freeway overpass (or underpass).

Chapter 4, Integration Framework and Empirical Evaluation, is an introduction to computer simulation integration platforms and their use in the transportation systems sector. It provides, in addition to an overview of the framework architectures, an introduction into ontology-based modeling and its integration into transportation; the workflow-based application integration in transportation; and detailed case studies for a marine terminal traffic network simulation, an airport operation simulation, a highway ramp control simulation, and vehicle tracking using the Internet of Things paradigm.

Chapter 5, Simulation Tools in Transportation, gives an overview of transportation simulation tools including continuous systems simulation tools, such as block-oriented and equation-oriented simulation tools, and discrete-event simulation tools. Some of the many available simulation software packages are described with a focus on those used for the case studies in this book. Finally, a ProModel-based case study for a maritime transportation analysis is introduced.

Chapter 6, Transportation Use Cases, introduces, from a general perspective, critical issues in the design, development, and use of simulation models of transportation systems. Case studies in Chap. 6 include real transportation projects such as the McDuffie Coal Terminal at the Alabama State Docks in Mobile, the Container Terminal at the Port of Mobile and its intermodal container handling facilities, and an analysis of the operation of an intermodal terminal center before the design of any planned expansion is finalized. A case study on port security inspection is included due to the increased security requirements for the operation of seaports. The objective of the simulation is to evaluate the impact of various inspection protocols on the operation of the container terminal at the Port of Mobile. The objective of the Interstate Traffic Congestion case study is to determine the congestion point as traffic increases and to evaluate adding additional lanes at congestion points. Tunnels are an important solution for the transportation infrastructure, e.g., for crossing a river such as the Mobile River in downtown Mobile, Alabama. Besides the maritime transportation sector, the aviation domain also calls for innovative solutions to optimize their operational needs. Thus the first of the two aviation case studies focuses on passenger and freight operations at Hamburg Airport to estimate the maximum numbers of passengers and freight that can be dispatched to identify potential opportunities in process optimization with regard to the expected growth in passenger and freight numbers. In the second aviation case study, an Italian airport transportation operation conducted by an international student team is introduced to demonstrate how an international group of students can be motivated to conduct an innovative, advanced project in a very complex area of concentration in modeling and simulation.

Besides the methodological and technical content, all chapters of the book contain comprehensive questions from the chapter-specific area to help students determine if they have gained the required knowledge, identifying possible knowledge gaps and conquering them. Moreover, all chapters include references and suggestions for further reading.

I would like to express my special thanks to Prof. Bernard Schroer, Ph.D., University of Alabama in Huntsville, USA, for our long collaboration on research-oriented projects in transportation analysis and modeling and simulation; to Prof. Jerry Hudgins, Ph.D., University of Nebraska-Lincoln, USA, for his great support in working at the University of Nebraska-Lincoln on computer modeling and simulation; to Prof. Dr. Thomas Hanschke, Chairman of the Simulation Science Center at Clausthal-Göttingen (SWZ), Germany, for electing me as a member of the Simulation Science Center Clausthal-Göttingen; to Prof. Louis G. Birta, Ph.D., University of Ottawa, Canada, for inviting me to contribute this book to the series, *Simulation Foundations, Methods and Applications*, he is editing; to Patricia Worster, University of Nebraska-Lincoln, for her excellent assistance in proofreading; and to Simon Rees, Springer Publ., for his help with the organizational procedures between the publishing house and the author.

For developing sample models and exercise problems and for executing prototype simulation models which appear in the book, I would like to thank my graduate students at Clausthal University of Technology (TUC), Germany, and University of Hamburg (UHH), Germany.

Finally, I would like to deeply thank my wife, Angelika, for her encouragement, patience, and understanding during the writing of this book.

This book is dedicated to my parents, Wilhelm and Hildegard Möller, whose hard work and belief in me made my dreams a reality.

Clausthal-Zellerfeld, Germany   Dietmar P.F. Möller

Contents

Computational Foundation in Transportation and Transportation Systems Modeling

This chapter begins with a brief overview of transportation systems. Section 1.1 covers the classification of models used for multimodal transportation systems. Section 1.2 introduces transportation and transportation systems for the movement of passengers or freight. Thereafter, Sect. 1.3 introduces model building in transportation by describing the main types of models and their representation through mathematical notation. Section 1.4 covers the important topic of modeling formalism in transportation systems to analyze their behavior and/or composite structure. Simulation models in the transportation sector are approximate imitations of the real-world phenomena which never exactly imitate the real-world phenomena. Therefore, models have to be verified and validated to the degree required for the models' intended purpose or application which is introduced in Sect. 1.5. Section 1.6 describes a transportation case study for planning a Seagate harbor expansion by a dry port. Section 1.7 contains comprehensive questions from the transportation system area, and a final section includes references and suggestions for further reading.

1.1 Introduction

The transportation systems sector—comprised of all modes of transportation, each with different operational structures and approaches to security—is a vast, open, interdependent network moving millions of tons of freight and millions of passengers. Every day, the transportation systems network connects cities, manufacturers, and retailers by moving large volumes of freight and passengers through a complex network of roads and highways, railways and train stations, sea ports and dry ports, and airports and hubs (Sammon and Caverly 2007). Thus, the transportation systems sector is the most important component of any modern economy's infrastructure in the globalized world. It is also a core component of daily human life with all of its essential interdependencies, such as demands for

© Springer-Verlag London 2014

D.P.F. Möller, *Introduction to Transportation Analysis, Modeling and Simulation,*
Simulation Foundations, Methods and Applications,
DOI 10.1007/978-1-4471-5637-6_1

travel within a given area and freight transportation in metropolitan areas, which require a comprehensive framework in which to integrate all aspects of the target system. The transportation systems sector also has significant interdependencies with other important infrastructure sectors (e.g., the energy sector). Transportation and energy are directly dependent on each other for the movement of vast quantities of fuel to a broad range of customers, thereby supplying fuel for all types of transportation. Moreover, cross-sector interdependencies and supply chain implications are among the various sectors and modalities in transportation that must be considered (Sammon and Caverly 2007).

The transportation system sector consists of physical and organizational objects interacting with each other to enable intelligent transportation. These objects include information and communication technology (ICT), the required infrastructure, vehicles and drivers, interfaces for the multiple modes of transportation, and more (Torin 2007). Advanced transportation systems are essential to the provision of innovative services via multiple modes of transportation interacting and affecting each other in a complex manner, which cannot be captured by a single existing model of transportation systems traffic and mobility management.

Transportation systems models enable transportation managers to run their daily businesses safely and more effectively through a smarter use of transportation networks. But the transportation systems sector in today's open, interdependent network encompassing urban and metropolitan areas requires optimization of all operating conditions. This can be successfully achieved if the interactions between transportation modes, the economy, land use, and the impact on natural resources are included in transportation systems planning strategies.

The proposed future of multimodal transportation systems cannot be measured through planning alone. Mathematical models of transportation systems and mobility management, incorporating both real and hypothetical scenarios, should be embedded in transportation systems analysis, including the evaluation and/or design of the traffic flows, determining the most reliable mode of operation of physical (e.g., a new road) and organizational (e.g., a new destination) objects, and the interaction between the objects and their impact on the environment. These mathematical models are fundamental to the analysis, planning, and evaluation of small-, medium-, and large-scale multimodal transportation systems (Cascetta 2009). The success of model-based scenario analysis can be evaluated by the resulting forecast or prediction of the transportation system response. An ideal design or operational methodology for a transportation system can be achieved using model-based analysis in conjunction with backcasting or backtracking. Thus, modeling and simulation can play a central role in planning, developing, and evaluating multimodal transportation systems, improving transportation efficiency, and keeping pace with the rising demands for optimizing multimodal transportation systems.

Various simulation models capture different aspects of a transportation system enabling evaluation of complex simulation scenarios where each one represents a certain aspect of a transportation system or a certain operational strategy. Multimodal transportation system models can be classified as:

- *Supply models*: representing the multimodal transportation systems sector services used to travel between different operating points within a given area
- *Demand models*: predicting the relevant aspects of travel demand as a function of system activity and level of service provided by the transportation system
- *Assignment models*: using the objects of the multimodal transportation system assignments

1.2 Transportation and Transportation Systems Sector

The importance of understanding and determining dynamic behavior in multimodal transportation in the transportation systems sector has been recognized for a long time, because without adequate modes of transportation, the globalized economy can neither grow nor survive. However, multimodal transportation of freight and passengers contributes to congestion, environmental pollution, and traffic accidents and has a tremendous impact, especially in metropolitan areas. Therefore, a prerequisite for any solution proposing to make the existing transportation systems sector with its multiple modalities more effective is a precise analysis and understanding of the traffic demands. This should take economic forecasts into account to keep pace with the growing demand. Thus, modeling and simulation-based analysis, along with holistic optimization, can help in studying complex transportation scenarios and identifying solutions without committing expensive and time-consuming resources to the implementation of various alternative strategies.

A holistic evaluation of the impact of different transportation policies in a complex transportation scenario requires a comprehensive simulation environment which integrates all aspects of the target transportation system. However, transportation systems analysis is a multidisciplinary field which draws on economics, engineering, logistics, management, operations research, political science, psychology, traffic engineering, transportation planning, and other disciplines (Manheim 1979). Therefore, numerous details must be considered in determining how major concepts can be applied in practice to particular modes and problems of the transportation systems sector.

Transportation modes include:

- *Air transportation sector*: involves modeling and simulation of airport terminal operations, such as baggage handling, gate handling, security check handling, ramp operations (such as vehicle management on the apron), freight handling, and taxiway and runway operations.
- *Maritime transportation sector*: involves modeling and simulation of container terminal operations, including the logistics of efficient container handling, intermodal transport to and from dry ports which expands the sea port container yard capacity, as well as ferry and cruise ship operations.

- *Rail transportation sector*: involves modeling and simulation of freight movement to determine operational efficiency and rationalize planning decisions. Freight simulation can include aspects such as commodity flow, corridor and system capacity, traffic assignment/network flow, and freight plans that involve travel demand forecasting. Rail transportation models also integrate passenger travel.
- *Roadway or ground transportation sector*: widely uses modeling and simulation for both passenger and freight movement. Simulation can be carried out at a corridor level or at a more complex roadway grid network level to analyze planning, design, and operations with regard to congestion, delay, and pollution. Ground transportation models are based mostly on all options of roadway travel, including bicycles, buses, cars, pedestrians, and trucks. In traditional road traffic models, an aggregate representation of traffic is typically used where all vehicles of a particular group obey the same rules of behavior; in microsimulation, driver behavior and network performance are included so that complete traffic problems can be examined (URL 1).

In this book, we study the presence of multiple modes of transportation with regard to transportation systems analysis, modeling, and simulation use cases, developed as part of independent grant projects. In addition to simulating individual modes, it is often more important to simulate a multimodal network, since in reality modes are integrated and represent more complexities that can be overlooked when simulating modes. Intermodal network simulation can also help in gaining a more comprehensive understanding of the impact of a certain network and its policy implications. Transportation simulations can also be integrated with urban environmental simulations, where a large urban area, including roadway networks, is simulated to better understand land use and other planning implications of the traffic network on the urban environment (Ioannou et al. 2007). Hence, manifold transportation applications can be analyzed by modeling and simulation to holistically evaluate the impact of different policies in complex scenarios, such as freight transportation in metropolitan areas, congestion problems in transportation systems traffic, and mobility management. In general, these types of analyses are too complicated or difficult for traditional analytical or numerical methods, which means they require a comprehensive simulation model which consider all aspects of the target system.

Simulation in the transportation systems sector, based on comprehensive simulation models of transportation systems and using specific simulation software, enables better planning, design, and operation of transportation systems. This new approach helps to solve transportation problems (e.g., evaluating the impact of terminal handling strategies on local traffic conditions and terminal throughput) (Ioannou et al. 2007), through real-world demonstrations of present and/or future scenarios (e.g., in traffic engineering and transportation planning). It also assists with difficult obstacles, such as cost requirements in planning and building new infrastructure.

1.3   Models and Their Mathematical Notation

Depending on the nature of the problems in the transportation systems sector, activity-based models, demand-based models, discrete choice analysis, dynamic traffic assignment, and public transportation models as well as traffic flow models, and their different representative forms as macroscopic, microscopic, and mesoscopic models, are used for in-depth, state-of-the-art study in the transportation systems sector.

The advantage of macroscopic traffic models is that they simply capture the general relationships between flow, density, and speed. Thus, macroscopic models abstract traffic to traffic streams that pass along traffic pipes which can be represented as continuous flow, often using formulations based on hydrodynamic flow theories, and by network theory, described as voltage and current in an electrical network. Hence, for segments of the network, the edges in the traffic graph, indicating traffic densities, can be calculated. The advantages of the macroscopic approach are its simplicity and the possibility of solving the traffic model equations in accordance with the rules of network theory, even analytically. Numerical solutions are necessary if a dynamic adaptation of the segment element's resistance and capacity is required. It is obvious that macroscopic models can be easily parameterized, because the level of information required can be compared with the information provided by measurements from traffic censuses. The main disadvantage to using the macroscopic approach is its lack of information, because the traffic streams in macroscopic models are based on statistical information. This means that there is no information about individual traffic events, because the model cannot predict information about departure points and destinations of traffic events passing a certain segment in the traffic graph. However, in most transportation applications, this information is essential to interpret changes (e.g., in the road net or the traffic load) and their implications for individual vehicle driving time and/or the distribution of individual traffic within the traffic network.

Microscopic traffic models are used to capture the behavior of individual objects (vehicles and drivers) in great detail, including interactions among vehicles, lane changing, and behavior at merge points. Therefore, they provide information that macroscopic models are not able to reproduce. Microscopic models generate information at the individual object level, describing acceleration and deceleration at the process level, separately for each object, and can also include sophisticated strategies for routing individual objects. The main disadvantage to using the microscopic approach is the simulation time required, caused by the large number of objects and the resulting interactions which must be handled by the model. However, the difficulty in properly parameterizing microscopic models by defining attributes of individual model objects adequately, completely, and consistently is another main disadvantage.

Mesoscopic traffic models incorporate objects such as the microscopic modeling paradigm, but at an aggregate level, usually by speed-density relationships and queuing theory approaches. These models require more effort in specification and interpretation, because the mesoscopic approach does not give any constructive advice on how to build the model. However, the mesoscopic approach offers a wide range of adaptability to meet a user's needs, the information available, and the objectives for simulation. Therefore, the mesoscopic approach can be introduced as a pragmatic modeling level for specifying a traffic simulation with a minimum of information required.

The use of these types of models in the transportation systems sector entails specific challenges (i.e., how to adequately analyze transportation systems scenarios which match the essential demands of growth in transportation). This helps in selecting the optimal solutions for development of the particular transportation systems sector infrastructure essential to the specific modality and/or multimodality or the transportation systems operation. Thus, transportation systems analysis, modeling, and simulation identify concepts that are truly fundamental to the planning, design, and/or management of the transportation systems sector.

Modeling and simulation itself is an iterative process consisting of successive mathematical model building, a description using the mathematical concepts and language, and computer-based simulation steps. The advantage of a mathematical model is that it can be manipulated in accordance with the scope of the simulation study by changing the input variables, output variables, parameters, and structure of the model to accurately match the transportation system behavior to improve transportation systems efficiency. Thus, the transportation system representation describes a set of variables containing information to specify the evolution of the system over time. A system, in general, can either be continuous or discrete in time. Thus, the transportation systems sector models can be classified according to continuous and discrete time, state, and space.

The structural concept of a transportation system can be given in the state space notation. Henceforth, a transportation system represented by the state vector $x(t_1)$ is determined through the initial state $x(t_0)$, and the input function $u(\cdot)$ over the time interval (t_0, t_1) for all $t_1 > t_0$ is said to be a dynamic system. A dynamic system is of finite order if the state vector $x(t_1)$ has a finite number of objects.

Model building of transportation systems consists of the following steps:

- Make decisions about the following information:
  - What is already known?
  - What sources of relevant data are available?
  - What are the major assumptions?
  - What are the major constraints?
  - What is intended to be forecast and/or predicted with the model?
  - What is intended to be backcast and/or backtracked with the model?
  - What approach can verify that the model is built correctly?
  - What methods are able to validate the correctness of the model?
  - What type of simulation system/tool is required?
- Identify the most important model objects and how they are connected to each other. As for any complex task, visualization helps a lot!
- Conduct a thorough literature review if a model has already been developed and published that suits the area of concentration. There is no need to reinvent the

wheel. However, it is essential to fully understand all assumptions, constraints, and applicability of a published model before using it.

- Conduct a thorough review of the data that are planned to be used, and identify discrepancies and inconsistencies between and within the data sets. Often data are missed and/or are not measurable. This requires a plan for dealing with missed and/or nonmeasurable data if feasible uncertainties associated with the data should be quantified.
- In any case, start with a simple model, because there is a simple trade-off between complexity and accuracy (see Fig. 1.4).
- Decide what are important variables and constants and determine how they relate to each other. Input and output variables are important. Depending on the type of model decision variables, random variables, state variables, etc., are important.
- Compile mathematical equations that relate variables to each other.
- Identify the parameters of the mathematical equations and compile a plan on how to estimate the parameters from data. This can be done by fitting the equations to the data. Complex models require sophisticated parameter identification methods.
- Validate the model against a data set that has not been used for model building.
- Test the model and update the equations based on new data and information.

Following the aforementioned steps helps to design simulation scenarios and evaluate the transportation system-wide impact of local changes in the transportation system network. It should be noted that a simulation model developed cannot capture the entire transportation systems behavior as every single simulation model will be developed based on specific boundaries and constraints. For example, a general road traffic simulation model cannot simulate the traffic flow inside a container terminal or at an airport ramp, while the container terminal or the airport ramp simulation model cannot capture the traffic flows outside the terminal or the airport.

In conclusion, it should be noted that the scope for using simulation models in various disciplines can be different. In control systems engineering, they are concerned with understanding and controlling segments of systems to provide useful economic products for society. Engineering in this way deals with system synthesis and optimization because engineers are primarily interested in a simulation model which can be executed under normal operating conditions. Engineering scopes using simulation models control systems optimally or, at least, keep them in relatively close vicinity of conditions that avoid danger due to a possible drift out of system margins of safe operation. In contrast, in the life sciences, biomedical scientists are not solely interested in simulation models of biomedical systems that are executed under normal operating conditions. Life science researchers prefer to develop simulation models that adequately describe system behavior outside the normal operating range, which can be interpreted, in medical terms, as a disease use case notation. This represents a system model that is operated outside of normal operating set points. In the transportation systems sector, two tasks must be performed: model building of transportation systems scenarios which optimally fit with the real-world system behavior and solve the model's response through simulation for evaluation purposes. Simulation itself can be introduced in general terms as mockup of a transportation process through a model to gain expertise and knowledge, which is transferable to reality. The combination of these steps is referred to as transportation systems analysis. The modeling formalisms in transportation are discussed in detail in the following section.

1.4   Modeling Formalisms

Mathematical models are used for many different purposes; to explain behavior and data, to provide a compact representation of data, and more (Janssens et al. 2014). Therefore, model building is a method that can be used to solve complex problems in the transportation systems sector, including traffic engineering and transportation planning. However, success in analyzing transportation systems based on transportation system models depends upon whether or not the model is properly chosen, because a mathematical model has achieved its purpose when an optimal match between simulation results, based on the mathematical model and data sets gathered through real measurements and experiments, is obtained. Therefore, a mathematical model helps to explain a system's behavior; and studying the effects of the different model objects through simulation allows predictions of the model scenario behavior. Simulation itself can be introduced as imitation, mimicking, or mockup of a dynamic process within a model to gain expertise which is transferable to reality.

In order to develop suitable models for the transportation systems sector, a thorough understanding of the transportation system and its operating conditions is essential. Since a model is an abstraction of the overall system's behavior, it will only capture some properties of the whole transportation system. Therefore, it is often necessary to use several different models or model concepts. These concepts depend on a priori knowledge based on:

- Inputs
- Outputs
- System states

for the decision of the unknown, as shown in Fig. 1.1.

1.4.1   General Formalisms

If a model structure is determined for the transportation systems sector, the next step is to describe the model formalisms in terms of mathematical equations. They describe the dynamics of the transportation system at a more abstract level of detail. The mathematical notation eligible to describe the systems behavior is based on different theories, functions, matrices, and principles, such as:

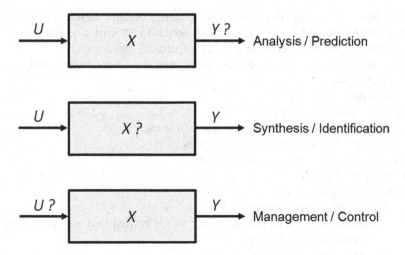

Fig. 1.1 Types of systems concepts that depend on a priori knowledge of inputs, outputs, or system state: analysis/prediction, synthesis/identification, management/control

- *Bellman optimization principle*: says that in some optimization problems, any optimal solution is composed of optimal partial solutions. An example is the calculation of the shortest path of a road network. A shortest path P between the nodes A and B which passes through the nodes X and Y must use a shortest path between these two nodes between X and Y. That is not the case, P could be shortened, using a shorter part way between X and Y, and then any shortest path between A and B would have been P, contrary to the assumption. The so-called Bellman-Ford Algorithm used to calculate shortest paths based on dynamic programming takes advantage of this principle. It determines the value of a decision problem at a certain point in time in terms of payoff from some initial choices and the value of the remaining decision problem that results from those initial choices. This allows breaking a dynamic optimization problem into simpler subproblems, described as Bellman's principle of optimality.
- *Bond graph theory*: it is a graphical representation of a physical dynamic system. A bond graph is composed of bonds which link together single port, double port, and multi-port elements. Each bond represents the instantaneous flow of energy $\frac{dE}{dt}$ or power. The flow in each bond is denoted by a pair of variables called power variables whose product is the instantaneous power of the bond.
- *Conditional path theory*: study the problem of locating a median path of limited length under the condition that some existing facilities are already located. The existing facilities may be located at any subset of vertices. This approach makes use of the Dijkstra algorithm to find the shortest paths from a point in a graph (the so-called source) to a destination. It turns out that one can find the shortest paths from a given source to all points in a graph in the same time. This problem is related to the spanning tree one. The graph representing all paths from one vertex to all others must be a spanning tree, which means it must include

all vertices. There will also be no cycles assumed because a cycle would define more than one path from the selected vertex to at least one other vertex. Dijkstra's algorithm itself takes two sets of vertices into account:

- **S**: set of vertices whose shortest paths from the source have already been determined
- **V-S**: remaining vertices

The other data structures required are:

- **d**: array of best estimates of shortest path to each vertex
- **pi**: array of predecessors for each vertex

The basic mode of operation is (URL 2):

1. Initialize **d** and **pi**.
2. Set **S** to empty.
3. While there are still vertices in **V-S**:
   - Sort the vertices in **V-S** according to the current best estimate of their distance from the source.
   - Add **u**, the closest vertex in **V-S**, to **S**.
   - Relax all the vertices still in **V-S** connected to **u**.

- *Cost function*: a mathematical equation used to predict the cost associated with a certain action or a certain level of output. Cost functions are often used to forecast the expenses associated with a specific activity, in order to determine what pricing strategies to use in order to achieve a desired result. Hence, a cost function for all possible output vectors y and all positive input cost vectors i is achievable if y belongs to the domain $V(y)$

$$DomV = \left\{y \in R_+^m : V(y) \neq 0\right\}$$

with input correspondence V which maps output vectors y in R_+^m into subsets of $2^{R_+^n}$ (URL 3). These subsets are vectors of inputs that will generate the given output vectors. Thus, V is the correspondence, given by $V = R_+^m \rightarrow 2^{R_+^n}$. Thus, the cost function $C(y,w)$ can be given as:

$$C(y,w) = \min\{wx : x \in V(y)\}; \quad y \in DomV, w > 0.$$

- *Differential equation*: mathematical equation that involves an unknown function as derivatives, i.e., $y' + y = x + 4$. Here the unknown function is y. Differential equations arise whenever a deterministic relation involving some continuously varying quantities (expressed by functions) and their rates of change in space and/or time (expressed as derivatives) is known or postulated. Differential equations play a prominent role in many scientific and engineering disciplines. Differential equations are mathematically studied from different perspectives, mostly concerned with their solutions which are their set of functions that satisfy the equation. Only simple forms of differential equations are solvable by explicit formulas; however, some properties of solutions of a given differential equation may be determined without finding their exact form. In general an ordinary differential equation (ODE) is a differential equation in which the unknown function (also known as dependent variable) is a function of a single independent variable. Ordinary differential equations are classified according to the order of the highest derivative of the dependent variable with respect to the independent variable appearing in the equation. The most important cases for applications are first-order and second-order differential equations. Some methods in numerical partial differential equations (PDE) convert the partial differential equation into an ordinary differential equation, which must then be solved. Both ordinary and partial differential equations are furthermore classified as linear and nonlinear. In this context a differential equation is called linear if the unknown function and its derivatives appear to the power of 1 and nonlinear otherwise.

- *Game theory*: is the study of strategic decision making. It is often used in studying mathematical models of conflict and cooperation of decision makers. It also applies to a wide range of behavioral relations, and today it is used as an umbrella term for the logical side of decision science.
- *Graph theory*: is the study of graphs, which are mathematical structures used to model pairwise relations between objects. A graph can be undirected, meaning that there is no distinction between the two vertices associated with each edge or its edges can be directed from one vertex to another. More specifically a graph can be introduced as an ordered pair

$$G = (V, E)$$

comprising a set V of vertices or nodes together with a set E of edges or lines, which are 2-element subsets of V, that connect them. V and E are usually taken to be finite. The order of a graph is $|V|$, the number of vertices. The graph's size is $|E|$, the number of edges. The degree of a vertex is the number of edges that connect to it.

- *Markov process*: is a stochastic process that satisfies the Markov property. The Markov process can be used to model a random system that changes states according to a transition rule that only depends on the current state. A Markov process is called memoryless if the process satisfies the Markov property so far one can make predictions for the future of the process based solely on its present state as well as knowing the process's full history. Thus, a stochastic process $x(t)$ is called a Markov process if for every n and $t_1 < t_2 \ldots < t_n$, we have

$$P(x(t_n) \leq x_n | x(t_{n-1}), \ldots, x(t_i)) = P(x(t_n) \leq x_n | x(t_{n-1}));$$

thus, the future probabilities of a random process can be determined by its most recent values.

- *Petri net theory*: is a mathematical modeling language for the description of distributed systems consisting of places, transitions, and arcs. Arcs run from a place to a transition or vice versa, but never between places or between transitions. Places from which an arc runs to a transition are called input places of the transition; places to which arcs run from a transition are called the output places of the transition. Thus, in general terms, a Petri net is a directed bipartite graph in which nodes represent transitions (i.e., events that can occur) and places (i.e., conditions). The directed arcs describe which places are pre- and/or postconditions for which transitions (signified by arrows). Furthermore, Petri nets contain a discrete number of marks, so-called token. From a general perspective, a transition of a Petri net can fire if it is enabled, i.e., there are sufficient tokens in all of its input places; when the transition fires, it consumes the required input tokens and creates tokens in its output places.

- *Queuing theory*: is a mathematical approach studying waiting lines, or queues. In queueing theory, a model is build so that queue lengths and waiting times can be predicted. Single queueing nodes can be described using Kendall's notation in the form

$$A/B/c/N/k$$

where A represents the interarrival time distribution, B is the service time distribution, c is the number of parallel servers of a station ($c \geq 1$), N represents the system capacity, and k is the size of the population. Many theorems in queuing theory can be proved by reducing queues to mathematical systems known as Markov chains.

- *Stochastic system*: is a system and/or model containing a random element wherefore its state is nondeterministic, meaning it is unpredictable without a stable pattern or order. The term stochastic is often taken to be synonymous with probabilistic, but, strictly speaking, stochastic conveys the idea of actual or apparent randomness, whereas probabilistic is directly related to probabilities and is therefore only indirectly associated with randomness. Thus, any system or process being analyzed using probability theory is stochastic at least in part.

- *Systems theory*: is an interdisciplinary study of systems with the goal of elucidating principles that can be applied to all types of systems at all nesting levels in all fields of research. The term originates from Bertalanffy's "General Systems Theory" (von Bertalanffy 1976). The term system itself can be introduced as a set of two or more interrelated elements with the following properties (Ackoff 1981):

1. Each element has an effect on the functioning of the whole.
2. Each element is affected by at least one other element in the system.
3. All possible subgroups of elements also have the first two properties.

By substituting the concept of "element" for that of "component," it is possible to arrive at a definition that pertains to systems of any kind, whether formal like mathematics or language, existential like real-world, or affective like aesthetic, emotional, or imaginative. In each case, a whole made up of interdependent components in interaction is introduced as a system. Therefore, a system is a group of interacting components that conserves some identifiable set of relations with the sum of the components plus their relations (i.e., the system itself) conserving some identifiable set of relations to other entities which can include other systems (Laszlo and Krippner 1998).

These and other types of theories, functions, matrices, and principles can overlap with a given model, involving a variety of different structural objects. Mathematical models of the transportation systems sector can be derived by considering physical laws and basic relationships characterizing the behavior of the transportation system, such as the equation of motion, and existing boundary conditions.

Let us consider, as a simple example, the dynamics of a vehicle in a transportation traffic system which moves with a varying velocity v. If the vehicle moves an arbitrary distance x in a given time t, the average velocity of the vehicle can be simply expressed by the equation

$$v = \frac{x}{t}. \tag{1.1}$$

If the vehicle is moving linearly, the movement of the vehicle can be calculated over a small period of time Δt. If the velocity of the vehicle has no possibility of changing within this period, the equation

$$v = \frac{\Delta x}{\Delta t} \tag{1.2}$$

is an approximation of the instantaneous velocity of the vehicle moving over small distances Δx. If we continue to decrease the period of time, we find that as Δt approaches zero, the above equation becomes an exact expression of the instantaneous velocity of the vehicle at any instant of time t', as follows:

$$v(t') = \lim_{\Delta t \to 0} \frac{\Delta x}{\Delta t}\bigg|_{t=t'}, \tag{1.3}$$

which can be rewritten as

$$v(t) = \frac{dx(t)}{dt}, \tag{1.4}$$

which means that both velocity v and arbitrary distance x are functions of time t, and v is a measure of the instantaneous rate of change of the distance x with respect to time t.

Let us consider, in this simple example, a change of velocity with respect to time, which is the case when the vehicle speeds up or slows down and the vehicle is said to accelerate or decelerate. Therefore, the motion can be described as follows:

$$a(t) = \frac{dv(t)}{dt},\tag{1.5}$$

where $a(t)$ is the acceleration, a function of time t. Substituting $v(t)$ in this equation, we obtain

$$a(t) = \frac{d\left(\dfrac{dx(t)}{dt}\right)}{dt} = \frac{d^2x(t)}{dt^2},\tag{1.6}$$

which is the second derivative of distance with respect to time.

The most important step in model building is translating an understanding of the transportation systems sector into the mathematical notation of systems theory, a method that originates from Bertalanffy's General System Theory, which states that:

> ... there exist models, principles, and laws that apply to generalized systems or their subclasses, irrespective of their particular kind, the nature of their component objects, and the relationships between them. It seems legitimate to ask for a theory, not of systems of a more or less special kind, but of universal principles applying to systems in general. (von Bertalanffy 1976)

As can be seen from this citation, the term "system theory" does not yet have a well-established, precise meaning; however, systems theory can reasonably be considered as a specialization of systems thinking, a generalization of systems science, and a systems approach. Bertalanffy divides systems inquiry into three major domains: Philosophy, Science, and Technology. In his work with the Primer Group, Bánáthy generalized the domains into four domains of systemic inquiry (von Bertalanffy 1976), as shown in Table 1.1, which operates in a recursive relationship. Integrating Philosophy and Theory as Knowledge, and Method and Application as Action, systems inquiry then is the knowledgeable action.

Applying system inquiries to the transportation systems sector requires transforming the characteristics of the transportation systems in such a way that a particular system notation can be found out of the set of possible system descriptions from the systems theory approach. In any case, a set of mathematical

Table 1.1 Domains of systems inquiries

Domain	Description
Philosophy	Ontology, epistemology, and axiology of systems
Theory	Set of interrelated concepts and principles applying to all systems
Methodology	Set of models, strategies, methods, and tools that instrument systems theory and philosophy
Application	Application and interaction of the domains

equations are usually obtained that describe the important variables of the transportation system. Describing a transportation system, based on the translation of systems knowledge into the language of a mathematical model, depends on the adequate form of its representation. In the case of a time-invariant, continuous-time system, the mathematical model MM_{TICTS} is based on ordinary differential equations as a set of dynamical equations of the form

$$MM_{TICTS} : (U, X, Y, f, g, T) \qquad (1.7)$$

with $u \in U$: set of inputs, $x \in X$: set of states, $y \in Y$: set of outputs, f : rate of the change function, g : output function, T: time domain, and

$$\begin{aligned} x' &= f(x, u) \\ y &= g(x, u) \end{aligned} \qquad (1.8)$$

It can be convenient for some applications to transform these mathematical equations into a standard or normal form. The important point is to decide on the state variables which are essential to characterize the system. Such model formalism is a specific set structure.

Let $MM = \Sigma$ (e.g., three forms of representation can be used: input, output, and state). Correspondingly are a state X, a set U of input values, and a set Y of output values. Thus, a mathematical model of a transportation system is called dynamic if it can be defined as set structure Σ:

$$\sum := (X, Y, U, v, T, a, b), \qquad (1.9)$$

with state variable X, set of output values Y, set of input values U, set of admissible controls v, time domain T, state transition map a, and read-out map b.

In some cases, it can be necessary to specify unmeasurable and/or random inputs. These system disturbances can be described as impacts of uncontrollability and/or unobservability of the dynamic system to be modeled, which can mathematically be described by stochastic, continuous-time models, as follows:

$$MM_{SCT} : (U, V, W, X, T, f, g), \qquad (1.10)$$

with

$$\begin{aligned} x' &= f(x, u, w, t) \\ y &= g(x, v, t) \end{aligned} \qquad (1.11)$$

The vectors v and w are random model disturbances. In a case where v and w are random or stochastic vector processes, meaning that the stochastic properties of these vectors are not related to the model specification, then x and y will be the same process.

In many cases, and especially in management and operational research, the dynamic system can be thought of as being built of a collection of events.

Even the state variables change at specific time instants. A mathematical description which can be based on the notation of a mathematical discrete-event model yields

$$MM_{DEVS} : (V, S, Y, \delta, \lambda, \tau, T) \tag{1.12}$$

with V: set of external events, S: sequence of states, Y: set of outputs, δ: transition function, λ: output function, and τ: time function.

Many dynamic systems have properties that vary continuously in space, which can be described based on distributed models. The mathematical expression for distributed models is based on partial differential equations, which result in the following mathematical model description:

$$MM_{PDE} : (U, \Theta, Y, F, r, g, z, T) \tag{1.13}$$

with

$$0 = f\left(\Theta, \frac{\partial \Theta}{\partial t \partial z}, u, z, t\right) z \in Z; \ 0 = r(\Theta, z, t) z \ dom \ Z; \ y = g(\Theta, z, t) z \in Z. \tag{1.14}$$

Apart from the independent variable t, the space coordinate z is introduced. Vector Θ of the dependent variables can vary in space and time. The equation(s) hold(s) in a spatial domain Z, while conditions given by r are provided on the boundary of the domain $domZ$. There is input u and output y.

1.4.2   State Models

Next we introduce a class of systems whose states are chosen to consist of a finite number of variables. The state of such a system can be represented by a finite-dimensional column vector x, which is called a state vector; and the components of x are called state variables. The state at time t_0 is, by definition, the required information at t_0 that, together with input u, uniquely determines the behavior of the system for all $t \geq t_0$. This can be interpreted as translating the equations of the dynamic system variables into state-variable equations in their standard form. Thus, the modeling procedure in the transportation systems sector consists of the definition of dynamical equations representing the model of the transportation system and translating it into a set of descriptive state equations that can be expressed in the following state-variable model dynamical equations:

$$\begin{aligned} x'(t) &= f(x(t), u(t), t) \\ y(t) &= g(x(t), u(t), t) \end{aligned} \tag{1.15}$$

with $x(t)$ as n-dimensional state vector, $x'(t)$ as its derivative, $y(t)$ as q-dimensional output vector, and $u(t)$ as p-dimensional input vector which have the general form

$$x = \begin{bmatrix} x_1 \\ x_2 \\ \vdots \\ x_n \end{bmatrix}, u = \begin{bmatrix} u_1 \\ u_2 \\ \vdots \\ u_m \end{bmatrix}, y = \begin{bmatrix} y_1 \\ y_2 \\ \vdots \\ y_k \end{bmatrix}.$$

The dynamic system is specified by the vector-valued functions f and g as follows:

$$f(x, u, t)0 \begin{bmatrix} f_1(x, u, t) \\ f_2(x, u, t) \\ \vdots \\ f_n(x, u, t) \end{bmatrix}, g(x, u, t)0 \begin{bmatrix} g_1(x, u, t) \\ g_2(x, u, t) \\ \vdots \\ g_k(x, u, t) \end{bmatrix}.$$

If the system is linear in x and u, a set of n first-order differential equations is

$$x_1' = a_{11}(t)x_1 + a_{12}(t)x_2 + \ldots + a_{1n}(t)x_n + b_{11}(t)u_1 + \ldots + b_{1m}(t)u_m$$
$$x_2' = a_{21}(t)x_1 + a_{22}(t)x_2 + \ldots + a_{2n}(t)x_n + b_{21}(t)u_1 + \ldots + b_{2m}(t)u_m \quad (1.16)$$
$$\ldots$$
$$x_n' = a_{n1}(t)x_1 + a_{n2}(t)x_2 + \ldots + a_{nn}(t)x_n + b_{n1}(t)u_1 + \ldots + b_{nm}(t)u_m$$

as well as m algebraic equations relating output variables, state variables, and control variables, which can be written as follows:

$$y_1 = c_{11}(t)x_1 + c_{12}(t)x_2 + \ldots + c_{1n}(t)x_n + d_{11}(t)u_1 + \ldots + d_{1m}(t)u_m$$
$$y_2 = c_{21}(t)x_1 + c_{22}(t)x_2 + \ldots + c_{2n}(t)x_n + d_{21}(t)u_1 + \ldots + d_{2m}(t)u_m \quad (1.17)$$
$$\ldots$$
$$y_k = c_{k1}(t)x_1 + c_{k2}(t)x_2 + \ldots + c_{kn}(t)x_n + d_{k1}(t)u_1 + \ldots + d_{km}(t)u_m$$

The linear continuous-time system can be rewritten in terms of a vector matrix

$$x' = A(t) \cdot x + B(t) \cdot u; \quad x \in \mathfrak{R}^n; u \in \mathfrak{R}^R; t > 0$$
$$y = C(t) \cdot x + D(t) \cdot u; \quad y \in \mathfrak{R}^R; t > 0. \quad (1.18)$$

The mathematical model, given in (1.18), is called linear. The matrixes of $A(t)$, $B(t)$, $C(t)$, and $D(t)$ are the transforms on the respective vector space, which is as follows:

$A(t): \mathfrak{R}^n \to \mathfrak{R}^n$ as a (n, n)-matrix called the system matrix
$B(t): \mathfrak{R}^m \to \mathfrak{R}^n$ as a (n, m)-matrix called the input matrix
$C(t): \mathfrak{R}^n \to \mathfrak{R}^k$ as a (k, n)-matrix called the output matrix
$D(t): \mathfrak{R}^r \to \mathfrak{R}^p$ as a (p, r)-matrix called the transition matrix

The state variables in (1.15) are said to be related by a nonlinear transformation, and hence they are called nonlinear state equations. A block diagram of the state-variable model dynamical equations is shown in Fig. 1.2.

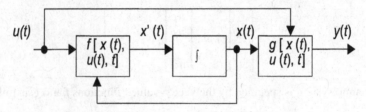

Fig. 1.2 Structural representation of the state-variable model of (1.15)

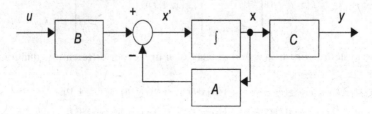

Fig. 1.3 Structural representation of the linear state-variable model with $D \equiv 0$

The state equations given in (1.18) are said to be related by a linear transformation. For a given initial state values $x_1(0), \ldots, x_n(0)$ and a given input function $u(t)$ defined for $t > 0$, there exists a unique solution of the state equations $x_1(t), \ldots, x_n(t)$ defined for all $t > 0$ (i.e., the functions $x_1(t), \ldots, x_n(t)$) which satisfy the state equations exactly for all $t > 0$, and hence there exists a unique output function $y(t)$ defined for $t > 0$.

In case that element of matrices A, B, C, and D is time dependent, the state variables given in (1.18) are said to describe a linear time-variant system.

Assuming that the influence of $u(t)$ on $y(t)$ is normally indirect, we may neglect D, writing $D \equiv 0$, which results in a structural diagram of a linear state-variable model, as shown in Fig. 1.3.

A sufficient condition for (1.18) to have a unique solution is that every entry of $A(\cdot)$ is a continuous function of t defined over $(-\infty, \infty)$. The entries of B, C, and D are also assumed to be continuous in $(-\infty, \infty)$. Since the values of A, B, C, and D change with time, the dynamical equation (1.18) is called a linear time-varying dynamical equation. If the matrices A, B, C, and D are independent of t, then the equation is called a linear time-invariant dynamical equation. Hence, an n-dimensional, linear, time-invariant dynamical equation has the form

$$x' = A \cdot x + B \cdot u \tag{1.19a}$$

$$y = C \cdot x + D \cdot u \tag{1.19b}$$

where A, B, C, and D are, respectively, $n \times n$, $n \times p$, $q \times n$, and $q \times p$ matrices. In the time-invariant case, the characteristics of (1.19a, b) do not change with time; hence there is no loss of generality in choosing the initial time t_0 to be 0. The time interval of interest then becomes $[0, \infty]$.

1.4.3 Methodological Principles

With reference to the spectrum of available models, the variety of levels of conceptual and mathematical representations is evident, which depend on the goals and purposes for which the model usage was intended, the extent of a priori knowledge available, data gathered through experimentation and measurements on the system, estimates of systems parameters, as well as system states. Hence, a transportation system can be seen as a system that is deconstructed of a certain level of detail.

1.4.3.1 Continuous-Time Principle

From a more general point of view, the mathematical representation of transportation systems is based on the foundations of deconstructed systems of continuous time at any required level.

- *Behavior level*: at which one can describe the dynamic system as a black box, in which we record measurements in a chronological manner based on a set of trajectories that characterize the system behavior. The behavior level is of importance because experimentation with dynamic system addresses this level due to the input–output relationship, which can be expressed for a black-box system as

$$\underline{y}(t) = F(\underline{u}, t), \qquad (1.20)$$

with $u(t)$ as an input set, $y(t)$ as an output set, and F as a transfer function for a state structure level, at which one can describe the dynamic system taking into account the system state structure that results by iteration over time in a set of trajectories, the so-called transient behavior. The internal state sets represent the state-transition function that provides the rules for computing the future states, depending on the current states:

$$\underline{y}(t) = G(\underline{u}(t), \underline{x}(t), t). \qquad (1.21)$$

A state of a dynamic system represents the smallest collection of numbers, specified at time $t = t_0$ in order to uniquely predict the behavior of the system for any time $t \geq t_0$ for any input belonging to the given input set, provided that every element of the input set is known for $t \geq t_0$. Such numbers are the so-called state variables.

- *Composite structure level*: at which one can describe the dynamic system by connecting elementary black boxes which can be introduced as a network description. The elementary black boxes are the components, and each one must be described by a system representation at the state level. Moreover, each component must define input and output variables as well as a specification determining the interconnection of the components and interfacing the input and output variables.

Difficulties in developing mathematical models may arise because transportation systems are, in general, extremely complex. In addition, a sufficient number of operating data are often unavailable. Hence, developing a mathematical model for a real-world transportation system first requires the selection of the type of model structure and, thereafter, the respective essential model parameter values for simulating the model. In a case such that the required model parameters are not available, a more simplified model can be developed eliminating intrinsic characteristics of the transportation system, because an overcomplicated mathematical model can cause mathematical difficulties.

Two major facts are important when developing mathematical models of real-world transportation systems:

1. A model is always a simplification of reality but should never be so simple that its answers are not true.
2. A model has to be simple enough that studying and working with it is easy.

Hence, a suitable model is a compromise between mathematical difficulties caused by equations that are too complicated and the accuracy of the final result. The corresponding relationships are shown in Fig. 1.4.

From Fig. 1.4, one can conclude that there is no reason to develop expensive models because the increment of quality is less than the increase in cost. This point is important because a mathematical model is a very compact way to describe transportation systems. But a complex model not only describes the relationships between the transportation system inputs and outputs, it also provides detailed insight into the transportation system structure and internal relationships. This is due to the fact that the main relationships between the variables of the transportation system modeled are mapped into appropriate mathematical equations.

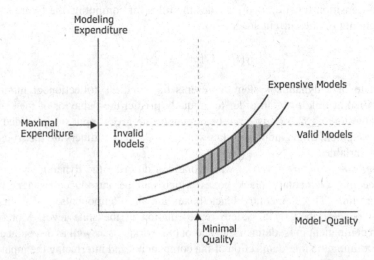

Fig. 1.4 Dependence of the modeling expenditure (costs) versus the degree of accuracy (model quality)

In principle, there are two different approaches to obtaining mathematical models of transportation systems: (1) the deductive or theoretical approach, based on the derivation of the essential relations of the transportation system, and (2) the empirical approach, based on experiments on the transportation system itself. It should be noted that practical approaches often use a combination of both, which might be the most advantageous way. The two methods result in the:

- *Empirical method of experimental modeling*: based on measures available on the inputs and outputs of a transportation system. Based on these measurements, the empirical model allows model building of the transportation system, as shown in Fig. 1.5. The characteristic signal-flow sequence of the experimental

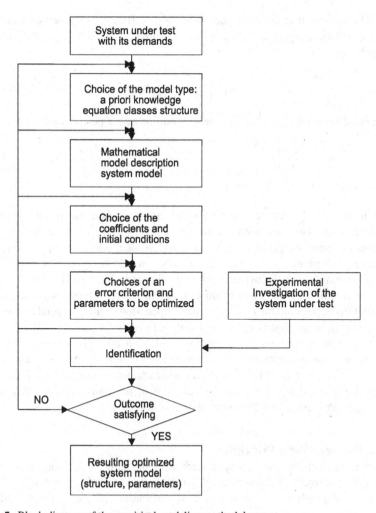

Fig. 1.5 Block diagram of the empirical modeling methodology

modeling process is used to determine the model structure for the mathematical description, based on a priori knowledge, which has to fit with the used error criterion, which is chosen in the same way as the performance criterion for the deductive-modeling method.

- *Deductive method of theoretical or axiomatic modeling*: represents a bottom-up approach starting at a high level of well-established a priori knowledge of transportation system objects, representing the mathematical model. In real-world situations, problems occur in assessing the range of applicability of these models. The deductive-modeling methodology is supplemented by an empirical model validation proof step. Afterwards, the model can be validated by comparing the simulation results with the data known from the transportation system whether they match an error criterion or not.

Let e be an error margin, which depends on the difference between measures on the real-world system y_{RWS} and data from the simulation of the mathematical model y_{MM}, as follows:

$$e := e\left(\underline{y}_{RWS}(t), \underline{y}_{MM}(t)\right). \tag{1.22}$$

The error criterion can be determined by minimizing a performance criterion

$$J = \int_0^t e^2 \cdot dt \rightarrow Min+. \tag{1.23}$$

The model fits the chosen performance criterion when the results obtained from simulation compared with the results from the transportation system data are within the error margin of the error criterion. If the model developed did not fit the chosen performance criterion, a modification would be necessary at different levels, as shown for the deductive-modeling scheme in Fig. 1.6. The result of the modification, which can be understood as a specific form of model validation, is a model that fits better than the previously developed model. It is essential to mention that a transportation system model not only describes relationships between its inputs and outputs, such as for black-box models, it also gives insight into the transportation system structure and into intrinsic and internal systems relationships at the respective level of representation of non-black-box models. This is due to the fact that the relationships between the variables of the transportation system are mapped into appropriate mathematical expressions.

1.4.3.2 Discrete-Time Principle

In comparison with the previously described model building of continuous-time systems, the treatment of discrete-time systems follows a different modeling paradigm which depends on the appearance of trajectories of the system variables. In Fig. 1.7, the typical time-dependent trajectories of a continuous-time and a

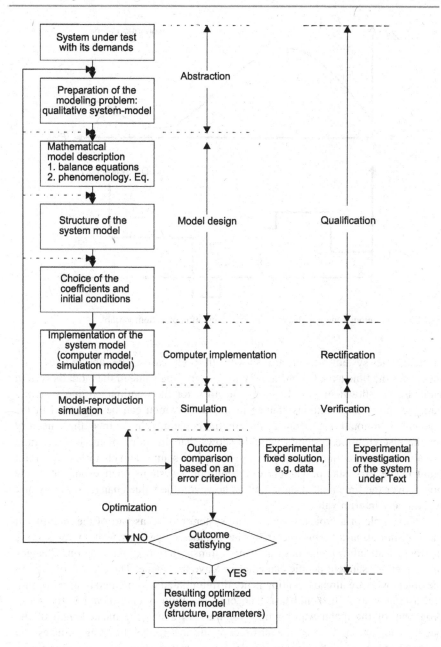

Fig. 1.6 Block diagram of the empirical modeling expanded by the deductive-modeling methodology

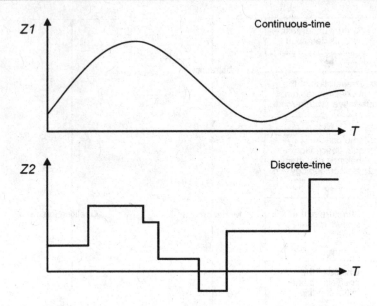

Fig. 1.7 Continuous-time and discrete-time representation of system variables

discrete-time system model are shown. In both cases, the abscissa (x-axis) represents the time axis; the ordinate (y-axis) marks the value of the model's system variable. As shown in Fig. 1.7, the trajectory for the continuous-time system's variable shows a continuous change in its value, which can be expressed in the mathematical notation of ordinary differential equations. In contrast, the trajectory for the discrete-time system's variable shows that the value of the discrete-time variable is constant in between the respective time stamps and changes at a certain point in the time stamp. Therefore, discrete-time systems change only at a few points over the time axis at these points; however, the value changes abruptly and without any interim value.

An example of a typical discrete-time system model, as part of the mobility in the transportation systems sector, is the number of persons waiting for service in front of an information desk at a mass transportation system station. Changes in number are sudden: people can enter or leave the queue. The process of joining the others who are already waiting is not differentiated in more detail: approaching and asking who is first and last. The only intention of this model is to give a prognosis of the mean waiting time for the customers, the mean length of the queue, and so on. Hence, the abstraction during model building reduces the dynamic behavior of the system model upon sudden changes in the number of people waiting. The number of people in the queue is a classical discrete-time model variable. With this simple example, the two basic principles of model building of discrete-time systems are introduced: (1) the definition of an event in the course of a model variable and (2) the condition for its dynamic behavior between the events which describe the two essential principles of discrete-time systems:

- An event is an instantaneous occurrence that changes the state of a system.
- The value of a time-discrete model quantity stays constant during the time interval defined by two consecutive events.

. Based on these principles, systems for different application areas in the transportation systems sector can be modeled sufficiently. When discussing the modeling of transportation systems by means of discrete events, two problems have to be mentioned:

- The term *event* implies a resolution of time which goes to infinitely short time periods. An event happens without any consumption of time. However, in reality, the execution of an event may take very little time but it does take some time, which results in the problem of the resolution of the time axis.
- The problem of what happens if two or more events occur simultaneously is caused by the definition of the event itself; and the problem has to be solved by simulation algorithms, which execute a system with two or more events.

Once the definition of the event and the description of its semantics are known, the way to model discrete-event systems is obvious. The description of the system dynamics consists of a chronologically sorted list of events which occur between the start time and the end time of the observation. All of the knowledge about the system is represented in this list. As in continuous models, an initial value for the model quantities influenced by the events must be given.

In practice, the modeler has to specify these events and put them into the correct order. Looking at the events used to model the very simple system shown in Fig. 1.8, lots of very similar events can be found. The example shows a single serving unit with a queue for the waiting customers. The customers are created randomly and receive a varying service time. After service, the customers leave the system. This system is the simplest example of discrete-event simulation and is called a single-server system.

Let us look at the events for the single-server system shown in Fig. 1.9: Object e1 enters the queue at time t1; Object e2 enters the queue at time t3; Object e3 enters the queue at time t4. ... These are the events which describe the arrivals

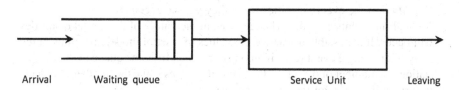

Arrival   Waiting queue   Service Unit   Leaving

Fig. 1.8 Discrete-time representation of a system concept

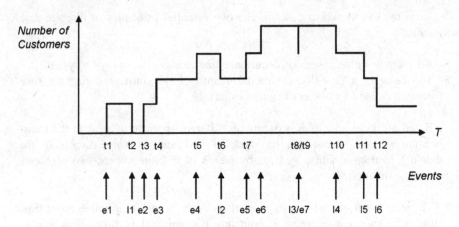

Fig. 1.9 Time events for a single-server system

of customers. On the other hand, there are the events describing departures. Elapsed time for customers' service: Object e1 finishes service at time t2, Object e2 finishes service at time t6, and Object e3 finishes service at time t9. ... To simplify the task of specifying all of these events, a more general specification scheme is offered using model description languages and the corresponding simulation systems.

The idea is to build classes of events which describe the dynamics on a more abstract level, such as the particular events introduced above. The main event classes could be:

- Arrival of a customer
- Customer enters queue
- Start service
- End of service
- Customer leaves system

Using these more abstract event classes, all arrivals, all entrances into the queue, all service starts, and more can be modeled by a single piece of model code. Therefore, the general syntax of an event in a model description language consists of two defining parts:

1. *Condition* of the events specifies when they will be executed.
2. *Body* of the events specifies what changes in the values of the model quantities will occur. It is possible to change the values of a set of model quantities in the body of one single event (e.g., if an object is taken from the queue to the service station, the number of objects in the queue decreases while the number of objects in service increases for the same amount).

In respect to the event condition, a further classification of events can be made as follows:

1. *Time events*: whose event condition exclusively uses the simulation time T and whose execution depends on the course of T exclusively
2. *State events*: whose condition is a free Boolean expression which can include any model variable and whose execution depends on the state of the model variables as they change—or even on the values of any other variables in the model

Example 1
(time event)
```
WHENEVER   T > T_enter
BEGIN_BODY
      number_customers := number_customers + 1;
END_BODY
```

Example 2
(state event)
```
WHENEVER   number_customers / number_service_units > 5
BEGIN_BODY
      number_serverice_units := number_service_units + 1;
END_BODY
```

If the dynamic of the system follows predetermined rules, such as iteration in time and interdependencies in certain states of the model, the user has the option to specify classes of events in which more than one activity in the real world is represented by a single event in the model description.

Example 3
(state event)
Whenever the value of a fuel level in a tank reaches its upper limit, 25% is taken out by the controller.
```
WHENEVER   tank_level >= level_max
BEGIN_BODY
      tank_level := tank_level * 0.75;
END_BODY
```

Independently from the way the tank is filled (e.g., time discrete by buckets or time continuous by a fuel flow from a fuel bunker), this event ensures that the level will not exceed the given limit.

Example 4
(time event)
The following event represents activities which model the arrival of customers at a service station. The event is triggered by setting the next time, and the event

will be active inside the body of the event itself. Therefore the variable InterArrivelTime can have a fixed value or can be represented by a random number to model a random arrival process.

```
WHENEVER   T >= T_NextArrival
BEGIN_BODY
        customers_in_queue := customers_in_queue + 1;
        T_NextArrival := T_NextArrival + InterArrivelTime;
END_BODY
```

For the single-server system, the transient behavior can be modeled by the following set of event classes with an implicit time condition for their next activation that is set by the procedures schedule_arrival_event and schedule_departure_event.

Example 5
(complete set of events to simulate the single-server system)

```
    WHENEVER Arrival_event
        IF number_in_server = 1
                THEN ( number_in_queue := number_in_queue +1;        )
                ELSE ( number_in_server := number_in_server +1;
                        schedule_departure_event (T + T_service_time ); )
        schedule_arrival_event (T + T_interarrival_time );
        protocol_state_changes ();
    END Arrival_event
                        WHENEVER Departure_event
        IF number_in_queue == 0
            THEN ( number_in_server := 0;    )
            ELSE ( number_in_queue := number_in_queue -1;
                    schedule_departure_event (T + T_service_time ); )
        protocol_state_changes ();
    END Departure_event
```

To run discrete-time/event model simulations, algorithms have been established based on the following specifications:

- Execute the events which happen in the simulation period between T_start and T_end completely.
- Execute events exactly at the point in time their condition becomes true.
- Execute the events in the right order.
- Execute the events without consumption of simulation time.

Assume the simulation interval is given by the start time and the end time for the run and that the resolution *deltaT* of the time axis is determined (e.g., by the next event simulation). The most simple simulation algorithm then is

```
Set ActTime := StartTime;
WHILE ActTime < EndTime
      DO
      WHILE NOT (all event conditions are false)
            DO
            <find an event_condition in model description which is true>
            <execute the corresponding event>
            END
      ActTime := ActTime + deltaT;
      END
```

This algorithm executes the simulation correctly but consumes a lot of CPU time as it checks all event conditions at every time step *deltaT*. Due to the characteristics of discrete time/event models, almost nothing happens at the point in time under observation. It is typical to hold a given value constant for a certain period of time until the next event changes it. Checking the event conditions at every point in time can usually be dispensed with as it causes an enormous consumption of calculation time. On the other hand, the algorithm is very simple and requires nothing concerning the formulation of the events.

Because of its run-time behavior, the algorithm is refined; and the result is the so-called next event algorithm. Its data structure consists of two elements:

1. Current time
2. Future event list, an ordered list of events which are to be executed in the future

Each of these events has a time stamp which shows the point in time that its condition becomes true. The list is ordered by a time stamp. By doing so, the event to be executed next moves to the top of the list.

The advantage of this event list is that there will not be other events between two entries in the list. So the algorithm does not need to check all of the conditions between two events and knows exactly when the next change in value of a model quantity will happen. The algorithmic version is as follows:

```
# initialize
          <set start time>
          <set end time>
          <put an initial set of events into the event list>
          T := T_start;
# simulation loop
          WHILE T < T_end
          DO
                current_event := first entry of the event list;
                T := current_event.time_stamp;
                execute (current_event);
                delete_from_event_list (current_event);
                current_event : <first entry of the event list>;
                T := current_event.time_stamp;
          END.
```

The disadvantage of this approach is that it needs the help of the modeler: somebody has to put new entries in the next event list. After the initialization at start time, this is done by expanding the body of the events. Within the event specification, the modeler has to specify when the active event will be active again or if there is another event which is triggered by the active event and when it is set for execution. These are the two types mentioned previously: self-triggered events (e.g., by interarrival time) or condition-triggered events at the same point in time (e.g., customer enters empty queue and is transmitted to the service unit at the same point in time).

All other more sophisticated solutions for discrete-event simulation algorithms are based on these two basic approaches. They modify the search for the next event in the list, they allow parallelism by distributing the event list, and they integrate continuous model elements in the processing of the simulation algorithm.

1.5   Model Validation

Modeling in the transportation systems sector is a complex procedure that is comprised of at least three steps:

1. *Qualification*: represent the level of model details, meaning what aspects of the complex transportation system actually need to be incorporated into the model and of which level of detail, and what aspects can be safely ignored. Therefore, the process of model building focused on the respective attributes, objects, and relations to describe the transportation system in an abstract manner, which is the so-called abstract model.
2. *Verification*: is the follow-up process of qualification, which means the abstract transportation system model will be transformed into a computer program, the executable so-called simulation model of the transportation system. Therefore, for a 10,000 lines of code simulation model, verification will show a poor programming if the program is written before attempting any debugging. Using a commercial simulation package reduces the amount of programming required. Moreover, simulation packages contain powerful high-level macro programming statement.
3. *Validation*: is the follow-up process beyond verification, focused on fit or no fit of the model in accordance with the respective dynamic behavior of the real-world transportation system under study. Validation of the model refers to the quality of the model in terms of best fit; and falsification, the opposite of validation of the model, means low quality of the model in terms of not fitting with information and data on the system.

Validation of models as introduced in this book is a procedure that involves the extent to which the system model is built, its tractability, and its credibility with regard to the fulfillment of the purpose for which the model has been built. However, model validation in general is a multidimensional procedure reflecting the model purpose, current theories, and experimental test data relating to the particular system under study, together with other essential sources of knowledge such as comparison with expert opinion and/or comparison with another model as well as statistical procedures for comparing real-world observation on the transportation system or the confidence interval approach based on independent data. Therefore, model validation can be referred to as a complex procedure that integrates several levels of detail:

- *Behavioral level*: the model output reproduces the behavior of the dynamic system under study which is a comparison with an existing system.
- *State structure level*: the model can be synchronized with a system state for the prediction of future system behavior.
- *Composite-structure level*: the model represents the internal interactions of the dynamic system which can be proved using a confidence interval rather than a hypothesis test to validate a simulation model (Law 2007).

A more straightforward validation method is deductive analysis to show the validity of the model representation reflecting the model's purpose, depending on the validity of the a priori knowledge used. Validation through deduction can be achieved in two ways:

- Investigation of the exactness of the premises that validates the model
- Checking other consequences of the premises that validate (e.g., a priori information, confidence-interval approach based on independent data, comparison with another model, comparison with expert opinion, etc.)

Besides the deductive analysis, the inductive analysis can also be used as a straightforward validation, whether or not the inductive model building procedure has been carried out in a mathematically and logically correct way. Assuming a model represents a source of data, and then a valid model at a certain point in time has to have these signs specified.

From a more practical point of view, a model is sufficiently valid if it fits the concept reflecting the model building purpose. Hence, a truly valid model is a model that fits the internal and external model building criteria:

- *Internal criteria*: enabling conditions within the model itself to be judged without external reference to the model purpose
  - *Consistency*: requiring that the model building contain no logical, mathematical, or conceptual contradictions
  - *Algorithmic validity*: requiring that the algorithm for analytical solution or numerical simulation is appropriate and leads to accurate solutions
- *External criteria*: referring to the model itself, like the model purpose, theory, and/or data, which are:
  - *Empirical validity*: requiring the model building to correspond to the available data

- *Theoretical validity*: requiring the model to be consistent with accepted theories and/or models
- *Pragmatic validity*: requiring testing that determines the extent to which the model satisfies the objectives for which it has been developed
- *Heuristic validity*: requiring tests that are associated with the assessment of the heuristic potential of the model (e.g., for scientific explanation, discovery, and/or hypothesis testing)

Considerations of validity are required from the very beginning of model building. Empirical and theoretical validity can be used by examining whether the respective validation criteria are met or not, which can then be used as a performance index. A performance index is a quantitative measure of the performance of a model of a dynamic system and is chosen so that emphasis is given to the important real-world constraints.

A suitable performance index *PI* is the integral of the square of the error

$$PI = \int_0^T e^2(t)dt, \tag{1.24}$$

where the upper limit T is a finite time chosen somewhat arbitrarily so that the integral approaches a steady-state value of the transient behavior of the system model; and e is a measure of the error between the real-world system and the system model.

Another possible performance criterion is the integral of the absolute magnitude of the error, which can be written as:

$$PI = \int_0^T |e(t)|dt. \tag{1.25}$$

This performance index is particularly useful for computer simulation studies. In order to reduce the contribution of the large initial error to the value of the performance integral and to place an emphasis on errors occurring later in the response, another performance index has been proposed

$$PI = \int_0^T t \cdot |e(t)|dt. \tag{1.26}$$

This performance index is designated as the integral of the time multiplied by the absolute error. Another similar performance index is the integral of time multiplied by the squared error, which is:

$$PI = \int_0^T t \cdot e^2(t)dt. \tag{1.27}$$

The general form of the performance index is

$$PI = \int_0^T f[e(t), u(t), y(t), t]\,dt,\qquad (1.28)$$

where f is a function of the error, input, output, and time.

1.6 Case Study in Transportation Systems Analysis

In Sect. 1.3, the advantages and disadvantages of the three types of traffic flow models represented by performance functions were introduced. The respective simulation models are classified as: (1) aggregated space discrete or space continuous macroscopic simulation models, (2) aggregated discrete space continuous flow mesoscopic simulation models, and (3) disaggregated microscopic simulation models, all used for in-depth state-of-the-art study in the transportation systems sector. These models are used to analyze traffic performance at an aggregated level as discrete space, continuous flow traffic models. The aggregated variables used are capacity, flow, and occupancy. Traffic, however, is represented discretely by tracing trips of individual packets—a group of vehicles is referred to as a packet—each of which is characterized by a departure time and a destination path (Cascetta 2009).

Discrete flow models are based on time discretization, a division of the reference period into interval [k], which is below the equal duration ΔT. Hence, trips in discrete flow models begin at a representative time instant t_k in interval [k]. Assuming that the time instant of intervals in discrete flow models is a single-interval case, then $t_k = [k] \cdot \Delta T$. Discrete flow models often assume that relevant flow variables are averaged over time intervals.

The variables of discrete models and their relationships must be defined first. Some of the variables are (Cascetta 2009):

- *Time*: introduced in relation to the absolute time τ.
- *Topological*: the same as in continuous flow, continuous-time models.
- *Flow*: follow the same definition as continuous flow models, but in discrete flow models, they represent counts which represent the number of users in the interval [k] rather than flows.
- *Travel time*: continuous variables that vary with time τ in the continuous case. In the discrete case, however, not all instants τ are meaningful because not all correspond to the arrival or departure of a packet.

Demonstrating the proposed concepts in this transportation system case study of discrete traffic flow, which can be either individual vehicles or groups of vehicles moving together on a network and experiencing the same conditions (Cascetta 2009), a traffic simulation model for metropolitan Hamburg, Germany, will be used for the transportation scenario analysis. Scenario analysis, as well as scenario planning, is an improved decision-making approach which includes the consideration of outcomes and their implications on transportation. Thus, specifying a simulation model implies interdependencies between the level of detail for the traffic network in relation to the traffic load at the links, the routing, and the treatment of the required simulation time. Thus, the significance of simulation models developed presupposes the appropriate selection of the level of detail and the real data available for model parameterization.

Metropolitan Hamburg, Germany's second largest city (with approximately 2.5 million inhabitants in the metropolitan area), has to cope with significant urban traffic for which an appropriate infrastructure must be in place to avoid serious bottlenecks. A bottleneck in the transportation systems sector is defined as a shortage of infrastructure resources of the intermodal and multimodal transportation systems network. Thus, bottleneck identification is concerned with the analysis of infrastructure resource plans, optimization of intermodal and multimodal transportation chains, timely and concurrent use of resources, transaction analysis, multicriteria approach, and more. The major bottlenecks in metropolitan Hamburg are the two links that traverse the Elbe River, which are the Elbe River tunnels connecting with interstate highway A7 in the direction of the Danish border and Hannover, as shown in Fig. 1.10, and the Köhlbrand Bridge, connecting with interstate highways A1 and A255. Since no further crossings exist along the Elbe River, the city naturally has to cope with most of the transit traffic, e.g., from Scandinavia to Western or Central Europe.

Moreover, Hamburg is the world's eighth largest and Europe's second largest container harbor, handling more than 4.6 million twenty foot equivalent units

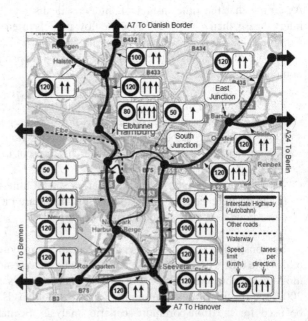

Fig. 1.10 Metropolitan Hamburg traffic network used for transportation modeling and simulation

(TEU) of inbound containers and 4.4 million TEU of outbound containers in 2011. It is assumed that this container load will double within the next 10 years. The port of Hamburg operates four container terminals:

1. Container Terminal Altenwerder (CTA): one of the most modern container handling facilities in the world. Container handling is almost entirely automated using autonomous guided driverless vehicles (AGVs) to transport the containers between the quay and the container yards. Fifteen container gantry cranes are operated at four berths for large vessels at CTA (see Fig. 1.10).
2. Container Terminal Burchardkai (CTB): the largest sea freight handling facility in the port of Hamburg. More than 5,000 ships per year are loaded and unloaded at 10 berths with 27 container gantries. It is assumed that the capacity will increase to 5.2 million TEU (nearly double the present capacity) in the coming years with twin 40-foot container cranes helping to achieve this goal by enabling the loading/unloading of two 40-foot containers in one move.
3. Container Terminal Tellerort (CTT): provides four berths with eight container gantries with the ability to handle post-Panamax size ships. The terminal has its own container rail station with 720 meters of track and three new Transtainer® cranes capable of handling block trains quickly without shunting. The Transtainer crane, especially the rubber-tired Transtainer crane (RTTC), is the key element invented by PACECO® Corp. for handling containers in a terminal. The RTTC is a rigid gantry crane mounted on large rubber tires which travels in a straight line on a horizontal path, the same as if it were running on rails.
4. Container Terminal Eurogate (CTE): connects directly to interstate highway A7. Six large-ship berths with 21 container cranes (of which 19 are post-Panamax types) and more than 140 van carriers ensure rapid handling. Handling 2.7 million TEU in 2008, the Eurogate Container Terminal in Hamburg is the second largest terminal operated by the EUROGATE Group in Germany.

Containers arriving and departing at seaport terminals require intermodal transfers. The most frequently used are vessel to truck and vice versa, closely followed by vessel to train and vice versa. Both of these transfer modes require their own infrastructure resources. Planning for these resources can be achieved using scenario analysis and an appropriate modeling and simulation environment. The selection of such an environment is based on the following criteria:

- Simulation and performance evaluation of traffic flows is essential; functionalities for traffic prediction are not required.
- The transportation network investigation can be customized in terms of topology (i.e., nodes/links), flow offered per origin–destination route, link speed, lane numbers, and capacity.
- Uncomplicated customization of these parameters for scenario analysis (e.g., increasing the number of lanes or the speed limit).
- Multimode support.
- Traffic flow visualization is desirable but not necessary.

These requirements are met by the Virtual Intermodal Transportation System (VITS) based on the ProModel® Visualize, Analyze, and Optimize (VAO) discrete-event simulator. VITS is a discrete-event traffic simulator which covers road, rail, and water mode; its methodology combines aspects of the so-called microscopic and macroscopic traffic simulation. Readers that are interested in a complete and detailed description are referred to Tan and Bowden (2004).

In transportation analysis, truck road traffic can be modeled individually, which means that for each truck, attributes denoting speed and destination can be assigned. However, for computational simplification, each of the model's truck entities can be parameterized to represent more than one truck for the purpose of road utilization and speed calculation. Given a network topology consisting of nodes (e.g., interstate highway junctions and exits as well as plants and ports and other locations important for freight traffic) and links (road, rail, or waterway segments each of which connect two nodes), trucks continuously appear at any node; and their interarrival time is assumed to be exponentially distributed with a higher mean during the daytime than at night. Each truck traverses a fixed route, i.e., a sequence of road links that depend on the origin–destination node pair assigned to the vehicle. A truck that eventually reaches its destination node is removed from the system (Wittmann et al. 2007).

Therefore, a traffic network model was developed based on a modified version of the VITS ProModel application that supports intermodal traffic. As previously mentioned, this type of traffic network simulation model is based on a number of nodes and edges describing the overall traffic network lane capacity, performance with regard to travel time, load, and more. A general model of a traffic network composed of nodes and edges is shown in Fig. 1.11.

The nodes in Fig. 1.11 are traffic chain components. Together with edges, they represent an intermodal cluster formation in transportation systems. The gray marked node in Fig. 1.11 represents the four Elbe River tunnel system (GSP) which, under specific circumstances, can become bottlenecks. The black and the white nodes represent two different types of transportation core areas as part of the assumed intermodal and multimodal transportation concepts. DP1 represents a transportation core area load center node, while DP2 characterizes a complex

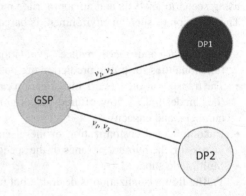

Fig. 1.11 Simple
transportation network model
based on nodes and edges

seaport freight distribution terminal node. The nodes indicate the capacity of inbound and/or outbound container load units. The edges, the lines in between the two nodes, e.g., GSP and DP1 and GSP and DP2, indicate the intermodal transportation weighting factors. Let v_1 indicate train transportation capacity from GSP to the transportation core areas, DP1 and DP2; v_2 indicate lane capacity from GSP to DP1; and v_3 characterize lane capacity from GSP to DP2. The mode and edge representation applied to the maritime transportation chains shown in Fig. 1.11 originate from the mathematical graph theory, whereby a graph is an ordered pair $G = (V, E)$ where:

- V is the vertex set whose elements are the vertices or nodes of the graph. This set is often denoted $V(G)$ or just V.
- E is the edge set whose elements are the edges, or connections between vertices, of the graph. This set is often denoted $E(G)$ or just E. Individual edges are ordered pairs (u, v) where u and v are vertices in V.
- Two graphs, G and H, are considered equal when $V(G) = V(H)$ and $E(G) = E(H)$.

The order of a graph is the number of vertices in it, usually denoted by $|V|$ or sometimes n. The size of a graph is the number of edges in it, denoted by $|E|$ or sometimes m. If $n = 0$ or $m = 0$, the graph is called empty or null. If $n = 1$, the graph is considered trivial.

This theoretical approach to graphs can be used for planning transportation traffic network infrastructure, e.g., building new traffic roads and/or new port facilities modeled through nodes, for which high-capacity transport modalities must be available for the majority of transportation chains represented by the nodes' edges. As a result of the constraints of nodes and edges, a modality shift from trucks to trains, and vice versa, can occur. Vehicles in the transportation traffic network can be modeled individually by attributes, including current location, load, speed, and final destination. Vehicles stochastically appear at any node because the interarrival time is assumed to be exponentially distributed and traverse to fixed routes, e.g., a sequence of road links represented by the nodes and edges to reach the final destination.

Bottleneck analysis is a process-related approach to identifying the dependencies in a sequence of actions to predict their impact on the transportation traffic network infrastructure and, if possible, to immediately decide how to rectify the bottleneck and optimize more transportation chains. Bottlenecks are identified using a scenario planning and analysis approach, which identifies the various impacts on optimal and/or suboptimal behavior of the transportation traffic network. Thus, bottleneck analysis is helpful in accomplishing a combination of best, worst, or real case scenario. Therefore, the results obtained through bottleneck analysis can be classified into different categories. Assuming bottleneck analysis deals with the calculation of adequate availability of resources, the resulting best case, worst case, and real case scenarios are shown in Table 1.2 (Moeller et al. 2012).

Due to complexity and constrains, such as transportation flow, time, and cost, identification of bottlenecks is not a trivial task. Identification and elimination of shortages is generally the first step in finding a possible and/or optimal solution.

Table 1.2 Different scenarios for the availability of resources

Best case scenario	Worst case scenario	Real case scenario
Resources for intermodal transportation chains are available; no shortage appears	Resources for intermodal transportation chains are not available in the required amount; and/or if worst comes to worst, only one resource is available but several are needed, i.e., a shortage appears	Resources available for intermodal transport chains are well balanced because only minor shortages occur. The achievable solution is in between the best case and worst case categories
Result: High-cost solution. Resources available cannot be used in an optimal way because more resources are available than required, meaning that this category is not an optimal solution for the effective use of required resources. This problem exists especially for railways, in comparison to trucks, as railways are not available on short notice	*Result:* Low-cost solution; resources available are not adequate	*Result:* Suboptimal solution

When a shortage is identified and rectified, it is sometimes discovered that the criteria-based function achieved is suboptimal. This is due to another, so far hidden, shortage that is identified after the first one is eliminated, which is the case with the so-called multishortage bottleneck analysis. Solving multishortage bottlenecks requires intelligent algorithms to determine the desired optimal transportation sequence.

Besides the aforementioned description, it is obvious that a bottleneck analysis can also be used if the transportation traffic network is described based on vague or fuzzy data, including relative or linguistic statements like "too congested," "not enough load," "not as fast," and "insufficient transportation traffic network," which require a specific algorithmic description. Such a description can be based on the fuzzy set theory (Kaufmann and Gupta 1991). In the case of the fuzzy set approach, the fuzzified input data are processed by the fuzzy inference machine; and the result obtained will be defuzzified. The inference machine includes the rule base, consisting of rules expressed in the form of IF . . .. THEN clauses such as

IF x1 (load capacity) is small AND x2 (congestion) is low,
THEN Y (on time delivery) is very high
....
IF x1 (load capacity) is small AND x2 (congestion) is medium,
THEN Y (on time delivery) is high.

Whenever it is necessary to change assumptions and/or parameters of the transportation traffic network, scenario analysis is the state-of-the-art method to be used to overcome problems resulting from the identification of potential shortages affecting the transportation traffic network. Scenario analysis for the Elbe River tunnels contemplates the possible transportation traffic outcome in metropolitan Hamburg while the Elbe River tunnels are closed for:

- Periodic maintenance
- Long period reconstruction work
- Demolition through an assumed fuel truck accident and resulting fire

Scenario analysis necessitates:

- Thinking about how to respond to each scenario
- Thinking about the realistic consequences
- Simulating scenarios with different responses to consider the likelihood of each scenario

Based on what is perceived to be the outcome of each scenario and basis of the plan for the next step, which requires testing to perceive a good scenario that is:

- Challenging
- Internally consistent
- Plausible to respective decision makers/stakeholders
- Recognizable from data/signals
- Relevant to topics or issues of interest
- Set in time and scope of the analysis
- Surprising or novel

For this reason, scenario planning is necessary to understand future trends in strategic decision making based on an analysis of the consequences of the most likely scenario. In doing so, at least the following facts have to be taken into account, such as:

- Driving forces
- Extremes of possible outcomes
- Key questions
- Stakeholders
- Time and scope
- Transportation flow
- Uncertainties

Assuming the transportation traffic flow can be described by trucks with containers that have been loaded at the container terminals in Hamburg harbor and driven to the final destination, the calculation of the speed of a truck along a road segment abstracts from microscopic vehicle interaction: Applying the Bureau of Public Roads (BPR) equation, speed depends on the (macroscopic) parameters of road capacity and utilization. The speed assigned to the link (from which truck speeds are derived by sampling a normal distribution) during the next period (1 h) is set such that the expected travel time $\hat{t}_i$ i required for traversing the link amounts to

$$\hat{t}_i = t_i \left[1 + a\left(\frac{x_i}{C_i}\right)^{\beta} \right]$$

subject to free-flow travel time ti (constrained, e.g., by the relevant speed limits only), link capacity Ci, and flow during the last period xi. Parameters α and β are set to 0.45 and 7.5, respectively, as suggested in Grady (2002). The flow xi is measured in terms of passenger cars and by counting trucks entering the link since the last speed update, applying an equivalence factor of 2.5 passenger cars per truck. Passenger car traffic is not modeled explicitly; the flow xi is chosen such that trucks account for 25 % of the overall traffic. Tan and Bowden (2004) intend to replace this estimation for the truck equivalence factor and passenger-car-to-truck ratio estimation with more accurate numbers (e.g., the equivalence factor depending on road (interstate highway, US highway, state highway) and terrain type (flat, rolling, mountainous)) in the future. Link capacity depends on road type, speed limit, and number of lanes and varies between 2,200 and 2,400 passenger car units per hour per lane, as suggested by Roess et al. (1990).

In contrast to road traffic, rail and water modes abstract from traffic density influencing travel times. Trains and barges appear at nodes connected to rail or waterway links and traverse each link on their route at a constant speed that is assigned to each link individually. Rail track and river capacities are assumed to suffice for any rail and barge traffic offered, thus always traversing the relevant links at the desired (maximum) speed. Despite single-mode transportation, in which trucks, rails, and barges, which appear with respect to an exponential interarrival time, traverse links on different single-mode routes. VITS also provides intermodal transfers in which routes served by different modes may be linked. For example, the freight delivered to a port by one or more trucks is loaded onto a barge so that barge departures are not sampled from a random distribution but depend on truck arrivals at the port as well as the barge-to-truck capacity ratio and the duration required for loading. Thus, interdependencies between the different modes of transportation can be traced and bottlenecks influencing the intermodal network's overall performance can be identified (Wittmann et al. 2007).

The results shown in Table 1.3 describe the network performance indicators, including the average travel time (TT) of all vehicles and of those on two chosen routes (namely, interstate highway A1 from Bremen westbound to the A24 highway to Berlin and the A7 from Hanover northbound to the Danish border, the latter including the Elbe River tunnel) as well as the average vehicle speed on two selected links (namely, the Elbe River tunnel northbound and interstate highway A1 northeast bound from South Junction to East Junction). Confidence interval half-widths, as stated in Table 1.3, refer to a confidence level of 95 %. For comparison, Table 1.3 also quotes the optimal performance indicator values

Table 1.3 Simulation results for metropolitan Hamburg VITS' speed update policy (i.e., vehicle velocity adjusted on entering new links only) versus the suggestion of instant speed updates of vehicles in between two nodes (Wittmann et al. 2007)

Speed update policy/ scenario	Average TT all vehicles (h)	Average TT A/north-bound (h)	Average TT A21/A24 westbound (h)	Average speed link river Elbe tunnel N (km/h)	Average speed link Junction SE (km/h)
VITS/1	0.58 ± 0.01	0.79 ± 0.03	0.62 ± 0.01	79.3 ± 0.1	115.4 ± 0.2
VITS/2	0.64 ± 0.02	0.93 ± 0.05	0.62 ± 0.01	70.2 ± 0.2	115.4 ± 0.2
VITS/3	0.75 ± 0.01	0.75 ± 0.03	1.00 ± 0.04	79.3 ± 0.1	108.9 ± 0.3
Inst/1	0.46 ± 0.01	0.52 ± 0.01	0.60 ± 0.01	78.4 ± 0.1	115.0 ± 0.3
Inst/2	0.47 ± 0.01	0.54 ± 0.01	0.60 ± 0.01	70.9 ± 0.2	115.1 ± 0.3
Inst/3	0.53 ± 0.01	0.52 ± 0.01	0.71 ± 0.01	78.4 ± 0.1	108.7 ± 0.3
Freeflow	0.39 ± 0.00	0.40 ± 0.00	0.54 ± 0.00	80.0 ± 0.0	120.0 ± 0.0

resulting from the assumption of all vehicles always driving at the speed limit, regardless of traffic density.

The results in Table 1.3 show that the closure of four of the eight Elbe River tunnel lanes (Scenario 2) or doubling of the west–east traffic from Bremen to Berlin (Scenario 3) deteriorates the network's performance. However, travel durations are significantly worse for the VITS speed policy than for instant speed updates (e.g., travel time from the southern to the northern model border 0.79 ± 0.03 h instead of 0.52 ± 0.01 h in Scenario 1). Since Scenario 1 traffic demand was set such that traffic density was low and that the driving speed at each link was very close to the speed limit, the VITS speed update policy (with vehicle low speeds always persisting at least until the next node is reached) has yielded a result surprisingly far from the free-flow travel time of 0.40 h. Accordingly, reducing the capacity of the Elbe River tunnel from eight to four lanes (Scenario 2), yielding an average speed declining from 79.3 ± 0.1 km/h to 70.2 ± 0.2 km/h on a link of 5 km on an otherwise unchanged route of a total length of 43.5 km results in travel time increasing yet further to 0.93 ± 0.05 h.

The traffic simulator application was developed for, but not limited to, metropolitan Hamburg to provide a tool for scenario analysis evaluating the impact of shortages due to the real-world situation involving the Elbe River tunnels. Such investigation typically includes performance measures, such as vehicle travel times, edge speeds, or throughput, resulting in a valuable decision support tool offering judgment as to whether solutions, part of the scenario analyzed, are sufficient with respect to the given target performance measures.

1.7 Exercises

1. Explain what is meant by the term modeling.
2. List and define three main characteristics in modeling a real transportation problem.

3. Explain what is meant by the term behavioral level of modeling.
4. Give an example for the behavioral level modeling.
5. Explain what is meant by the term composite-structural level of modeling.
6. Give an example for the composite-structure level modeling.
7. Explain what is meant by the term empirical modeling.
8. Give an example for an empirical modeling.
9. Explain what is meant by the term deductive modeling.
10. Give an example for a deductive modeling.
11. Explain for what test signals can be used.
12. Give the mathematical description for a unit step.
13. Give the mathematical description for a ramp function.
14. Explain what is meant by the term simulation?
15. Give an example of a simulation model of first order.
16. Why differential equations are important modeling transportation systems?
17. Give the mathematical description for a second-order differential equation and explain their coefficients and derivatives.
18. Explain the structural representation of the state-variable model in Fig. 1.3.
19. Give the block diagram of structural representation of the linear state-variable model in Fig. 1.3 with $D \neq 0$.
20. Explain what is meant by the term queuing system.
21. Give an example of a queuing system in transportation.
22. Define the five main characteristics in modeling discrete-event system.
23. Give an example for a time event.
24. Give an example for a state event.
25. A crucial part of modeling building is the quality of the model builds which describe whether or not a mathematical model fits a system accurately.
26. Give an example for proving the model quality.
27. Explain what is meant by the term verification in modeling.
28. Based on Naylor and Finger, Banks et al. formulated a three-step approach to model validation. Describe the three-step approach for model validation.

References and Further Readings

Ackoff RL (1981) Creating the corporate future. Wiley, New York
Banks J, Carson JS II, Nelson BL, Nicol DM (2001) Discrete-event system simulation. Prentice Hall International, Upper Saddle River
Carson JS (2002) Model verification and validation. In: Proceedings of the winter simulation conference, pp 42–58
Cascetta E (2009) Transportation systems analysis. Springer, New York
Grady B (2002) Review of the regional transportation commission of southern Nevada's draft 2003–2025 regional transportation plan & 2003–2005 transportation improvement plan conformity finding. Smart Mobility Inc., Norwich
Hoover SV, Perry RF (1990) Simulation—a problem-solving approach. Addison-Wesley, Reading
Ioannou P, Chassiakos A, Valencia G, Hwan C (2007) Simulation test-bed and evaluation of truck movement concepts on terminal efficiency and traffic flow, Metrans project 05–11 final report

Janssens D, Yasar A-U-H, Knapen L (eds) (2014) Data science and simulation and transportation research. IGI Global Publ., Hershey

Kaufmann A, Gupta MA (1991) Introduction to fuzzy arithmetic: theory and applications. Van Nostrand Publ., New York

Kheir NA (1995) Systems modeling and computer simulation. Marcel Dekker, Inc., New York

Kitamura R, Kuwahara M (eds) (2005) Simulation approaches in transportation analysis: recent advances and challenges. Springer, New York

Law AM (2007) Simulation modeling and analysis. McGraw-Hill International, Boston

Laszlo A, Krippner S (1998) Systems theories: their origins, foundations, and development. In: Jordan JS (ed) Systems theory and a priori aspects of perception, Chapter 3. Elsevier, New York, pp 47–74

Manheim ML (1979) Fundamentals of transportation systems analysis, vol 1. MIT Press Classic, Boston

McClamroch NH (1980) State models of dynamic systems—a case study approach. Springer, New York

Moeller DPF, Froese J, Schroer B, Anderson M (2012) Scenario planning and bottleneck analysis on intermodal maritime transportation chains in metropolitan Hamburg. In: Weber R, Waite B, Gauthier J (eds) Proceedings of AlaSim international 2012 modeling and simulation conference & exhibition. AlaSim Publ, Huntsville

Naylor TH, Finger JM (1967) Verification of computer simulation models. Manage Sci 14:92–101

Roess RP, McShane WR, Orassas EE (1990) Traffic engineering, 2nd edn. Prentice-Hall, Englewood Cliffs

Sammon JP, Caverly RJ (2007) Transportation systems. US Department of Homeland Security, USA

Sokolowski JA, Banks C (eds) (2009) Principles of modeling and simulation—a multidisciplinary approach. Wiely, Hoboken

Tan AC, Bowden RO (2004) The Virtual Transport System (VITS)—final report, Department of Industrial Engineering, Mississippi State University, MS, http://www.ie.msstate.edu/ncit/Research/VITS% 20Project.htm

Torin M (2007) "War Rooms" of the street: surveillance practices in transportation control centers. Commun Rev 10(4):367–389

Treiber M, Kesting A (2013) Traffic flow dynamics: data, models, simulation. Springer, New York

von Bertalanffy L (1976) General systems theory: foundations, development, applications. Harper, New York

Wittmann J, Göbel J, Möller DPF (2007) Refinement of the Virtual Intermodal Transportation System (VITS) and adoption for metropolitan area traffic simulation. In: Wainer GA, Vakilzadian H (eds) Proceedings SCSC 07: moving towards the unified simulation approach. SCS Publ., San Diego, pp 411–415

Links

(URL 1) www.microsimulation.drfox.org.uk

(URL 2) https://www.cs.auckland.ac.nz/software/AlgAnim/dijkstra.html

(URL 3) http://www2.econ.iastate.edu/classes/econ501/hallam/documents/CostFunctions.pdf

Transportation Models

2

This chapter begins, in Sect. 2.1, with a brief overview of the use of models in the transportation sector, several types of models used in transportation planning, and the specific evaluation methods used. Thereafter, the theory of traffic flow is introduced which enables investigation of the dynamic properties of traffic on road sections with regard to the respective variables defined at each point in space and time. Based on the mathematical equations derived, different transportation system scenarios are investigated in Sect. 2.2. Section 2.3 examines queuing theory, the mathematical study of waiting in lines or queues. Transportation system models incorporate queuing theory to predict, for example, queuing lengths and waiting times. Section 2.4 analyzes transportation systems with regard to existing demand and the potential impact of changes resulting from transportation planning and development projects. Traffic management has become a critical issue as the number of vehicles in metropolitan areas is nearing the existing road capacity, resulting in traffic congestion. In some areas, the volume of vehicles has met and/or exceeded road capacity. The methodological background of congestion is described in Sect. 2.5. Graph theory is introduced in Sect. 2.6. It is widely used to model and study transportation networks. Section 2.7 focuses on shortages occurring in transportation systems, so-called bottlenecks. The main consequence of a bottleneck is an immediate reduction in the capacity of the transportation system infrastructure. Section 2.8 describes a ProModel-based case study for a four-arm road intersection. Section 2.9 contains comprehensive questions from the transportation model area of concentration, and the final section includes references and suggestions for further reading.

2.1 Introduction

A model can be introduced as a schematic description of a real-world system, theory, or phenomenon that accounts for its known or inferred properties used for further study of its characteristics or to predict or evaluate the intrinsic

© Springer-Verlag London 2014   45
D.P.F. Möller, *Introduction to Transportation Analysis, Modeling and Simulation*,
Simulation Foundations, Methods and Applications,
DOI 10.1007/978-1-4471-5637-6_2

dynamic behavior. In the transportation system sector, models or systems of models (so-called models of models (MOM)) are used to simulate traffic performance and traffic flow. Incorporating traffic requirements, as defined by technical and/or organizational constraints of real-world transportation systems, these models are used to predict impacts and/or to evaluate possible options for transportation planning and evaluation.

Transportation planning is defined as the process of making decisions about transportation resource needs, preferences, and values. Planning occurs at many different levels from day-to-day decisions to more general major decisions to strategic decisions with long-term impacts. Best practices in transportation planning can be achieved by coordinating short-term decisions in support of strategic long-term goals. An example of such comprehensive planning is transportation infrastructure planning with regard to land use, economic development, and social planning. Another example would be when manifold potential options exist to reduce traffic congestion, and some of these solutions may also help to overcome other traffic problems, such as finding parking spots and minimizing pollution emissions. A comprehensive transportation planning process will result in the prioritization of transportation activities and the efficient allocation of resources.

In the evaluation of transportation planning activities, the evaluation itself can be used as a method of determining the value of a potential planning option in order to support decision making. That is why evaluation in transportation planning is often applied when it comes to decision making (Small 1998; Litman 2006; USDOT 2003; CUTR 2007), and there are specific evaluation methods used, such as:

- *Cost-effectiveness (CE)*: This method compares the costs of different potential options for achieving a specific objective, such as building a particular highway or delivering a particular amount of airfreight, etc. The quantity of benefits (outputs) are held constant, so there is only one variable, the cost of inputs.
- *Cost-benefit analysis (CBA)*: This method compares the total incremental benefits with the total incremental costs for each of the potential options. This analysis is not limited to a single benefit or objective, such as potential highway routes which can differ in construction costs as well as quality of the services offered.
- *Lifecycle cost analysis (LCA)*: This method incorporates, in addition to CBA, the value of investments at the respective schedule, which allows a comparison of projects with regard to their cost and benefit milestones.
- *Multiple accounts evaluation (MAE)*: This method considers quantitative and qualitative evaluation criteria and can be used in cases where some impacts cannot be financially benchmarked. Using this evaluation method each potential option is rated for each potential criterion.

In general, transportation planning is involved with the evaluation, assessment, design, and siting of transportation facilities and is based on specific transportation planning models for which the respective environmental goals and objectives are defined. Problems which call for transportation planning or are identified during the implementation of a solution require potential alternatives for development which have to be evaluated with regard to the existing budget. In this sense, the role of transportation planning is shifting from a purely technical analysis, including environmental aspects and sustainability, to a more integrated transportation framework which also embeds behavioral psychological aspects, e.g., persuading automobile drivers to use public transportation rather than their personal automobiles.

Several of the models used in transportation planning are the so-called travel demand models (TDMs) which have been developed to evaluate transportation demands in terms of the numbers of traveling individuals who may search for specific prices, transport services, modes, etc., in order to predict the corresponding traffic volumes and their potential impacts, such as congestion, pollution emissions, etc. Most TDMs are four-step models which follow these steps (TDM Encyclopedia 2013):

1. *Trip generation*: this approach predicts the total trips that start and end in a particular area of interest, the traffic analysis zone (TAZ), based on factors such as the zone's land use patterns; number of residents and jobs; demographic factors; transportation system features, such as number of roads, quality of transit service, etc.; and the distance between two zones.
2. *Trip distribution*: this approach focuses on trips that are distributed between pairs of zones, based on the distance between them.
3. *Mode split*: this approach focuses on trips that are allocated among the available travel modes.
4. *Route assignment*: this approach focuses on trips that are assigned to specific facilities included in the highway and transit transportation networks.

These models make use of travel surveys and census data to determine transportation demands, establish baseline conditions, and identify future trends. The trips used as a basis in these models are often predicted separately by purpose, i.e., work, shopping, etc., and thereafter aggregated into total trips on the respective network. This modeling approach allows the prediction of congestion problems because they mainly focus on measures of peak-period motor vehicle trips on major roadways. As a result of these predictions, a so-called level-of-service (LOS) roadway report is available with a letter grade from A (best) to F (worst) which indicates vehicle traffic speed and delay. As mentioned in TRB (2007), these models often incorporate several types of bias favoring automobile transport over other modes and undervaluing travel demand model (TDM) strategies. Because the travel surveys they are based on tend to ignore or undercount nonmotorized travel, they undervalue nonmotorized transportation improvements for achieving transportation planning objectives (Stopher and Greaves 2007). Moreover, they do not accurately account for the tendency of traffic to maintain equilibrium and the effects of traffic generated by roadway capacity expansion, thereby exaggerating future congestion problems and the benefits of roadway capacity expansion.

A number of recent studies have examined ways to better predict how smart growth locations and demand management programs can affect trip and parking generation (Lee et al. 2012). Based on the assumption that a standard application of trip rates for an area with many smart growth characteristics will result in an overestimation of the number of trips generated, this study identifies eight available methodologies, five of which are candidate methods which are compared with the traditional trip generation method in a two-part assessment.

Economic models are used to evaluate and compare the value of particular transportation improvements, such as widening a roadway, improving public transit, or implementing a TDM strategy. The models compare the various categories of benefits and costs. They tend to consider a relatively limited set of benefits, since most of these models were originally developed to evaluate roadway improvement options. They generally assume that total vehicle mileage is constant and so is not well designed to evaluate the full benefits of TDM strategies that reduce automobile trips. For example, these models often ignore parking and vehicle ownership cost savings that result when travelers shift from automobile travel to alternative modes; and they generally ignore the safety benefits that result from reductions in total vehicle mileage (Ellis et al. 2012).

Integrated Transportation and Land Use Models are designed to predict how transportation improvements will affect land use patterns, e.g., the location and type of development that will occur if a highway or transit service is improved. They are often integrated with traffic models. These are considered the best tools for evaluating transportation policies and programs because they can measure accessibility rather than just mobility, but they are costly to develop, are complex, and may be difficult to apply, particularly for evaluating individual, small-scale projects (Dong et al. 2006). Some models predict how particular land use factors, such as density and mix, affect travel behavior and their impacts on congestion and pollution emissions (Donoso et al. 2006; Scheurer et al. 2009; Bartholomew and Ewing 2009). The Smart Growth Area Planning (SmartGAP) tool synthesizes households and firms in a region and determines their travel demand characteristics based on their built environment and transportation policies affecting their travel behavior (TRB 2012).

Transportation simulation models are a newer approach to modeling the behavior and needs of individual transport users (so-called agents), rather than aggregate groups. This improves the consideration of modes such as walking and cycling; the transport demands of nondrivers, cyclists, and the disabled; and the effects of factors such as parking supply and price, transit service quality, and local land use. Simulation models can provide a bridge between other types of models, since they can incorporate elements from the conventional traffic, economic, and land use models. Simulation models have been used for many years in individual projects and are increasingly used for area-wide analysis. Transportation simulation models allow traffic flow and network flow aspects to be combined for investigation of transportation systems with continuous services, such as road systems and transportation systems with discrete services, such as airplanes, buses, ships, and trains.

To conclude, the biases in current models tend to exaggerate the benefits of roadway capacity expansion and understate the value of alternative modes and TDM solutions. More accurate and comprehensive modeling is, therefore, a key step in developing more optimal transport planning and the implementation of specific TDM strategies. Therefore, in TDM Encyclopedia (2013), the various problems common with current models and how they can be corrected are described. These deficiencies are not necessarily intrinsic; significant improvements can be made to existing models and how they are applied. For example, many problems could be reduced by simply educating planners and decisions makers about modeling assumptions, biases, and weaknesses so that they can take these factors into account.

2.2 Traffic Flow Models

The theory of traffic flow investigates the dynamic properties of traffic on road sections. Dynamic models of traffic flow date from the 1950s, representing traffic flow based on an analogy with lines of water flows in rivers, an approach that allows to treat individual vehicles as "continuous fluid." Against this background, macroscopic traffic flow theory relates on variables declaring the dynamic properties of traffic which are:

- Density k
- Flow rate q
- Speed v

These result in the fundamental statement that flow q equals density k multiplied by speed v. These variables are defined at each point in space and time which means that the discrete nature of traffic is transferred into continuous variables. The evolution in time of these state variables can be modeled by partial differential equations (PDEs) comprising the conservation of mass (vehicles) and an experimental relation between flow rate q and density k. Using this approach, traffic flow models can be formulated for density k by the number of vehicles n at time t_0 occupying a given length x of a road or, more in general, on the location interval Δx of a roadway at a particular instant, as follows:

$$k = \frac{n}{\Delta x}. \tag{2.1}$$

The total space s of the n vehicles can be set equal to Δx, and thus we can write

$$k = \frac{n}{\sum_i s_i} = \frac{1}{s}, \tag{2.2}$$

where the mean space occupancy in the interval s_i is defined as

$$\bar{s} = \frac{1}{n}\sum_n s_i. \tag{2.3}$$

From (2.3), it can be seen that density k depends on the designated roadway point x_0, the time t_0, and the measurement interval, defined as an area in the t-x space. As introduced in Immers and Logghe (2002), for a location x_1, we can take the center of the measurement interval Δx. Thus, (2.1) can be rewritten in order to include these factors:

$$k(x_1, t_1, s_1) = \frac{n}{\Delta x}.\tag{2.4}$$

2.2.1 Uncongested Traffic Conditions

For uncongested traffic conditions, freeway traffic data suggests that desired speeds are relatively constant and chosen by the drivers. Under stationary conditions, the flow-rate-versus-density ratio can be expressed as mean speed v, which appears to be nearly constant for uncongested traffic flow. Introducing the flow-rate-versus-density ratio under congested conditions causes driver behavior to become an important factor. Assuming drivers can no longer choose free-flow speed under congestion, a simple classification can define driver types: aggressive drivers T_{AD} and nonaggressive drivers T_{NAD}. Assuming that each driver type drives at his/her desired speed, the uncongested flow-rate-versus-density relationship is a weighted average of the desired speeds. With regard to such behavior, a regression of traffic flow on total density interacts with proportions of distinct driver/vehicle types T_I with $I = AD$ or $I = NAD$. This results in estimates of free-flow speeds for these drivers, described for the uncongested flow rate by the following equation (Kockelman 2001):

$$q_u = \sum_{T_I} v_{free, T_I} p_{T_I} k,\tag{2.5}$$

where q_u is the total uncongested traffic flow rate, v_{free, T_I} is the free traffic flow speed of driver/vehicle type T_I, $p_{T_I} k$ is the density of driver/vehicle type T_I, and p_{T_I} is the proportion of vehicles on the road of driver/vehicles type T_I.

2.2.2 Congested Traffic Conditions

In the case of a congested condition, the driving situation is different because speed is no longer constant for tumescent densities. Drivers can no longer choose free-flow speeds because they have to be aware of the spacing at which they follow the car in front of them. For this situation, the behavioral assumption is of selected spacing d as a linear function of congested speed v_C. Since total vehicle density k is the inverse of average spacing of vehicles on the roadway and average spacing is a proportion-weighted sum of type densities, one can solve for the total density k_T as a function of speed as shown in the following equation from Kockelman (2001):

$$k_T = \frac{1}{\sum_{T_l} p_{T_l} d_{T_l}} = \frac{1}{\sum_{T_l} p_{T_l}(a_{T_l} + b_{T_l} v)}, \qquad (2.6)$$

where d_{T_l} stands for intervehicle spacing (front-to-front) of the lth driver type, v is mean speed, and a_{T_l} and b_{T_l} are constants defining the lth driver type behavior.

2.2.3 Flow-Density and Speed-Flow Graphs

Based on the foregoing specifications and definitions, the following graphs can be introduced (Muench 2004), showing the congested and uncongested flow rate q versus density k (Fig. 2.1) and the speed v versus flow rate q (Fig. 2.2). As indicated in Fig. 2.1, the optimal traffic flow capacity q_m correlates with the inflection point k_m at which the uncongested flow rate changes into the congested flow rate, meaning the more density k increases, the more traffic flow q decreases, which can be expressed by the equation of flow rate q with v_f as free space mean speed shown in Fig. 2.1.

In Fig. 2.2, it is shown that the optimal traffic flow capacity q_m correlates with the inflection point v_m at which the uncongested free-flow speed changes into the congested flow speed. In other words, the more the flow rate q increases, the more the mean speed v decreases, which results in the equation of flow rate q with v_f as free space mean speed shown in Fig. 2.2.

Flow rate q is the interaction of density k and mean speed (stationary traffic conditions) u. Thus, flow rate q represents the number of vehicles n passing some

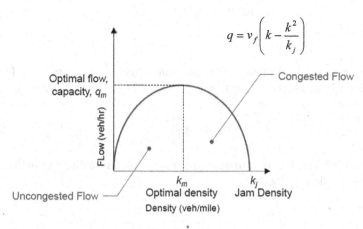

$$q = v_f\left(k - \frac{k^2}{k_j}\right)$$

Optimal flow, capacity, q_m

Congested Flow

FLow (veh/hr)

Uncongested Flow

k_m
Optimal density

k_j
Jam Density

Density (veh/mile)

Fig. 2.1 Flow rate versus density graph (Muench 2004)

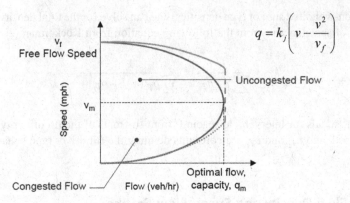

Fig. 2.2 Speed versus flow rate graph (Muench 2004)

designated roadway point x_0 in a given time interval Δt. For time interval Δt at any location x, such as measurement interval S, flow rate q be calculated as follows:

$$q(x, t, S) = \frac{n}{\Delta t}. \qquad (2.7)$$

The time interval Δt is the sum of headways h between vehicles as their bumpers pass a given point x_0:

$$\Delta t = \sum_{i=1}^{n} h_i. \qquad (2.8)$$

Introducing a mean headway $\bar{h}$, we find the following expression for the traffic flow rate q:

$$q_u = \frac{n}{\displaystyle\sum_{i=1}^{n} h_i} = \frac{1}{\bar{h}}. \qquad (2.9)$$

Mean speed v is the quotient of flow rate q and density k. Mean speed is a function of location x, time interval Δt, and measurement interval S which results in:

$$v(x, t, S) = \frac{q(x, t, S)}{k(x, t, S)}. \qquad (2.10)$$

In another form, this definition of mean speed is also called the fundamental relation of traffic flow theory:

$$q = k \cdot v, \qquad (2.11)$$

This relation links flow rate q, density k, and mean speed v. Knowing two of these variables immediately leads to the remaining third variable.

2.2.4 Traffic Flow Scenarios

Based on the mathematical equations above, we can work out some traffic scenarios as case study examples.

Scenario 1

Problem: Let us assume that a vehicle is traveling in uncongested conditions for a total distance D of 100 miles. For the first 60 miles of this distance D_1, the vehicle travels $v_1 = 55$ mph; and for the next 40 miles D_2 of the total distance, it travels $v_2 = 65$ mph. For this scenario, the weighted average speed over the time spent traveling those 100 miles is of interest to know.

Solution: Intuitively, driving at 55 mph will take longer than driving at 65 mph. Hence, the weighted average speed v_{wa} for the entire trip is less than the arithmetic mean speed v_{am} of 60 mph. Thus, we will demonstrate that this is true by the following calculations:

- 60 miles at 55 mph $= t_1 = 65.45$ min
- 40 miles at 65 mph $= t_2 = 36.92$ min

Now we can calculate the weighted average speed:

$$v_{wa} = \frac{v_1 \cdot t_1 + v_2 \cdot t_2}{t_1 + t_2} = \frac{(55\,\text{mph} * 65.45\,\text{min} + 65\,\text{mph} * 36.92\,\text{min})}{65.45\,\text{min} + 36.92\,\text{min}} = 58.6\,\text{mph}.$$

Scenario 2

Problem: Let us assume that five vehicles with different driver types T_I are driving in uncongested conditions over a given distance D of 100 miles. For each vehicle, this distance requires a different time due to the different speeds chosen by the different driver types T_I as shown in Table 2.1.

Solution: To calculate the average speed v_a, we first have to calculate the average travel time t_a as follows:

$$t_a = \frac{t_1 + t_2 + t_3 + t_4 + t_5}{5veh} = \frac{533\,\text{min}}{5veh} = 106.6\,\text{min}.$$

Table 2.1 Driving time for a given distance by different driver types

Vehicle	Time required to drive 100 miles
1	$t_1 = 80$ min
2	$t_2 = 100$ min
3	$t_3 = 133$ min
4	$t_4 = 120$ min
5	$t_5 = 100$ min

Now we can calculate the average speed v_a based on the average travel time t_a

$$v_a = \frac{D}{t_a} = \frac{100\,miles * 60\,min}{106.6\,min * h} = \frac{6{,}000}{106.6} = 56.28\,mph.$$

Scenario 3

Problem: Let us assume that 20 vehicles pass a given point x_0 in 1 min and move a length of 1 mile. For this scenario flow rate q, density k, space mean speed, space headway h_s, and time headway h_t are of interest to know.

Solution: In general, the time (in seconds) between moving vehicles, as their front bumpers pass a given point x_0, can be calculated based on (2.8) as follows:

$$t = \sum_{i=1}^{n} h_i.$$

And, therefore, flow rate can be calculated based on (2.9) as follows:

$$q = \frac{n}{\sum_{i=1}^{n} h_i} = \frac{20\,veh \cdot 60\,min}{1\,min\quad h} = 1{,}200\,\frac{veh}{h},$$

which results for density k in

$$k = \frac{v}{\Delta x} = \frac{20\,veh}{1\,mile} = 20\,\frac{veh}{mile}.$$

Calculating space mean speed has to take into account that space mean speed of vehicles moving along and traversing a roadway segment of a known length l follows (2.11) assuming v is space mean speed:

$$v = \frac{q}{k} = \frac{1{,}200\,\frac{veh}{h}}{20\,\frac{veh}{h}} = 60\,\frac{mile}{h}.$$

Now we can calculate the space headway h_S and the time headway h_T. Space headway h_S can be calculated in an idealized manner taking into account the result for density k as follows:

$$k = \frac{1}{h_i}$$

$$\bar{h}_i = \frac{1}{k} = \frac{1}{40\,\frac{veh}{mile}} = 0.025\ mile.$$

Time headway h_T can be calculated in an idealized manner taking into account the result for space headway h_S as follows:

$$\bar{h}_S = v \cdot \bar{h}_T,$$

$$\bar{h}_T = \frac{\bar{h}_S}{v} = \frac{0.025\text{mile}}{60\frac{\text{mile}}{h}} = 1.5 \text{ s}.$$

In general, traffic measurements are executed at a fixed location x_f which allows an easy measure of occupancy o. As introduced in Immers and Logghe (2002), the relative occupancy o_R of a vehicle in measurement interval S and time interval Δt can be calculated as follows:

$$o_R(x, t, S) = \frac{1}{\Delta t} n \sum o. \tag{2.12}$$

Assuming all vehicles have the same length l_V, then the relative occupancy o_R and density k can be given as follows (Immers and Logghe 2002):

$$o_R(x, t, S) = l_V k(x, t, S). \tag{2.13}$$

Scenario 4
Problem: Let a traffic stream have a mean speed v of 50 mph and a flow rate q of 1,000 vehicles/h. All vehicles are assumed to be 5 m in length l_V. What is the relative occupancy?
Solution: From (2.11) we receive

$$k = \frac{q}{v} = \frac{1,000\frac{vehicle}{h}}{50\,\text{mph}} = 20\frac{vehicles}{\text{mile}}.$$

Given that density k is 20 vehicles/mile means that k corresponds with space occupancy o_S of 80,465 m per vehicle. With an assumed vehicle length l_V of 5 m, the corresponding relative occupancy o_R is 6.21 %. Calculating the relative occupancy o_R by using (2.13) gives

$$o_R(x, t, S) = l_V k = 0.005\,\text{km} \cdot 0.62137\frac{\text{mile}}{\text{km}} \cdot 20\frac{vehicles}{\text{mile}} = 6.21\,\%.$$

It should be noted that this formula cannot be used in real-world applications because a traffic stream is neither homogeneous nor stationary in reality. A possible solution calculating traffic density is to measure the traffic flow rate and traffic mean speed using the equations given in Immers and Logghe (2002) and then calculate traffic density by using the fundamental relation of traffic flow theory in (2.11).

2.2.5   Traffic Flow Behavior

With regard to the level of detail the models use to represent the traffic flow behavior, the models can be classified, as introduced in Chap. 1: macroscale as macroscopic traffic flow models, representing the traffic behavior at an aggregated level; microscale as microscopic traffic flow models, representing the movement of individual vehicles; and mesoscale models representing traffic flow at the level of detail of a single vehicle.

Macroscopic traffic flow models can be described as:

- *Space continuous models*: where state variables are defined at each point in space
- *Space discrete models*: where basic variables affecting link performance, such as density or speed, do not vary along the link

With regard to the control flow in the macroscale model, there is no consideration of detailed individual transportation units. Using the fundamental relation of traffic flow theory in (2.11) to describe the changes in time and location of the macroscopic variables along a road:

$$q(x, t) = k(x, t) \cdot v(x, t). \qquad (2.14)$$

Let the road to be modeled be divided into cells with length Δx, and the density of cell i at time tj is represented by $k(i,j)$; then the number of vehicles in cell i is $k(i,j).dx$. Then, one time interval Δt later, at $tj + 1$, density will change. Let us assume that a number n of vehicles have traveled from cell i-l into cell i which results in a traffic inflow of

$$q(i - 1, j) \cdot \Delta t, \qquad (2.15)$$

and a number n of vehicles have traveled from cell i to cell $i + 1$ which results in the traffic outflow

$$q(i, j) \cdot \Delta t. \qquad (2.16)$$

Let us also assume there are branching and exit roads which will enable in- and outflows in the form of

$$z(i, j) \cdot \Delta x \cdot \Delta t, \qquad (2.17)$$

where z is expressed per time and length of unit and is positive for an increase in the number n of vehicles.

Let the limit for time step Δt and cell length Δx approach zero, and we can write the partial differential equation representing the conservation law of traffic as follows:

$$z(x,t) = \frac{\partial k(x,t)}{\partial t} + \frac{\partial q(x,t)}{\partial x}. \tag{2.18}$$

Let the traffic flow be stochastic. This requires a stochastic model because the variables of the traffic flow cannot be described as a deterministic process but could be described as a stochastic process:

- Sequence of vehicle arrivals (arrival pattern)
- Sequence of service times at maintenance check of vehicles (service pattern)
- Queuing behavior

Let arrivals and services be independent, randomly distributed variables with time constant parameters. Let N be a random variable describing the queue length and n realizations of N. Let the queuing phenomena be defined by the following notation:

$$A/B/c(d,e)$$

where A denotes the type of arrival pattern variable describing time intervals between two successive arrivals, B denotes the type of service pattern, c is the number of service stations, d is the queue storage limit, and e denotes the queuing behavior, such as *FIFO* (First In First Out), *LIFO* (Last In First Out), etc., where d and e, if defined by ∞ (no constraint on maximum queue length) and by *FIFO*, are generally omitted. This representation in traffic flow models allows, with the help of queuing analysis, to determine how long it takes to complete a trip and/or how long it would have taken if there was no queuing, congestion, etc., which refers to the topic of Sect. 2.3, Queuing Models.

2.3   Queuing Models

Queuing theory is the mathematical study of waiting lines or queues. In queuing theory, a model is constructed so that queue lengths and waiting times can be predicted (Sundarapandian 2009). Queuing theory is generally considered a branch of operations research, a discipline which deals with the application of advanced analytical methods to help make better decisions, as the results are often used when making business decisions about the resources needed to provide a service. Thus, planning efficient transportation systems and networks is a crucial factor for urban insertion since it gives access to economic activity, facilitates family life, and more. Hence, the importance of transportation in human life and global economy cannot be overemphasized, and queuing theory can help to study traffic behavior near a certain section where demand exceeds available capacity. Queuing can be discovered in many common situations like boarding a bus or a train or a plane, freeway bottlenecks (see Sect. 2.7), etc. In transportation engineering, queuing can occur at red lights, stop signs, bottlenecks, or any traffic-based flow constriction. When not dealt with properly, queues can result in severe network congestion or gridlock conditions, therefore making them important to be studied and understood by engineers. For example, based on the departure and arrival pair data, the delay of every individual vehicle can be determined. Using an input-output queuing diagram, it is possible to determine the delay for every individual vehicle: the delay of the ith vehicle is time of departure-time of arrival $(t_d - t_a)$. Thus, the total delay is the sum of the delays of each vehicle.

Let us assume that the traffic flow q to and through an intersection is controlled by a traffic light sequencer. This can be accomplished by analyzing the cumulative flow of vehicles as a function of time. As it is known from traffic flow operation experience with traffic light sequencing, traffic lights change in the following sequence:

$$green \rightarrow amber \rightarrow red \rightarrow amber \rightarrow green$$

whenever, e.g., a person pushes a button. Let us assume the light is red and the traffic flow is stopped from time t_1 to t_2 during the red signal interval. At the start of the green interval (t_2), traffic begins to leave the intersection, with the so-called saturation traffic flow rate q_{Sat}, and continues until the queue is exhausted. Thereafter, the departure rate $D(t)$ equals the arrival rate $A(t)$ until t_3, which is the beginning of the next red signal. At this point, the process starts all over. The resulting type of traffic flow is called interrupted flow. Interrupted traffic flow is a flow regulated by an external means, such as a traffic signal. Under interrupted traffic flow conditions, vehicle-vehicle interactions and vehicle-roadway interactions play a secondary role in calculating the traffic flow. For interrupted traffic flow, the following impacts can be identified:

- Determining the optimal cycle length and phase length for traffic lights with regard to the daytime-dependent numbers of vehicles moving in different possible directions at the respective crossing
- Evaluating consequences, adding lanes, or changing the geometric configuration of an interstate highway on recurrent (peak period) and nonrecurrent (incident happens) delays
- Optimizing the frequency at which trucks should be dispatched along a route, taking cost of operation and service quality into account

Contrary to the interrupted traffic flow is the uninterrupted traffic flow which depends on vehicle-vehicle interactions and interactions between vehicles and the roadway situation. For example, vehicles traveling on an interstate highway are moving in an uninterrupted traffic flow so long as no congestion occurs as part of an accident on the interstate highway.

Thus, the dominant effect of queuing theory in transportation is the delay of a trip from an initial destination to a final destination, measured as

- Time in system
- Average speed
- Waiting time

Therefore, queuing analysis allows determining how long it will take to complete a trip assuming an uncongested situation and how long it would have taken if a queuing or congestion situation is assumed. For these cases, the performance measures, predicted with queuing models, are:

- Throughput rate at which vehicles proceed through the highway system, which means, in terms of transportation, how long a trip will take from the place of departure to the final destination.
- Crowding/congestion is the separation between or density of vehicles, which means, in terms of transportation, a specific number of cars and/or trucks are moving in their respective lanes on a highway from the place of departure to the final destination.
- Queue percentage refers to the number of vehicles that encounter a queue prior to traveling, which means, in terms of transportation, congestion happens where vehicles have to wait before they can drive on the highway from the place of congestion to the final destination.
- Transportation cost is the annual or per customer expense of providing transportation service, which means, in terms of transportation, the transportation ticket bill as part of public transportation from the place of departure to the final destination.
- Productivity of transportation depends on the amount of queuing and whether the transportation system is saturated, which means, in terms of transportation, a highway or roadway, which is congested every morning by vehicles commuting into a metropolitan area and leaving the same way in the evening to go back home, is saturated for peak traffic. The degree of saturation is:
  - Under saturated: $\lambda < \mu$.
  - Saturated: $\lambda = \mu$.
  - Oversaturated: $\lambda > \mu$.
  with the following notation:
  - Arrival rate (vehicles per unit time): λ.
  - Departure rate (vehicles per unit number): μ.

2.3.1 Little's Law

Let the average queue size (measured in vehicles) equal the arrival rate (vehicles per unit time) multiplied by the average waiting time (both delay time in queue and activity time (in units of time)); then the result is independent of particular arrival distributions, which is known as Little's Law (Little and Graves 2008). This law says that, under steady-state conditions, the average number of vehicles in a queuing system equals the average rate which vehicles arrive multiplied by the average time that a vehicle spends in the system. Letting

- L: average number of vehicles/customers in the queuing system
- W: average waiting time in the system for a vehicle/customer
- λ: average number of vehicles arriving per unit time

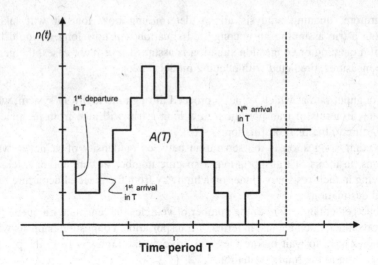

Fig. 2.3 Number of vehicles in a queuing system versus time (Little and Graves 2008)

the resulting law is called Little's Law:

$$L = \lambda W. \tag{2.19}$$

This equation is remarkably simple, extremely useful, and handy for back in the envelope calculations. The reason is that two of the terms in (2.19) may be easy to estimate but not the third. Thus, Little's Law provides the missing value (Little and Graves 2008).

In Fig. 2.3 we follow an example given in Little and Graves (2008) which shows a possible realization of a queuing system. With regard to Little's Law, one can make a heuristic argument interpreting the area under the curve in Fig. 2.3 in two different ways:

Let

- $n(t)$: number of vehicles in the queuing system at time t
- T: long period time
- $A(T)$: area under the curve $n(t)$ over the time period T
- $N(T)$: number of arrivals in the time period T

On the one hand, a vehicle in the queuing system is simply there. The number of items can be counted at any instant of time t to give $n(t)$. Its average value over T is the integral of $n(t)$ over T, meaning $A(T)$, divided by T. On the other hand, at time t, each of the vehicles is waiting and is accumulating waiting time. By integrating $n(t)$ over the time period T, we obtain a cumulative measure of the waiting time, again equal to $A(T)$. Furthermore, the arrivals are countable too and given by $N(T)$. Therefore, from Fig. 2.3, we can define

- $C = \frac{N(T)}{T}$: arrival rate during the time period T
- $L(T) = \frac{A(T)}{T}$: average queue length during time period T, indicating the number of customers in the system at time T
- $W(T) = \frac{A(T)}{N(T)}$: average waiting time in the system per arrival during T

A slight manipulation of Little's Law in (2.19) gives

$$L(T) = \lambda(T)W(T). \tag{2.20}$$

All of these quantities wiggle around a little as T increase because of the stochastic nature of the queuing process and because of end effects. End effects refer to the inclusion in $W(T)$ of some waiting by vehicles/customers which joined the system prior to the start of T and the exclusion of some waiting by vehicles/customers who arrived during T but have not left yet. As T increases, $L(T)$ and $\lambda(T)$ go up and down somewhat as vehicles/customers arrive and later leave.

Under appropriate mathematical assumptions about the stationarity of the underlying stochastic processes, the end effects at the start and finish of T become negligible compared to the main area under the curve. Thus, as T increases, these stochastic wiggles in $L(T)$, $\lambda(T)$, and $W(T)$ become smaller and smaller percentages of their eventual values so that $L(T)$, $\lambda(T)$, and $W(T)$ each go to a limit as T increase to infinity. Then, using the obvious symbols for the limits, we receive

$$\lim_{T \to \infty} L(T) = L; \quad \lim_{T \to \infty} \lambda(T) = \lambda; \quad \lim_{T \to \infty} W(T) = W$$

from which we get the desired result for (2.19). It is important to note the equation holds for each realization of the queuing system over time. This was argued by Little, in his original paper in 1961, noting that (2.19) held for each evolution of the time series of a particular queuing system (Little and Graves 2008).

2.3.2 Queuing Systems Attributes and Disciplines

Since the key elements of a transportation queuing system are vehicles and arrivals and/or departures, a queuing system can be described by the following attributes:

- Calling population, which represents the population of potential vehicles who may have called for an arrival and/or departure
- System capacity, which is the limit in numbers of vehicles that the queuing model can accommodate at any time
- Composition of arrivals and/or departures, which can occur at scheduled times or at random times

- Queuing discipline, which is the behavior of the queue in reaction to its current state
- Service mechanism, which means that service times may be constant or of some random duration.

Therefore, queuing models gain information about characteristic quantities that describe the workload of the transportation system or the time the activity needs to pass through the system. Against this background, the intention of using queuing models in transportation is to gain information about characteristic quantities that describe traffic flow, traffic density, etc., or the time a traffic flow needs to pass a distance, e.g., turnaround of an aircraft at an airport, across a flow interruption point, etc. The ways activities are processed through queues are based on specific queue disciplines which refer to the rule that a server uses to choose the next customer from the queue (if any) when the server completes the service of the current customer. Commonly used queue disciplines are:

- First come, first served (FCFS) or first in, first out (FIFO): means that customers (vehicles, passengers, etc.) are served one at a time and that the customer that has been waiting the longest is served first.
- Last come, first serve (LCFS) or last in, first out (LIFO): means it also serves customers (vehicles, passengers, etc.) one at a time; however, the customer with the shortest waiting time will be served first.
- Sharing: means activity capacity is shared equally between customers (vehicles, passengers, etc.).
- Priority: means customers (vehicles, passengers, etc.) with high priority are served first. Priority queues can be of two types: non-preemptive (activity in service cannot be interrupted) and preemptive (activity in service can be interrupted by a higher priority activity).
- Shortest activity first: means the next activity to be served is the one with the smallest size.
- Preemptive shortest activity first: means the next activity to be served is the one with the original smallest size.
- Shortest remaining processing time: means the next activity to be served is the one with the smallest remaining processing requirement.
- Round robin scheduling (RRS): means time slices are assigned to each activity in equal portions and in circular order, handling them all without priority; also known as cyclic executive.
- Multilevel feedback: means a scheduling algorithm which meets the following design requirements for multimode systems:
  - Gives preference to short activities.
  - Gives preference to I/O bound processes. This means it refers to a condition in which the time it takes to complete an activity is determined principally by the period spent waiting for input or output services to be completed.
  - Separates processes into categories based on their need for services.
- Service in random number: means random numbers are generated in a predictable fashion using a mathematical formula announcing the sequence of services.

Scenario 5

Problem: Let a transportation system have the following elements: a calling population, a waiting line, and services. Let calling population be infinite, i.e., if a vehicle leaves the calling population and joins the waiting line or enters service, there is no change in the arrival rate of other vehicles that may need service. Arrivals for service occur one at a time using a randomized schedule; once they join the waiting line, they are eventually served. In this transportation model, service times are assumed to be of some random length according to a probability distribution that does not change over time. Assume that system capacity has no limit, meaning that any number of vehicles can wait in line. Furthermore, the vehicles should be served in the order of their arrival by a single server, which results in the first come, first served (FCFS) or first in, first out (FIFO) service schedule.

Let arrivals and services be defined by the distribution of the time between arrivals and the distribution of the service times, respectively. For any simple transportation queue, the overall effective arrival time has to be less than the total service rate; otherwise, the waiting line will grow without bounds. If queues grow without bounds, they are called explosive or unstable. In cases where the arrival time will be for short terms greater than the service rate, there is a need for queuing networks with routing capabilities.

Queuing systems can be represented by terms such as state, event, simulation clock, etc. Hence, the state of the queuing system is represented by its number of vehicles as well as the state of the activity (server), which can be busy or idle. An event then represents a set of circumstances that causes an instantaneous change in the state of the system. There are only two possible events that can affect the state of the transportation system: the arrival event, which means the entry of a vehicle into the system, and the departure event, meaning the completion of and activity (service) on a vehicle. Furthermore, a simulation clock is used to track simulated time.

Solution: If a vehicle enters a discrete-event transportation system, the vehicle can find activity (server) either busy or idle, which results in two possible cases:

1. Vehicle begins with activity (service) immediately if the server is idle.
2. Vehicle enters queue for activity (server) immediately if server is busy.

It is not possible for the server to be idle and the queue to be empty, which can be interpreted as a third case. The results of which can be expressed in a matrix form for the potential unit actions upon arrival, as shown in Table 2.2.

After completing a service, as shown in Table 2.2, the server can become idle or remain busy with the next unit. The relationship of these two outcomes of the state of the queue is shown in Table 2.3. If the queue is not empty, another unit can enter

	Queue status	
Server status	Not empty	Empty
Busy	2	2
Idle	3	1

Table 2.2 Cases of unit actions upon arrival (for details see text)

Table 2.3 Server outcomes of Table 2.2 after service completion (for details, see text)	Server status	Queue status	
		Not empty	Empty
	Busy	1 or 2	3
	Idle	3	1 or 2

the server keeping him busy; or if the queue is empty, the server will be idle after a service is completed, which is indicated by the disjunctive indication of case 1 or 2. Again, it is impossible for the server to become busy if the queue is empty when a service is completed, which is indicated by case 3.

Simulating queuing systems requires the stipulation of an event list for determining what will be next. This event list tracks the future times at which different types of events occur. Hence the simulation system is able to calculate the respective simulation clock time, e.g., for arrivals and departures. If events occur at random times, the randomness needed can be realized through random numbers. A random number is a number generated by a process, whose outcome is unpredictable and which cannot be subsequentially reliably reproduced. This definition works fine provided that one has some kind of a black box that fulfills this task. Random numbers have the following properties:

• The set of random numbers is uniformly distributed between 0 and 1.
• Successive random numbers are independent.

When used without specific meaning, the word random usually means random with uniform distribution. A uniform distribution also known as a rectangular distribution is a distribution that has constant probability. A transformation which transforms from a two-dimensional continuous uniform distribution to a two-dimensional bivariate normal distribution or complex normal distribution is the Box-Muller transformation which allows pairs of uniform random numbers to be transformed to corresponding random numbers having a two-dimensional normal distribution. Random numbers can be generated with the respective queuing system simulation tools. When generating random numbers over some specified boundary, it is often necessary to normalize the distributions so that each differential area is equally populated.

Scenario 6
Problem: Let the transportation system in Scenario 5 have interarrival times and service times that can be generated from the distribution of random variables. Consider having seven vehicles with the interarrival times 0, 2, 6, 4, 3, 1, 2. Based on the interarrival times, the arrival times of the seven vehicles in the queuing systems result in 0, 2, 8, 12, 15, 16, 18.

Solution: Due to these boundaries, the first vehicle arrives at clock time 0, which sets the simulation clock in operation. The second vehicle arrives two time units later at clock time 2, the third vehicle arrives six time units later at clock time 8, etc. The second time values of interest in Scenario 5 are activity (service) times that are generated at random from a distribution of activity (service) times. Let the possible activity (service) times be one, two, three, and four time units. Hence, we are able to mesh the interarrival times and the activity (service) times, simulating the simple transportation queuing system.

In this queuing model, the first vehicle arrives at clock time 0 and activity (service) starts immediately, which requires four time units. The second vehicle arrived at clock time 2, but activity (service) could not begin until clock time 4. This occurred because vehicle 1 did not finish activity (service) until clock time 4. The third vehicle arrives at clock time 8 and is finished at clock time 10, etc. The strategy that serves vehicles in Scenario 6 is based on the *first come, first served (FCFS) or first in, first out (FIFO)* basis, which keeps track of the clock time at which each event occurs.

Furthermore, the chronological ordering of events can be determined as records of the clock times of each arrival event and of each departure event, depending on the vehicle number. The chronological ordering of events is needed as a base concept for the realization of discrete-event simulation systems.

2.3.3 Queuing Systems Parameters and Performance Measures

Further interesting parameters for queuing systems are the:

- Workload, which represents the percentage of the simulation time a resource was working
- Throughput, which is the number of vehicles per time unit that leave the system
- Mean waiting time
- Mean time in system
- Queue length
- Mean number of waiting vehicles

Moreover, knowledge of the layout of the queuing networks is of importance for the use of discrete-event simulation systems. The layout depends on:

- Open-queuing systems, which have sources and sinks. The jobs pass through the queuing net and leave it when all demands are satisfied. Typical examples of open-queuing systems are production lines, where the jobs are the raw materials that have to be processed using certain operations and leave the system as ready-made products.
- Closed-queuing systems, which are identified by a closed loop in which the jobs move through the queuing net. The number of jobs is fixed for the whole simulation time. A typical example of a closed-queuing system is a multiuser system with n terminals and a single central processing unit (CPU). The jobs circle between the terminals and the CPU; their number stays constant during the simulation time.

Simulating queuing systems generally requires maintaining data specifying the dynamic behavior of the discrete-event system, which can be done using simulation tables designed for the problem being investigated. Hence, the content of the simulation table depends on the system and can give answers such as:

- The average waiting time of a vehicle is determined by the total time the vehicles wait in the queue divided by the total number of vehicles.
- The average time a vehicle spends in the queuing system is determined by the total time the vehicles spend in the queuing system divided by the total number of vehicles.
- The average service time is determined by the total service time divided by the total number of customers.
- The average time between arrivals is determined by the sum of all times between arrivals divided by the number of arrivals − 1.
- The probability a vehicle has to wait in the queue is determined by the number of vehicles who wait in queue divided by the total number of vehicles.
- The fraction of idle time of the server is determined by the total idle time of the server divided by the total runtime of the simulation.

Moreover, it has to be decided whether:

- It is possible to leave the queue without being served at all.
- The number of jobs in the queue is limited.
- There are priorities for the jobs (static and/or dynamic).
- It is possible for a job with high priority to interrupt the service for a low-priority job and to occupy the service station immediately when entering the queue.

There are measures of performance for queuing systems available, but, with regard to the complexity of the queuing system investigated, some of which are not well defined.

Let D be the delay in queue of the ith customer, $W_i = D_i + S_i$ be the waiting time in system of the ith customer, $Q(t)$ be the number of customers in queue at time t, and $S(t)$ be the number of customers in system at time t. Then the measures

$$d = \lim_{n \to \infty} \frac{\sum_{i=1}^{n} D_i}{n}$$

and

$$w = \lim_{n \to \infty} \frac{\sum_{i=1}^{n} W_i}{n}$$

are called steady-state average delay d and steady-state average waiting time w. Similarly the measures result in

$$Q = \lim_{T \to \infty} \frac{\int_0^T Q(t)dt}{T}$$

and

$$L = \lim_{T \to \infty} \frac{\int_0^T L(t)dt}{T}$$

and are called steady-state time-average number in queue Q and steady-state time-average number or queue length L.

The most important equations for queuing systems among others are

$$Q = \lambda D,$$

and

$$L = \lambda W. \qquad (2.19)$$

These equations hold for every queuing system for which D and W exist. Equation 2.19 is the Little formula.

2.3.4 Kendall's Notation

Queuing systems offer a standard notation which can hold the following characteristics:

- s servers in parallel and one FIFO queue feeding all servers.
- $A_1, A_2, \ldots$ are random variables.
- $S_1, S_2, \ldots$ are random variables.
- A_i and S_i are independent.

Such a queue is called $GI/G/s$ queue, where GI (general independent) refers to the distribution of the A_i,s and G (general) refers to the distribution S_i,s.

If specific distributions are given for the $A_i s$ and $S_i s$, symbols denoting these distributions are used in place of GI and G. Thus, e.g., symbol M is used for the exponential distribution because of the Markovian, i.e., memory loss, property of the exponential distribution, the symbol E_k for a k-Erlang distribution, and D for deterministic (or constant) times. For any $GI/G/s$ queue, the quantity

$\rho = \lambda/sw$—with **sw** as service rate of the system when all servers are busy—is called utilization factor of the queuing system. Thus a single-server queuing system with exponential interarrival times and service times and a FIFO queue discipline is called M/M/1 queue, following Kendall's notation which was introduced to standardize the description of queuing models. Kendall introduced a notation for queuing systems, which includes information about the processes, such as job arrivals, and the distribution of the time that is needed in the server. This standard notation is based on a five-character code

$$A/B/c/N/k, \tag{2.21}$$

where A represents the interarrival time distribution, B is the service time distribution, c is the number of parallel servers of a station ($c \geq 1$), N represents the system capacity, and k is the size of the population.

The elements of queues and servers are represented in the term "station." Hence, a station can be described using Kendall's notation as

$$A/B/c- < strategy > [pre - emptive]\,[\text{max } imal\, queue - length] \tag{2.22}$$

The short forms for the mostly used distributions of queuing systems are:

- G: general (no limitation concerning the distribution)
- D: deterministic
- M: exponential distribution

It should be mentioned that the aforegoing discussed performance measures can also be analytically computed for $A/B/c$ queues with $c \geq 1$.

Scenario 7
Let Kendall's notation be used as follows:

1. $M/D/1$: represents the simplest example, the FCFS/FIFO principle.
2. $M/G/2$: represents a so-called preemptive systems example, the LCFS/LIFO principle.
3. $MM/1/\infty/\infty$: indicates a single-server system with unlimited queue capacity and infinite calling population. Interarrival times and service times are exponentially distributed.

Queuing systems typically have two states of behavior, short-term or transient, followed by long-term or steady-state behavior. If a queuing system is started, it must operate for a period of time before reaching steady-state conditions. A discrete-event simulation model of a queuing system must run for a sufficiently long period of time to exceed the transient period before measures of steady-state performance can be determined, which results in a specific notation for queuing systems containing:

- Steady-state probability of having n vehicles in system
- Probability of n vehicles in system at time t

- Arrival state
- Effective arrival state
- Effective rate of one server
- Server utilization
- Interarrival time between vehicles $n-1$ and n

Based on this notation for the various classes of queuing system models, a performance analysis can be introduced based on steady-state parameters for:

1. $M/M/1$ queues
2. $M/G/1$ queues
3. $M/E_k/1$ queues
4. $M/D/1$ queues
5. $M/M/1/N$ queues

For the first three queues, the service times are exponentially distributed for M, generally distributed for G, and Erlang distributed for E. For the fourth case, D, the service times are constant. For $M/M/1/N$ queues, the system capacity is limited to N; and for $M/M/c$ queues, the channels c operate in parallel.

The exponential distribution can be characterized as follows: Let X be an absolute continuous random variable. Let its support—the set of values that the random variable can take—be the set of positive real numbers:

$$R_x = [0, \infty).$$

Let $\lambda \in \mathfrak{R}_{++}$. We say that X has an exponential distribution with parameter λ if its probability density function is:

$$f_x(x) = \begin{cases} \lambda \exp(-\lambda x) & \text{if } x \in \mathfrak{R}_x \\ 0 \end{cases} \tag{2.23}$$

where parameter λ is called rate parameter.

A random variable having an exponential distribution is also called an exponential random variable.

The Erlang distribution is a continuous probability distribution which was developed to examine the number of telephone calls which might be made at the same time to the operators of the switching stations. This work on telephone traffic engineering has been expanded to consider waiting times in queuing systems. Erlang-distributed random numbers can be generated from uniform distribution random numbers ($U \in (0,1)$) using the following formula:

$$E(k, \lambda) \approx -\frac{1}{\lambda} \ln \prod_{i=1}^{k} U_i. \tag{2.24}$$

Simulation of queuing systems is often done manually, based on simulation tables. One has to decide, comparing the difference between possible analytical and

Table 2.4 Advances and limitations for analytical and simulation solutions

Solution	Advantages	Limitations
Analytical	Results which are general for use with all possible parameterizations	Preconditions, concerning the distribution of the interarrival times and time to be served
		Substantial problems, to handle queuing strategies
		Numerical efforts, to solve the state equations
		Results only for the steady state
		Only mean values, no predictions about the minimum and the maximum or the history of individual jobs
Simulation based	No preconditions concerning the distributions	A single simulation run only corresponds to a single random sample, all simulation results are singular solutions for the given initial state, and they are not general results for the whole model
	Any strategy can be reproduced	
	Observation of the individual history for jobs and queue lengths possible	

simulative solutions, which of the two methods should be used. This comparison can be restricted, reflecting limitations and advances (see Table 2.4).

2.3.5 Inventory System

Another important class of simulation problems of queuing systems involves inventory systems. An inventory system has a periodic review of length at which time the inventory level is observed, and an order that is made to bring the inventory up to a specified level of amount in inventory. At the end of the review period, an order quantity is placed.

Problem: Let us consider an inventory problem that deals with the purchase and sale of parts. The part sellers may buy the parts for 30 US$ each and sell them for 50 US$ each. Parts not sold at the end of the month are sold as scrap for 5 US$ each.

Solution: The problem to be solved with this inventory system is to determine the optimal number of parts the part seller should purchase, which can be done by simulating the demands for a month and recording the profits from sales each day. The profit P can easily be calculated as follows:

$$P = \left(\begin{matrix} sales \\ revenue \end{matrix} \right) - \left(\begin{matrix} cost\ of \\ parts \end{matrix} \right) - \left(\begin{matrix} profit\ loss \\ excess\ demand \end{matrix} \right) + \left(\begin{matrix} salvage\ sale \\ scrap\ parts \end{matrix} \right).$$

$$(2.25)$$

Based on the aforegoing example, the primary measure of the effectiveness of inventory systems, which are total system costs, can be extracted. Contributing to total inventory cost are the following:

- Item cost which represents the actual costs of the Q items acquired.
- Order costs which are the costs of initiating a purchase or production setup.
- Holding costs which are the costs for maintaining items in inventory.
- Shortage costs represent the costs of failing to satisfy demand.

In general, inventory problems of the type discussed above are often easier to solve then queuing problems.

2.3.6 Simulation Languages

Furthermore, discrete-event simulation of queuing models is based on simulation languages, which use programming languages. Assume that a model consists of two events: customer arrival and service completion. The events can be modeled with event subroutines, which are *ARRIVE* and *DEPART*, respectively. These subroutines contain an *INCLUDE* statement and can be described with generalized statements as follows:

```
SUBROUTINE ARRIVE
INCLUDE ´mm1.dc1´
...
Schedule next arrival
....
IF (SERVER.EQ.BUSY) THEN
.....
END
SUBROUTINE DEPART
INCLUDE ´mm1.dc1´
...
Check whether the queue is empty or not
....
IF (NIQ.EQ.0) THEN
.....
SERVER = IDLE
....
ELSE
Queue is not empty
NIQ=NIQ+1
....
END
```

2.3.7 Probability in Queuing Systems

In simulating queuing systems, the modeler sees a probabilistic world. The time it takes a system to fail, e.g., a traffic light system at a road intersection, is a random variable, as is the time it takes maintenance to repair the road intersection traffic light system. Thus, modeling probabilistic problems requires skills in recognizing the random behavior of the various phenomena that must be incorporated into the model, analyzing the nature of these random processes, and providing appropriate mechanisms in the model to mimic the random processes.

If X is a variable that can assume any of several possible values over a range of such possible values, X is said to be a random variable.

Let X be a variable in which the range of possible values is finite or countable infinite. For $x_1, x_2, \ldots$, the probability mass function of X is

$$p(x_i) = P(X = x_i) \tag{2.26}$$

$$p(x_i) \geq 0 \text{ for all } i$$
$$\sum_i p(x_1) = 1. \tag{2.27}$$

Assume X is a continuous random variable in which the range of possible values is the set of real numbers $-\infty < x < \infty$. If $f(x)$ is the probability density function of X, then

$$P(a \leq X \leq b) = \int_a^b f(x)dx \tag{2.28}$$

$$f(x) \geq 0 \text{ for all } x \text{ in } \Re$$
$$\int_{\Re_x} f(x) = 1. \tag{2.29}$$

The expected value of the random variable X is given by

$$E(X) = \sum_i x_i p(x_i) \quad \text{if } X \text{ is discrete,} \tag{2.30}$$

and by

$$E(X) = \int_{-\infty}^{\infty} x(x)dx, \tag{2.31}$$

if X is continuous. The expected value is also called the mean, denoted by μ. Defining the nth moment of X results in the variance of the random variable X

$$V(X) = E\left[(X - E(X))^2\right] = E\left[(X - \mu)^2\right]. \qquad (2.32)$$

Random variables can be based on continuous distributions or discrete distributions that are used to describe random phenomena. The focus of using distribution functions is analyzing raw data and trying to fit the right distribution to that data by answering four basic questions about the data to help in its characterization:

1. First question: relates to whether the data can take on only discrete values or whether the data is continuous.
2. Second question: focuses at the symmetry of the data and if there is asymmetry, which direction it lies in. In other words, are positive and negative outliers equally likely or is one more likely than the other?
3. Third question: looks whether there are upper or lower limits on the data; there are some data items like revenues that cannot be lower than zero, whereas there are others like operating margins that cannot exceed a value (100 %).
4. Fourth question: relates to the likelihood of observing extreme values in the distribution; in some data, the extreme values occur very infrequently, whereas in others, they occur more often (URL 1).

For continuous distributions some of which one can use are the:

- Erlang distribution *erlang(p)*: is a continuous probability with wide applicability primarily due to its relation to the exponential and Gamma distributions. The Erlang distribution was developed to examine the number of telephone calls which might be made at the same time to the operators of the switching stations, and has been expanded to consider waiting times in queuing systems.
- Exponential distribution *expo(β)*: fit, evaluate, and generate random samples with regard to interarrival times of vehicles/customers to a system that occur at a constant rate and time to failure of a piece of a component. Parameter β is the scale parameter with $\beta > 0$.
- Uniform distribution *(U/a,b)*: also known as rectangular distribution is a distribution that has constant probability. Can be used as a first model for a quantity that is assumed to be randomly varying between parameters a and b but about which little is known. Thus, the uniform distribution is essential in generating random values from all other distribution. Parameters a and b are real numbers with $a < b$; a is the location parameter, and $b - a$ is the scale parameter.
- Normal (or Gaussian) distribution *N(μ, σ²)*: continuous probability distribution showing that the probability of any real observation will fall between any two real limits or real numbers as the graph approaches zero on either side. Normal distributions are very important in statistics and are often used for real-valued random variables whose distributions are not known. The parameter of the normal distribution is the location parameter $\mu \in (-\infty, \infty)$ and scale parameter σ with $\sigma > 0$.

- Weibull distribution **Weibull(α,β)**: continuous probability distribution used as a rough model in the absence of data like time to failure of a component, or time to complete a task, or to describe a particle size distribution, etc. Parameters α and β are so-called shape parameters with $\alpha > 1$ and $\beta > 0$.

For discrete distributions some of which one can use are the:

- Bernoulli distribution **Bernoulli(p)**: probability distribution of a random variable with two possible outcomes used to generate other discrete random variates, e.g., *binominal*, *geometric*, and *negative binominal*. Its outcomes can take value 1 with success probability p and value 0 with failure probability $q = 1 - p$. Thus, parameter p holds $p \in (0,1)$.
- Binomial distribution **bin(t,p)**: discrete probability distribution of the number of successes in t independent Bernoulli trials with probability p of success on each trial; number of defective components in a batch of size t, e.g., number of vehicles or passengers of a random size. Parameters are t and p whereby t is a positive integer, and p holds $p \in (0,1)$.
- Geometric distribution **geom(p)**: is either of two discrete probability distributions:
  - Probability distribution of number X of Bernoulli trials needed in finding one success, supported on the set $\{1, 2, 3, \ldots\}$
  - Probability distribution of number $Y = X - 1$ of failures before first success, supported on the set $\{0, 1, 2, 3, \ldots\}$
- Poisson distribution **Poisson(λ)**: discrete probability distribution expressing the probability of a given number of events that occur in an interval of time when the events are occurring at a constant rate; number of components in a batch of random size. Parameter λ holds $\lambda > 0$.

Example 2.1

Assume the number X of defective assemblies in the sample n of manufactured assemblies is binomially distributed. Let $n = 30$ and the probability of defective assembly $p = 0.02$ results in

$$P(X \leq 2) = \sum_{x=0}^{2} \binom{30}{x} (0.02)^x (0.98)^{30-x} = \tag{2.33}$$

$$0.5455 + 0.3340 + 0.0988 = 0.9783$$

The mean number of defectives in the sample is

$$E(X) = n \cdot p = 30 \cdot 0.02 = 0.6. \tag{2.34}$$

The variance of defectives in the sample is

$$V(X) = n \cdot p \cdot q = 30 \cdot 0.02 \cdot 0.98 = 0.588. \tag{2.35}$$

Example 2.2

Assume a class of vehicle has a time to failure that follows the Weibull distribution with $\alpha = 200$ h, $\beta = 0.333$, and $\nu = 0$. The mean time to failure yields for the mean Weibull distribution:

$$E(X) = \nu + \alpha \Gamma \left(\frac{1}{\beta} + 1 \right) = 200\Gamma(3+1) = 200(3!) = 1200 \ h, \qquad (2.36)$$

and for the variance Weibull distribution:

$$V(X) = \alpha^2 \Gamma \left(\frac{2}{\beta} + 1 \right) - \left[\Gamma \left(\frac{1}{\beta} + 1 \right) \right]^2. \qquad (2.37)$$

The probability that a vehicle fails before 200 h can be calculated based on the cumulative distribution function of the Weibull distribution as follows:

$$F(x) = 1 - e^{-\left(\frac{x-\nu}{a} \right)^{\beta}} = 1 - e^{-\left(\frac{2000}{200} \right)^{0.333}} = 1 - e^{-2.15} = 0.884 \qquad (2.38)$$

2.4 Traffic Demand Models

To analyze and design transportation systems, it is necessary to estimate the existing demand and predict the impact of changes which will result from the transportation planning and development projects considered. In doing so, traffic analysis incorporates a wide spectrum of topics as part of transportation planning and development activities. Thus, traffic analysis is conducted to assist decision makers in improving their transportation planning decisions. One strength of modern traffic demand forecasting is the ability to ask "what if" questions about proposed plans and policies. For this reason, a computerized travel demand forecasting model is used to estimate the relationship between travel demand flows and their characteristics and transportation supply systems and their characteristics.

Traffic demand flow is introduced as an aggregation of individual trips, whereby each trip can be the result of multiple choices made by the users of the transportation system. These users can be individual travelers in passenger transportation, ramp traffic controllers, and/or control tower operators at airports, freight transportation operators, etc. In Cascetta (2009), some classification criteria of travel demand models are introduced, as shown in Table 2.5.

Travel demand models have been designed to include a method for evaluating transport demands with regard to the amount of travel people may choose under specific conditions, e.g., price or transport services. This information is then used to predict roadway traffic volumes and impacts, such as congestion, pollution emissions, etc. Most of the models use a four-step approach (Virginiadot 2014) as shown in Table 2.6.

Table 2.5 Characteristics of travel demand models (Cascetta 2009)

Characteristic	Model
Type of choice	Mobility or context
	Travel
Sequence of choice	Trip-based demand
	Trip chaining
	Activity based
Level of detail	Disaggregate
	Aggregate
Basic assumptions	Behavioral
	Descriptive

Table 2.6 Four-step approach used in travel demand models

Step	Initial tasks
1	Trip generation (the number of trips to be made)
2	Trip distribution (where those trips go)
3	Mode choice (how the trips will be divided among the available modes of travel)
4	Trip assignment (predicting the route trips will take)

The constraints for the task steps in Table 2.6 are shown in Table 2.7.

Once the four steps have been completed, the travel demand forecasting model provides planners with data for existing travel patterns, which are validated and cross-checked to determine how well the model predicts current data, such as park-and-ride utilization, highway vehicle traffic counts, etc.

Besides actual travel surveys, travel demand models use census data to determine the transportation demands, establish baseline conditions, and identify trends. Thus, trips are often predicted separately by purpose (i.e., work, shopping, etc.) and then aggregated into total trips on the network. From this perspective, it can be concluded that travel demand models are designed primarily to identify congestion problems because they mainly measure peak-period motor vehicle trips on major roadways. They generally report roadway level of service (LOS), and a letter grade from A (best) to F (worst) indicates vehicle traffic speeds and delays.

As described in TDM Encyclopedia (2014), travel demand models often incorporate several types of bias favoring automobile transport over other modes and undervaluing travel demand modeling strategies (TRB 2007). The travel surveys they are based on tend to ignore or undercount nonmotorized travel and so undervalue nonmotorized transportation improvements for achieving transportation planning objectives (Stopher and Greaves 2007). Most do not accurately account for the tendency of traffic to maintain equilibrium (congestion causes travelers to shift time, route, mode, and destination) and the effects of generated traffic that result from roadway capacity expansion, and so tend to exaggerate future congestion problems and the benefits of expanding roadway capacity. They are not sensitive to the impacts many types of travel demand model strategies have on trip generation and traffic problems and so undervalue travel demand model benefits.

Table 2.7 Constraints of the four-step travel demand model approach

Step	Constraints
1	Takes into account area factors such as:
	Number and size of households
	Automobile ownership
	Types of activities (residential, commercial, industrial, etc.)
	Density of development, how much travel flows from or to a specific area within the region
	For simplicity, a geographic unit called a transportation analysis zone (TAZ) is used to create trip generation rates for the region
2	Takes into account a certain number of trips generated from each TAZ based on which trip distribution can be analyzed, leading to trip origin and destination points within the region and the number of trips between each pair of TAZ
3	Takes into account the mode of transportation used between trip origins and destinations, i.e., cars, carpools, public transportation, etc.
4	Determines the selected routes taken from origins to destination points, assuming a preference for the fastest route to a destination, based on all kinds of information, such as:
	Actual or predicted congestion and/or other incidents
	Road conditions
	Transit schedules and fares
	Traffic signal systems
	Uses a multicriteria approach to determine the optimal trip assignment

Different reports have been published summarizing information from numerous site surveys, such as:

- Trip and Parking Generation models (Lee et al. 2012).
- Economic Evaluation models (Ellis et al. 2012).
- Integrated Transportation and Land Use models (Dong et al. 2006; Donoso et al. 2006; Scheurer et al. 2009; Bartholomew and Ewing 2009; TRB 2012).
- Simulation models which model the behavior and needs of individual transport users (called agents), rather than aggregate groups, which improves consideration of modes such as walking and cycling; the transport demands of nondrivers, cyclists, and the disabled; and the effects of factors such as parking supply and price, transit service quality, and local land use accessibility factors. Simulation models can provide a bridge between other types of models, since they can incorporate elements from conventional traffic, economic, and land use models. Simulation models have been used for many years on individual projects and are increasingly used for area-wide analysis (TDM Encyclopedia 2014).
- Energy and Emission models (Litman 2013).

Assume a trip-based demand model predicts the average number of trips d with given characteristics executed for a given reference period resulting in the equation (Cascetta 2009)

$$d[C_1, C_2, \ldots] = d(SEV, L : D). \tag{2.39}$$

where the average travel demand flow between two transportation analysis zones has the characteristics $C_1, C_2, \ldots C_n$, which can be expressed as a function of vector SEV, a socioeconomic variable, related to the activity system and/or the decision makers, and of a vector L, level-of-service attributes of the transportation supply system. Demand functions also involve a vector D of coefficients or parameters.

Trip characteristics that are considered relevant in trip-based demand models include (Cascetta 2009):

- u: user's class—category of socioeconomic characteristics
- o, d: zones of trip origin and destination
- p: trip purpose
- t: time period which is the time band in which trips are undertaken
- m: mode used during the trip
- tp: trip path, that is, the series of links connecting centroids o and d over the network, representing the transportation service providers by mode m

With demand flow denoted by $d_{o,d}^u[p, t, m, tp]$, the demand model can be expressed as

$$d_{o,d}^u[p, t, m, tp] = d(SEV, L). \tag{2.40}$$

It is difficult to incorporate freight information into transportation models because freight data is proprietary, and the release of that data is considered to be detrimental to the company's competitive position. Due to the difficulty in acquiring freight data, the inclusion of freight in most transportation plans and models has either been limited in scope or based upon limited sample sizes without knowledge of the contents. In Harris (2008), the Freight Analysis Framework Database, developed and distributed by the Federal Highway Administration, contains freight flows for 114 zones at the national level. This allows the formulation of a travel demand model for truck trips, with vehicles moving through counties of the so-called freight flow zones (TPC_i). These zones can be used to calculate the zonal truck counts for each county as follows:

$$TPC_i = TCZ_{ab} \frac{WF * FLC_i}{\sum FLC_{ij}}, \tag{2.41}$$

2.5 Congested Network Models

Traffic management has become a critical issue as the number of vehicles in metropolitan areas is nearing the existing road capacity, resulting in traffic congestion. In some areas, the volume of vehicles has met and/or exceeded road capacity. However, many roads are constructed with less space than is needed to accommodate the ever-increasing traffic flow, resulting in congestion. Traffic congestion occurs when the volume of traffic generates a demand for space that is greater than the available road capacity, commonly termed saturation (see Sect. 2.3). There are a number of specific circumstances which cause congestion. The majority of those circumstances are the result of a reduction in road capacity at a given point due to roadwork, weather conditions, accidents, and/or other incidents or an increase in the number of vehicles required for a given transportation volume of people and/or freight. But traffic congestion in transportation is not limited to roads. It is also a problem at airports, at harbors, on railways, and for travelers on public transportation networks.

As introduced in Sect. 2.2, traffic congestion can be studied either at a microscopic level, where the motion of individual vehicles is tracked, or at a macroscopic level, where vehicles are treated as a fluidlike continuum. Therefore, both macroscopic and microscopic models are used to address various traffic flow and congestion phenomena, such as phase transitions, a phenomenon whereby free-flow traffic can spontaneously break down for no obvious reason and persist in a self-maintained congested state for long periods (Kerner and Rehborn 1997). The importance of modeling and controlling traffic congestion can also be seen by reviewing the projects funded by the European Research Council (ERC) for the period 2012–2017. Due to the manifold types of traffic phenomena, traffic flow modeling cannot fully predict under what conditions a traffic jam, defined as heavy but smoothly flowing traffic, may suddenly occur. The reason is that individual incident, such as accidents, a single car braking, an abrupt steering maneuver by a single vehicle, or a truck breakdown, in a previously smooth traffic flow may cause a so-called cascading failure. A cascading failure in a traffic flow system of vehicles means that the failure of one vehicle can trigger the failure of successive vehicles. A cascading failure usually begins when one vehicle of the traffic flow system fails and the effect spreads out and creates a sustained traffic jam. When this happens, nearby traffic nodes must then, if possible, take up the stagnancy caused by the traffic jam which can, in turn, overload those nodes, causing them to fail, resulting in serious congestion. As mentioned in Sect. 2.2, theoretical traffic flow models apply the rules of fluid dynamics to traffic flow, like a fluid flow in a pipe. In spite of the poor correlation of theoretical traffic flow models to actual traffic flow, empirical models have been chosen with the scope to forecast traffic flow. These traffic models use a combination of macro-, micro-, and mesoscopic modeling features, with the addition of entropy effects, by grouping vehicles and randomizing flow patterns within the node segments of the network. These models are then calibrated by measuring actual traffic flows on the links in the network, and the baseline traffic flows are adjusted accordingly (Lindsey and Verhoef 1999; Lindsey et al. 2012).

Traffic flow can be described by the variables density (k), speed (v), and flow (q), measured in vehicles per lane per mile, mile per hour, and vehicles per lane per hour. At the macroscopic level, these variables are defined under stationary conditions at each point in space and time and are expressed by the fundamental equation of traffic flow theory, given in (2.11). For safety reasons, speed usually declines as density increases which means that the less vehicles per lane per hour, the nearer the vehicles are to driving at free-flow speed v_f. At higher densities, the flow in the flow-density graph (see Fig. 2.1), as well as the speed-flow graph (see Fig. 2.2), drops more rapidly, reaching zero at the congestion density, k_j, where speed and flow are both zero. Thus we can say that the uphill branch of the graph in Figs. 2.1 and 2.2 is referred to as uncongested, unrestricted free traffic flow; and the downhill branch of the graph in Figs. 2.1 and 2.2 is referred to as congested, restricted, or queued. Thus in general mathematical terms of transportation supply models, the speed-traffic flow graph in Fig. 2.2 can be used to formulate relationships among performance, cost, and flow. Hence, interpreting traffic flow as quantity of trips supplied by the road per unit of time, a trip cost curve $C(q)$ can be generated in the form of

$$C(q) = c_0 + \frac{u_C D}{v(q)}, \qquad (2.42)$$

where c_0 denotes trip costs, u_C is the unit cost of travel time, D is trip distance, and $v(q)$ is speed expressed in terms of flow. Then the trip cost curve based on (2.42) shows a positively sloped portion corresponding to the congested branch of the speed-flow curve. Thus, $C(q)$ measures the cost of a trip taken by a vehicle. Therefore, the total cost of q trips is then

$$TC(q) = C(q)q, \qquad (2.43)$$

and the cost of an additional trip is

$$AC(q) = \frac{\partial TC(q)}{\partial q} = C(q) + q\frac{\partial C(q)}{\partial q}, \qquad (2.44)$$

Congestion simulations and real-time observations have shown that in heavy but free-flow traffic, jams can arise spontaneously, triggered by minor events, such as the so-called butterfly effect, e.g., an abrupt steering maneuver by a single motorist. The butterfly effect is the sensitive dependency on initial conditions in which a small change at one place in a deterministic nonlinear system can result in large differences in a later state.

A team of Massachusetts Institute of Technology (MIT) mathematicians (Flynn et al. 2009) has developed a model that describes the formation of "phantom jams," in which small disturbances (a driver hitting the brake too hard or getting too close to another car) in heavy traffic can become amplified into a full-blown, self-sustaining traffic jam. Key to the study is the realization that the mathematics of such jams, which the researchers call "jamitons," are strikingly similar to the equations that describe detonation waves produced by explosions, according to Aslan Kasimov, lecturer in MIT's Department of Mathematics. That discovery enabled the team to solve traffic jam equations that were first theorized in the 1950s.

2.6   Graph Models

In mathematics graph theory, graphs which are mathematical structures used to model pairwise relations between objects are studied. A graph in this context consists of a nonempty set of vertices (or nodes) and a set E of links called "edges" that connect (pairs of) nodes. Each edge has either one or two vertices associated with it, called its "endpoints." An edge is said to connect its endpoints. A graph can be undirected, meaning that there is no distinction between the two vertices associated with each edge, or its edges can be directed from one vertex to another, etc., which have different formal definitions, depending on what kinds of edges are allowed. In this context, a directed graph is defined as

$$G = (V, E) \tag{2.45}$$

where V consists of a nonempty set V of vertices (or nodes), each node represents a variable, and

$$E \subseteq V \times V. \tag{2.46}$$

where E is the set of directed edges (or arcs) and edges encode the dependencies. Each directed edge $(u, v) \in E$ has a start (tail) vertex u and an end (head) vertex v. Note: A directed graph $G = (V, E)$ is simply a set V together with a binary relation E on V.

In graph theory, the following terminology is of importance. In a simple graph, each edge connects two different vertices; and no two edges connect the same pair of vertices. Multigraphs can have multiple edges connecting the same two vertices. When m different edges connect vertices u and v, we say that $\{u, v\}$ is an edge of multiplicity m. An edge that connects a vertex to itself is called a loop. A pseudograph can include loops as well as multiple edges connecting the same pair of vertices.

For a set V, let $[V]k$ denote the set of k element subsets of V. Equivalently, $[V]k$ is the set of all k combinations of V.

An undirected graph, (2.45) consists of a nonempty set V of vertices (or nodes) and a set

$$E \subseteq [V]^2, \tag{2.47}$$

of undirected edges. Every edge $\{u, v\} \in E$ has two distinct vertices $u \neq v$ as endpoints, and vertices u and v are then said to be adjacent in graph G. Note: The above definitions allow for infinite graphs, where $|V| = 1$.

Table 2.8 shows the terminology of graphs.

Two undirected graphs, $G1 = (V1, E1)$ and $G2 = (V2, E2)$, are isomorphic if there is a bijection $f: V1 \rightarrow V2$ with the property that for all vertices $a, b \in V1$

$$\{a, b\} \in E1 \text{ if and only if} \{f(a), f(b)\} \in E2.$$

Such a function f is called an isomorphism.

An arbitrary undirected graph can be introduced as encoding a set of independencies. As an example, the following rule states when two sets of variables

$$U_1, U_2 \subseteq V, U_1 \cap U_2 = O$$

are separated in an undirected graph. Let us denote separation by $\perp$ and take it to mean independence in the joint distribution over V:

$$U_1 \perp U_2 | U_3 \Longleftarrow\Longrightarrow all\ paths\ between\ sets\ U_1\ and\ U_2\ pass\ through\ set\ U_3$$

Let U_3 blocks the paths between X and Y; which can be interpreted as blocking the flow of information. A consequence of this rule is the following Markov property for Markov networks, also called the local Markov property:

$$A \perp everything\ else | n(A)$$

where $n(A)$ are the neighbors of variable A.

Let a set of variables that separate node A from the rest of the graph be called a Markov blanket for A. Hence, set $n(A)$ is a Markov blanket, and it is the minimal Markov blanket of A. Adding a node to a Markov blanket preserves the Markov blanket property.

Graph theory is widely used to model and study transportation networks:

- Airline networks can be modeled using directed multigraphs where:
  - Airports are represented by vertices.
  - Each flight is represented by a directed edge from the vertex representing the departure airport to the vertex representing the destination airport.
- Road networks can be modeled using graphs where:
  - Vertices represent intersections.
  - Edges represent roads.
  - Unidirected edges represent two-way roads.
  - Directed edges represent one-way roads.

One of the most interesting and powerful features of graphs is their use in modeling structures. With this possibility, one can model relationships, flight schedules, etc. By building a graph model, we use the appropriate type of graph (see Table 2.8) to capture the important features of the application. In a graph-based airport network, vertices represent the airport destinations; and edges represent the airway links between the destinations, as shown in Fig. 2.4.

To model an airport network in which the number of links between the vertices (airports) is important, we can use a multigraph model, as shown in Fig. 2.5.

To model an airport network in which diagnostic links at the vertices (airports) is of importance, we can use pseudograph model where loops are allowed, as shown in Fig. 2.6.

Table 2.8 Graph terminology

Type	Edges	Multiple edges allowed	Loops allowed
Simple graph	Unidirected	No	No
Multigraph	Unidirected	Yes	No
Pseudograph	Undirected	Yes	Yes
Simple directed graph	Directed	No	No
Directed multigraph	Directed	Yes	Yes
Mixed graph	Directed and unidirected	Yes	Yes

Fig. 2.4 Airport network

Fig. 2.5 Multigraph airport network

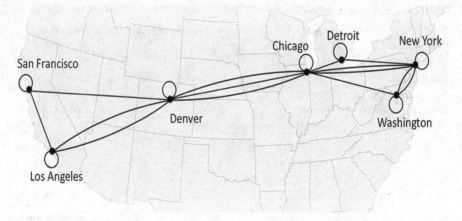

Fig. 2.6 Pseudograph airport network with loops

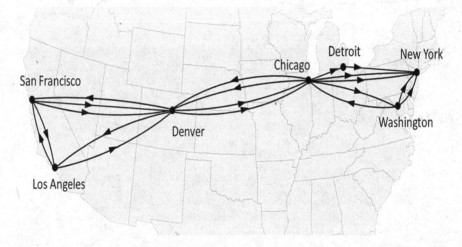

Fig. 2.7 Directed multigraph airport network

To model an airport network in which multiple one-way links between the vertices (airports) are of importance, we can use a directed multigraph model, as shown in Fig. 2.7. It should be noted we could also use a directed graph without multiple edges if we are only interested in whether there is at least one link from a vertex (airport) to another vertex (airport).

Simulations of the air transportation system with detailed models of terminal areas are often applied to optimize current concepts for managing air traffic and investigate future concepts. These simulations can include studies of operations at individual airports or wider regions covering several airports. All of them require detailed terminal and airspace models capable of representing both current and future operating conditions because no two airports are identical. To simulate the movements of aircraft on the ramp, taxiways, and runways, or passengers in the terminal from check-in through security to gates, or freight to the aircraft, it is convenient to use vertex (node) link (edges) graph models. Modeling aircraft, passenger, or freight movements on a node-link model makes determining the separation between aircraft, passengers, and freight straightforward. In addition, various levels of detail can be modeled depending on the nature of the simulation; and the models can be adapted to existing simulation tools, such as the Airport and Airspace Delay Simulation Model used by the Federal Aviation Administration (FAA) or the Total Airspace and Airport Modeler (Lee and Romer 2011).

The FAA's Airport and Airspace Delay Simulation Model (SIMMOD) is an event-step simulation model which traces the movement of individual aircraft and simulated air traffic control actions required to ensure aircraft operate within procedural rules. The FAA's SIMMOD uses a node-link structure to represent the gate/taxiway and runway/airspace route system. Input parameters depend on the type of aircraft and include permissible airborne speed ranges for use by Air Traffic Control (ATC), runway occupancy times, safety separations, landing roll and declaration characteristics, taxi speeds, and runway/taxiway utilization. Gate utilization depends on the aircraft type and airline (URL 2). With SIMMOD, testing and analyzing the impact of various air traffic scenarios is possible. For this purpose, SIMMOD computes aircraft travel times and delay statistics. SIMMOD can be downloaded from (URL 3). An airfield, based on SIMMOD, is given in (URL 4), showing the runways, terminal buildings, and ramps.

With the Total Airport and Airspace Modeler (TAAM), airports and airspace can be modeled to facilitate planning, analysis, and decision making and to evaluate the impact of changes to infrastructure, operations, and schedules. TAAM is recognized as a standard in the aviation industry and is widely used by Air Navigation Service Providers (ANSP), Civil Aviation Authorities (CAA), airspace planners, airport operators, and major air carriers. TAAM allows modeling of the complete ground operation of an aircraft, preparing scenarios for the airport, creating simulation for a baseline airspace configuration, etc. (URL 4).

2.7 Bottleneck Analysis

The capacity of a transportation system can be modeled as a series of pipes of varying capacity, with the smallest diameter or capacity holding back the entire system. Figure 2.8 illustrates a five-pipe system with different capacities (diameters).

Pipe 2 in Fig. 2.8 represents a bottleneck in the transportation system with regard to capacity. At location *pipe 1* before the bottleneck *pipe 2*, the arrival of vehicles follows a regular traffic flow. If the bottleneck is absent, the departure rate of vehicles at location *pipe 2* is essentially the same as the arrival rate at *pipe 1* at some later time, free-flow travel time T_{FF}. However, due to the bottleneck, the system at location *pipe 2* is now only able to have a departure rate of μ (see Sect. 2.3). The vehicle's arrival at location *pipe 3* takes into account the delay caused by the bottleneck of *pipe 2*. The reason is that output from one pipe

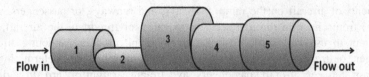
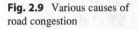

Fig. 2.8 *Bottleneck* in a transportation system pipe series

Fig. 2.9 Various causes of
road congestion

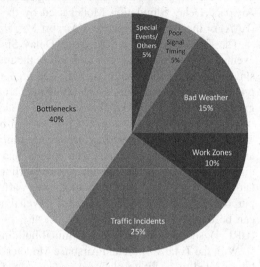

becomes the input to the next until the transportation vehicle exits *pipe 5*. As shown
in Fig. 2.9, *pipe 2* cannot handle the traffic flow that *pipe 1* can deliver; and,
therefore, it restricts the traffic flow. Because of *pipe 2*'s limited capacities, it
restricts the flow from upstream pipes and starves the downstream pipes. *Pipe
3, pipe 4, and pipe 5* can only work on what *pipe 2* delivers, meaning it determines
the transportation system's capacity. Therefore, bottlenecks are important
considerations because they impact the traffic flow and thereby the average speed
of vehicles. The main consequence of a bottleneck is an immediate reduction in
capacity of the transportation system roadway. The Federal Highway Authority has
stated that 40 % of all congestion is from bottlenecks, as shown in Fig. 2.9.

Bottlenecks are characterized with regard to their characteristic features, which
are stationary and moving bottlenecks. Stationary bottlenecks occur when a multi-
lane road is reduced by one or more lanes, which causes the vehicular traffic in the
ending lanes to merge into the other lanes.

Let us assume that at a certain location x_0, the highway narrows to one lane.
Thus, the maximum traffic flow rate is now limited to q_{cll}, since only one lane of the
two is available. The traffic flow rate is shared by q_{cll} and q_c, but its vehicle density
k_{cll} is higher.

As described for Fig. 2.9, we can state that before the first vehicles reach location
x_0, the traffic flow is unimpeded. However, downstream of x_0, the roadway narrows,
reducing the capacity by half. Thus, vehicles begin queuing upstream of x_0 which

Fig. 2.10 Slow tractor creating a moving bottleneck

results in a slower mean space speed v_{qu} of the vehicles compared with the free-flow speed v_f of vehicles. The vehicles driving in the one-lane queue will begin to clear, and the traffic jam can dissipate. But the free space mean speed of the vehicles driving on the now one-lane capacity road will be slower than the vehicles moving at free-flow speed v_f.

Moving bottlenecks are those caused due to slow-moving vehicles, such as trucks, that disrupt the traffic flow. Moving bottlenecks can be active or inactive bottlenecks. If the reduced traffic flow rate caused by a moving bottleneck is greater than the actual traffic flow rate downstream of the vehicle, then this bottleneck is said to be an active bottleneck. In Fig. 2.10, a moving bottleneck is represented by a slow-moving heavy tractor with a mean space speed v_t approaching a downstream location. If the reduced traffic flow rate of the tractor is less than the downstream traffic flow rate, then the tractor becomes an inactive bottleneck.

An analytical expression for capacity reductions caused by a moving bottleneck where each lane has an underperforming flow of traffic can be described in terms of the desired traffic flow rate and can be modeled by the flow conservation equation. This occurs when vehicles pass an observer moving with speed v when traffic is in a steady flow rate-density state (see Fig. 2.1 in Sect. 2.2):

$$q_r = q - kv \tag{2.48}$$

If v, q, and k are given, then q_r is the vertical separation between the corresponding steady-state point on the flow rate versus the density graph in Fig. 2.1. Equation 2.48 applies to an observer that either trails or precedes a moving bottleneck by a substantial but fixed distance, which means the steady-state traffic flow is on either side of the bottleneck. Such a bottleneck is said to be active when, as a result of its presence, the steady states upstream and downstream of it are different. This occurs in practice when the bottleneck holds back a queue, i.e., when a queue is detected behind it but no queue exists for a long sector of road downstream. Equation 2.48 implies that if a stable passing traffic flow rate q_r exists when an active bottleneck moves at speed v, then the two steady states before and after it must be somewhere on the red line of Fig. 2.1. Therefore, q_r will exclusively denote the passing rate when a bottleneck is active.

Identifying a bottleneck in a transportation system is critical; therefore, the importance of bottleneck analysis cannot be overstated because the results are used not only in determining capacity but also in planning and scheduling traffic flows.

Different methods for bottleneck analysis are known and applied in transportation analysis, such as:

• Capacity utilization
• Queuing time
• Elapsed time
• Shifting shortage

The capacity utilization method refers to the utilization of different resources and calculates the resource with the highest capacity utilization as shortage, which can be calculated after Wang et al. (2005) as follows:

$$B = \{i | p_i = \max(p_1, p_2, ..., p_n)\} \tag{2.49}$$

with p_i as capacity utilization of the ith resource. The advantage of this method is the intrinsic simplicity making it ideal for transportation planning applications, such as roadway capacity planning and design, congestion management, traffic impact studies, etc. The intersection capacity utilization method is also defined as the sum of ratios of the approach volume divided by the approach capacity for each part of the intersection which controls the overall traffic signal timing plus an allowance for clearance times (Crommelin 1974). Hence, it can be predicted how much reserve capacity is available and how much the intersection is over capacity but does not predict delay. Moreover, the capacity utilization method can be used to predict how often a roadway intersection will cause congestion. But for this, the method requires a specific set of data to be collected which includes traffic volume, number of lanes, saturated traffic flow rates, signal timings, reference cycle length, and lost times for an intersection. Then the method can sum the amount of time required to serve all movements at a saturation rate for a given cycle length and divide it by the reference cycle length. This means that the method is similar to summing critical volume to saturation flow ratios which allow consideration of minimum timings. Moreover, the concept of level of services (LOS) is used whereby LOS reports on the amount of reserve capacity or capacity deficits.

In order to calculate the LOS for intersection capacity utilization, the intersection capacity utilization (*ICU*) must be computed first, which can be achieved as follows:

$$ICU = \left(\max\left(t_{\text{Min}}, \frac{v}{si} \right) * RCL + \frac{t_{Li}}{RCL} \right) \qquad (2.50)$$

with t_{Min} as minimum green time, critical movement i, v/si as volume to saturation flow rate, RCL as reference cycle length, and t_{Li} as lost time for critical movement i (Husch 2003).

Once the *ICU* is fully calculated for an intersection, the *ICU* Level of Service for that intersection can be calculated based on the following criteria (Husch 2003):

A. *If ICU is less than or equal to 55%.*
B. *If ICU is greater than 55% but less than 64%.*
C. *If ICU is greater than 64% but less than 73%.*
D. *If ICU is greater than 73% but less than 82%.*
E. *If ICU is greater than 82% but less than 91%.*
F. *If ICU is greater than 91% but less than 100%.*
F. *If ICU is greater than 100% but less than 109%.*
H. *If ICU is greater than 109%.*

This grading criterion shows some specific details about the specific intersection (Husch 2003):

A. *Intersection has no congestion.*
B. *Intersection has very little congestion.*
C. *Intersection has no major congestion.*
D. *Intersection normally has no congestion.*
E. *Intersection is on the verge of congested conditions.*
F. *Intersection is over capacity and likely to experience congestion periods of 15 to 60 consecutive minutes.*
G. *Intersection is 9% over capacity and likely to experiences congestion periods of 60 to 120 consecutive minutes.*
H. *The intersection is 9% or greater over capacity and could experience congestion periods of over 120 minutes per day.*

To achieve an intersection capacity utilization level of service E or better is not always easy and, therefore, much care is given to the signal timings and geometric bottlenecks, such as lane drops, hard curves, hills, etc., in order to get the LOS to be better than E.

The queuing time method determines the shortage (bottleneck) in relation to the queuing time of the resources before loading and uploading containers for transportation within the supply chain, which can be calculated after Tan and Bowden (2004) as follows:

$$B = \{i | W_i = \max(W_1, W_2, ..., W_n)\} \tag{2.51}$$

with W_i as queuing time utilization of the ith resource. The advantage of the method is its easy implementation.

The elapsed time method is a traffic flow scheduling problem in which processing time is associated with their respective probabilities including the transportation time. Finding a good traffic flow schedule for a given set of activities helps transportation managers to effectively control traffic flows and provide solutions for activity sequencing. A traffic flow activity scheduling problem consists when determining the processing sequence for n vehicles on a road network. Therefore, the objective of the elapsed time method can be to minimize the time required to pass a bottleneck. The notation of the elapsed time method contains sequences, vehicles, activity processing times, and the probability associated with activity processing times which finally results in the expected processing time of an activity. The calculation of an elapsed time approach is based on the following criteria which have been mentioned previously:

S *Sequence of activities 1, 2, 3,...n*
V_j *Vehicle j, j = 1,2,.......k*
A_1 *Processing time of i*th *activity for vehicle V_1*
A_2 *Processing time of i*th *activity for vehicle V_2*
A_2 *Processing time of i*th *activity for vehicle V_3*

...

P_1 *Probability associated to processing time A_1 of i*th *activity for vehicle V_1*
P_2 *Probability associated to processing time A_2 of i*th *activity for vehicle V_2*
P_3 *Probability associated to processing time A_3 of i*th *activity for vehicle V_3*

...

T_1 *Transportation time of i*th *activity from vehicle1 to destination D_1*
T_2 *Transportation time of i*th *activity from vehicle2 to destination D_2*
T_3 *Transportation time of i*th *activity from vehicle3 to destination D_1*

...

PT_1 *Expected processing time of i*th *activity on vehicle V_1*
PT_2 *Expected processing time of i*th *activity on vehicle V_2*
PT_3 *Expected processing time of i*th *activity on vehicle V_3*

...

The sequence of activities is processed on the vehicles in the order $V_i, i = 1, 2, 3, ...$ with A_1, A_2, and A_3 as processing time of each activity on vehicle V_1, V_2, and V_3, respectively, assuming their respective probabilities P_i, $I = 1, 2, 3,\ldots$ such that $0 \leq P_i \leq 1$. T_i is the transportation time of the ith activities from vehicles V_i to destination D_i, $i = 1, 2, 3,\ldots$, respectively. The algorithm of the given problem is (shown only in part):

- Step 1: Define expected processing time PT_i on vehicle V_i, $i = 1, 2, 3$, as follows:

$$PT_i = A_i \times P_i \qquad (2.52)$$

- Step 2: Compute processing time by creating two fictitious vehicles, G and H, with their processing times G_i and H_i, respectively.
- Step 3: Define new reduced problem with processing times G_i and H_i as defined in Step 2.
- Step 4: Find the optimal sequence for two vehicles G and H with processing times of G_i and H_i obtained in Step 2.
- Step 5: Compute the in-out graph for the sequence obtained in Step 4 (Gupta et al. 2013).

The shifting shortage method, in contrast with the sole shortage approach, requested average active time steps of shortages based on which it will be possible to estimate the timeliness shortages are shifting. This allows identification of nonshortages too. This method differentiates between the probability of the existence of shortages and the existence of nonshortages. Moreover, this method allows separation between primary and secondary shortages due to the average shortage over time (Lima et al. 2008). But the primary methodological problem of this method is its implementation and the computing time required. Figure 2.11 shows the principle of moving shortages (with the shortages designated as S1 and S2). As shown in Fig. 2.11, at a specific time step, the shortage is caused by the active periods of tasks which may have the longest runtime. Therefore, the shifting shortage is based on the overlap of shortages.

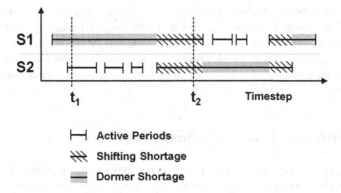

Fig. 2.11 Shifting shortage method

Primary delays as a result of shortages corroborate a belief in so-called distributions of:

- Shipping time.
- Arrival time.
- Quay time for uploading/loading.
- Accomplishable delay compensation through optimization of an objective function to maximize it; for each criterion a higher value will be preferred opposite a lower value.
- Accomplished improvement to compensate for delays.

In general, distribution assumptions can be summarized in a model which allows statistical data analysis (Stahl 2002). However, problems occur with secondary delays, the so-called domino effect, because they start with the distribution assumptions of the primary delays. Secondary delays in maritime transportation, as a consequence of primary delays, can, for example, result in delayed arrival of the following for uploading and loading the containers:

- Trucks
- Trains
- Feeders

The consequences of connection delays can be estimated using mathematical models which allow statistical calculations. The outcome is throughput estimation as a result of the delay, which can be compared with the original assumptions to show the implications of the delay from a general perspective, as well as for a single case study.

Based on the distribution graphs composed, shortages can be identified and rectified through a representative selection of objective functions following the multicriteria approach for simulation runs, which finally results in appropriate adjustments.

As can be seen in Fig. 2.11, short time delays are dominant for the maritime probability transportation chains model. For the shortage analysis, it is important to identify if the resources allocated for the transportation chains can be used without shortages. This means that the transportation job will be operated in an optimal manner. Otherwise, it has to be proven whether the transportation job can be operated with a restricted number of alternatives, meaning a nonempty set of alternatives.

2.8   ProModel Case Study: Road Intersection

Graph theory can be used to model road intersections with running and waiting links. Therefore, vertices (nodes) are usually located at the intersection between road segments included in a model for continuous service in transportation, such as an urban road network. Typical examples of road intersections in urban areas are individual modes, such as cars, buses, pedestrians, etc., using a road intersection network represented by nodes. Thus, links correspond to connections between nodes to constitute the urban road network. Figure 2.12 shows an example of an intersection in an urban road system in Clausthal-Zellerfeld, Germany, embedded in ProModel.

In this case study, two distinct types of links are considered: running links, which represent a vehicle's real movements as it moves along the road in the urban road section, and waiting or queuing links, representing queuing at intersections in the city road system, as shown in Fig. 2.13.

The level of detail of the road system depends on the purpose of the model. In this case, the road intersection to be studied is represented by nodes, where the accessed links converge into a four-arm road intersection. The graph model representation for this road intersection is shown in Fig. 2.14 for single node

Fig. 2.12 Screenshot of the four-arm road intersection in downtown Clausthal-Zellerfeld, Germany

Fig. 2.13 Representation of road intersection with running and waiting links

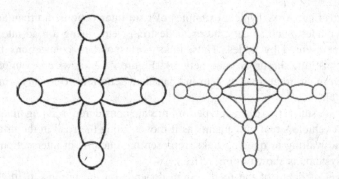

Fig. 2.14 Single node (*left*) model of road intersection in Fig. 2.13 and detailed representation (*right*). For details, see text

representation (left) and a detailed representation, including options for a driving maneuver in the different driving directions of the four-arm intersection (right).

In the single node representation of Fig. 2.13, a left turn can only be achieved if allowed. Moreover, different waiting times cannot be assigned to maneuvers with green-phase duration such as a right turn, for example, in the single node model. This will require a more detailed representation, as given in the right part of Fig. 2.14. In case one wants to expand the model with a parking supply representation, this type of an expanded network representation can be found in Cascetta (2009).

Link performance of a road intersection can be expressed by a cost function introducing the respective performance attributes, which can be travel time along a section, waiting time at the intersection, and monetary cost. In this case, a cost function can be obtained as the sum of the aforementioned performance functions which results in (Cascetta 2009)

$$c(f) = \alpha_1 tr_a(f) + \alpha_2 tw_a(f) + \alpha_3 tm_a(f) \tag{2.53}$$

with

- $tr_a(f)$: function relating to the running time on link a to the flow vector
- $tw_a(f)$: function relating to the waiting time on link a to the flow vector
- $tm_a(f)$: function relating to the monetary cost on link a to the flow vector

and α_i, $i = 1,2,3$, are weighting factors.

A more detailed diagram of the node in Fig. 2.14 has been developed by I. A. Jehle for the four-arm intersection in downtown Clausthal-Zellerfeld, as shown in Fig. 2.15:

In Fig. 2.15 the icon has the following meaning:

- Round and elliptic nodes represent the directions from which cars arrive at the junction → arriving points.
- Rectangular nodes show the traffic lights (or other traffic flow regulations) for different driving directions.

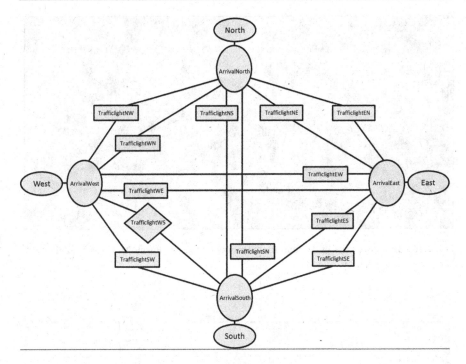

Fig. 2.15 Detailed four-arm road intersection graph model

To simulate the traffic movement at the four-arm intersection in downtown Clausthal-Zellerfeld, the detailed network in Fig. 2.15 has to be transferred into the basic logic components of the ProModel model. In Fig. 2.16, a bird's-eye view of the four-arm intersection is shown which serves as a background for the simulated scenarios in ProModel.

The representation in Fig. 2.13 is the background for a vehicle's movement simulation with ProModel. It refers to:

- Cars arriving at the road intersection from different cardinal directions represented by an icon in the simulation, as shown in Fig. 2.17 for entities
- Entities having a name to identify them and their speed

The node representation in Fig. 2.14 requires, in ProModel, the allocation of the respective locations which have specific attributes, as shown in Fig. 2.18:

- Locations have a name, a capacity describing the number of entities allowed at the location at one time, and rules by which entities enter the location.
- Units describe the number of locations with the same behavior.

The outermost nodes, North, East, South, and West, represent the link with other junctions from which cars arrive at the first simulated location of this direction. The

Fig. 2.16 Bird's-eye view of the four-arm intersection nodes in downtown Clausthal-Zellerfeld

Entities			
Icon	Name	Speed (mpm)	Stats
	CarNorth	50	Time Series
	CarEast	50	Time Series
	CarSouth	50	Time Series
	CarWest	50	Time Series

Fig. 2.17 Representation of *entities* of the four-arm intersection

Locations						
Icon	Name	Cap.	Units	DTs...	Stats	Rules...
	TrafficlightNE	INF	1	None	Time Series	Oldest, FIFO
	TrafficlightNS	INF	1	None	Time Series	Oldest, FIFO
	TrafficlightNW	INF	1	None	Time Series	Oldest, FIFO
	TrafficlightEN	INF	1	None	Time Series	Oldest, FIFO
	TrafficlightES	INF	1	None	Time Series	Oldest, FIFO
	TrafficlightEW	INF	1	None	Time Series	Oldest, FIFO
	TrafficlightSN	INF	1	None	Time Series	Oldest, FIFO
	TrafficlightSE	INF	1	None	Time Series	Oldest, FIFO
	TrafficlightSW	INF	1	None	Time Series	Oldest, FIFO
	TrafficlightWN	INF	1	None	Time Series	Oldest, FIFO
	TrafficlightWE	INF	1	None	Time Series	Oldest, FIFO
	TrafficlightWS	INF	1	None	Time Series	Oldest, FIFO
	North	INF	1	None	Time Series	Oldest
	ArrivalNorth	INF	1	None	Time Series	Oldest
	East	INF	1	None	Time Series	Oldest
	ArrivalEast	INF	1	None	Time Series	Oldest
	South	INF	1	None	Time Series	Oldest
	ArrivalSouth	INF	1	None	Time Series	Oldest
	West	INF	1	None	Time Series	Oldest
	ArrivalWest	INF	1	None	Time Series	Oldest

Fig. 2.18 Representation of *locations* of the four-arm intersection

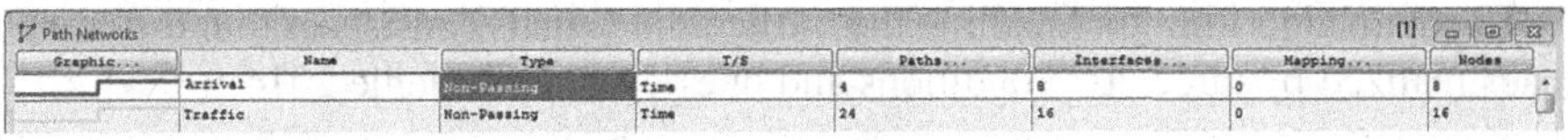

Graphic...	Name	Type	I/S	Paths...	Interfaces...	Mapping...	Nodes
	Arrival	Non-Passing	Time	4	8	0	8
	Traffic	Non-Passing	Time	24	16	0	16

Fig. 2.19 ProModel path network

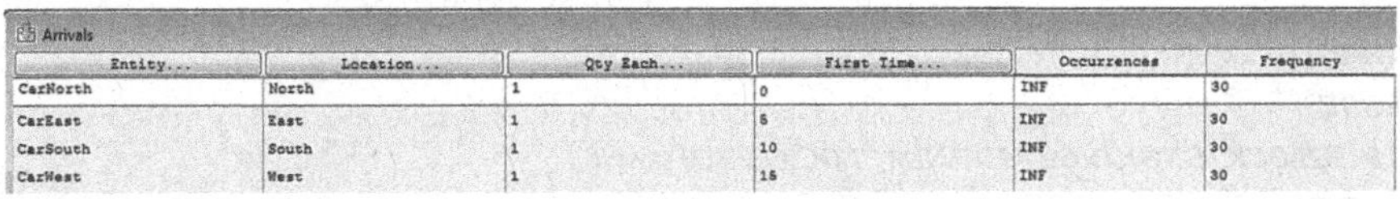

Entity...	Location...	Qty Each...	First Time...	Occurrences	Frequency
CarNorth	North	1	0	INF	30
CarEast	East	1	5	INF	30
CarSouth	South	1	10	INF	30
CarWest	West	1	15	INF	30

Fig. 2.20 Arrivals of entities at locations

Table 2.9 Global variables used in ProModel

ID	Type	Initial value	Stats
Drive 1	Integer	0	Time series, time
Drive 2			
Drive 3			
Drive 4			

nodes, *ArrivalNorth*, *ArrivalEast*, *ArrivalSouth*, and *ArrivalWest*, together with the underlying network, represent a complete digraph. To each of the edges of this graph, one node, representing a traffic light, is added. These traffic light nodes are also locations in the ProModel model. All locations of this node model have infinite capacity and process each entity as one unit.

The entities in Fig. 2.17 move on two different path networks, shown in Fig. 2.19, when running the simulation. The path network, Arrival, brings entities from the arrival locations to the junction nodes, ArrivalNorth, ArrivalEast, ArrivalSouth, and ArrivalWest. The inner network, Traffic, shows the graph structure.

The entities arrive at the locations, North, East, South, and West, individually at different times. The first entity which starts the simulation arrives at time 0:00 at the northern point. After 5 s, the next entity enters the traffic junction at the location East, and five and ten seconds later, entities at South and West arrive. These arrivals repeat themselves every 30 s until the end of the simulation. This scenario is shown in Fig. 2.20.

The global variables are shown in Table 2.9.

In this case study, driving phases for a green signal last 60 s. During this 60-s window, the traffic light for one direction shows red-yellow for 1 s, then green for 56 s, and yellow for 3 s. The signals for the other directions show a red light.

The yellow traffic light is divided into two subphases. The first one lasts 1 s and still allows cars to cross the intersection. The next phase, lasting 2 s, requires stopping at the traffic light.

The main algorithm for handling the traffic light signals consists of four different cases realized by three "if" conditions and one "else" case (Jehle 2014):

```
IF (CLOCK(SEC)/x@n=M) AND ((CLOCK(SEC)@x>0) AND (CLOCK(SEC)@x< (x=y
+1)))
                              //each direction has a green phase of x seconds
                              //signal for direction M green
THEN{DriveM=1}
ELSE {
IF (CLOCK(SEC)/x@n=M) AND (CLOCK(SEC)@x=0)
THEN {DriveM=4 DriveMAlt=22}
                              //signal for direction M red-yellow
                              //and yellow for the former M
ELSE {
IF (CLOCK(SEC)/x@n=M) AND (CLOCK(SEC)@x> (x-y))
THEN {DriveM=21}             //signal for direction M yellow
ELSE {DriveM=3}              //signal for direction M red
}
}
```

The algorithm of the above form can be used for every length x of the green phase and different numbers of driving directions. In this case study, there are four sets of driving directions. Each category combines some directions, which can be used without the need to cross the path of another car, as shown in Table 2.10.

The driving directions shown in Table 2.10 allow the logic macros of the ProModel simulation model to be specified as shown in Table 2.11.

The processing decides, for each entity, the way through the path network and the behavior during the simulation. The following algorithm shows the link between traffic light signals and the movement of the simulated cars for direction M.

```
IF (DriveM=1) OR (DriveM-21)        //drive if signal is green or yellow
THEN {MOVE ON Traffic}
ELSE {WAIT UNTIL (Drive1-1) OR (Drive1-21)}
                              //else wait until signal changes
```

This logic was implemented together with the macros in the processing table, showing the parts of the four entities in Tables 2.12, 2.13, 2.14, and 2.15.

Table 2.10 Driving directions

M	Driving directions
1	From North to South and West; from South to North and East
2	From North to East; from South to West
3	From East to West and North; from West to East and South
4	From West to North; from East to South

Table 2.11 Macros of the ProModel simulation model

ID	Macros	Operations
OpLogic1	IF (CLOCK(SEC)/60@4=1)AND	None
	((CLOCK(SEC)@60>0)AND(CLOCK(SEC)@60<58))	
	THEN{Dr ive1=1}	
	ELSE {	
	IF (CLOCK(SEC)/60@4=1)AND(CLOCK(SEC)@60=0)	
	THEN {Dr ive1=4 Dr ive4=22}	
	ELSE {	
	IF (CLOCK(SEC)/60@4=1)AND(CLOCK(SEC)@60>57)	
	THEN {Dr ive1=21}	
	ELSE {Dr ive1=3}	
	}	
	}	
OpLogic2	IF (CLOCK(SEC)/60@4=2)AND	None
	((CLOCK(SEC)@60>0)AND(CLOCK(SEC)@60<58))	
	THEN {Dr ive2=1}	
	ELSE {	
	IF (CLOCK(SEC)/60@4=2)AND(CLOCK(SEC)@60=0)	
	THEN {Dr ive2=4 Dr ive1=22}	
	ELSE {	
	IF (CLOCK(SEC)/60@4=2)AND(CLOCK(SEC)@60>57)	
	THEN {Dr ive2=21}	
	ELSE {Dr ive2=3}	
	}	
	}	
OpLogic3	IF (CLOCK(SEC)/60@4=3)AND	None
	((CLOCK(SEC)@60>0)AND(CLOCK(SEC)@60<58))	
	THEN {Dr ive3=1}	
	ELSE {	
	IF (CLOCK(SEC)/60@4=3)AND(CLOCK(SEC)@60=0)	
	THEN {Dr ive3=4 Dr ive2=22}	
	ELSE {	
	IF (CLOCK(SEC)/60@4=3)AND(CLOCK(SEC)@60>57)	
	THEN {Dr ive3=21}	
	ELSE {Dr ive3=3}	
	}	
	}	
OpLogic4	IF (CLOCK(SEC)/60@4=0)AND	None
	((CLOCK(SEC)@60>0)AND(CLOCK(SEC)@60<58))	
	THEN {Dr ive4=1}	
	ELSE {	
	IF(CLOCK(SEC)/60@4=0)AND(CLOCK(SEC)@60=0)	
	THEN {Dr ive4=4 Dr ive3=22}	
	ELSE {	
	IF (CLOCK(SEC)/60@4=0)AND(CLOCK(SEC)@60>57)	
	THEN {Dr ive4=21}	
	ELSE {Dr ive4=3}	
	}	
	}	

Table 2.12 Processing of entity CarNorth

Entity	Location	Operation	Output	Destination	Rule	Movelogic
CarNorth	TrafficlightNE	OpLogic1 OpLogic2	CarNorth	ArrivalEast		IF (Drive2=1) OR (Drive2=21) THEN {MOVE ON Traffic} ELSE {WAIT UNTIL (Drive2=1) OR (Drive2=21)}
	TrafficlightNS	OpLogic3 OpLogic4	CarNorth	ArrivalSouth		
	TrafficlightNW			ArrivalWest		
	ArrivalNorth			TrafficlightNE	0.4	MOVE ON Traffic
				TrafficlightNS	0.3	
				TrafficlightNW	0.31	
	North			ArrivalNorth	First1	MOVE ON Arrival
	ArrivalEast			EXIT	First1	
	ArrivalSouth					
	ArrivalWest					

Table 2.13 Processing of entity CarEast

Entity	Location	Operation	Output	Destination	Rule	Movelogic
CarEast	TrafficlightEN	OpLogic1 OpLogic2 OpLogic3 OpLogic4	CarEast	ArrivalNorth		IF (Drive3=1) OR (Drive3=21) THEN {MOVE ON Traffic} ELSE {WAIT UNTIL (Drive3=1) OR (Drive3=21)}
	TrafficlightES			ArrivalSouth		
	TrafficlightEW			ArrivalWest		
	ArrivalNorth			EXIT		
	ArrivalEast			TrafficlightEN	0.4	MOVE ON Traffic
				TrafficlightES	0.3 1	
				TrafficlightEW	0.3	
	East			ArrivalEast	First1	MOVE ON Arrival
	ArrivalSouth			EXIT		
	ArrivalWest					

Table 2.14 Processing of entity CarSouth

Entity	Location	Operation	Output	Destination	Rule	Movelogic
CarSouth	TrafficlightSN	OpLogic1 OpLogic2 OpLogic3 OpLogic4	CarSouth	ArrivalNorth		IF (Drive1=1) OR (Drive1=21) THEN {MOVE ON Traffic} ELSE {WAIT UNTIL (Drive1=1) OR (Drive1=21)}
	TrafficlightSE			ArrivalEast		
	TrafficlightSW			ArrivalWest		
	ArrivalNorth			EXIT	First1	
	ArrivalEast					
	ArrivalSouth			TrafficlightSN	0.3 1	MOVE ON Traffic
				TrafficlightSE	0.4	
				TrafficlightSW	0.3	
	South			ArrivalSouth	First1	MOVE ON Arrival
	ArrivalWest			EXIT	First1	

Table 2.15 Processing of entity CarWest

Entity	Location	Operation	Output	Destination	Rule	Movelogic
CarWest	TrafficlightWN	OpLogic1 OpLogic2	CarWest	ArrivalNorth		IF (Drive3=1) OR (Drive3=21) THEN {MOVE ON Traffic} ELSE {WAIT UNTIL (Drive3=1) OR (Drive3=21)}
	TrafficlightWE	OpLogic3 OpLogic4		ArrivalEast		
	TrafficlightWS			ArrivalSouth		
	ArrivalNorth			EXIT	FIRST 1	
	ArrivalEast					
	ArrivalSouth					
	ArrivalWest			TrafficlightWN	0.3 1	MOVE ON Traffic
				TrafficlightWE	0.3	
				TrafficlightWS	0.4	
	West			ArrivalWest	FIRST 1	MOVE ON Arrival

Fig. 2.21 Introducing the Internet of Things paradigm into transportation system analysis

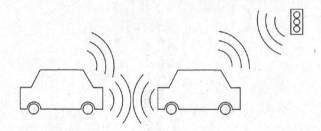

With regard to the advent of Internet and communication technologies and especially the new paradigm of the Internet of Things, vehicles and traffic lights at road intersections will be equipped with radio frequency identification (RFID) technology in the near future, enabling them to communicate wirelessly with each other and with traffic light systems. With the Internet of Things, it will be possible to send wireless radio signals from a traffic light system to vehicles approaching the traffic light system at a road intersection. By monitoring the actual speed of the vehicles, their arrival time at the road intersection traffic light system will be calculated. Based on this calculation, the vehicle will arrive at the traffic light system when it has changed from red to green. Thus the traffic light management system will send information to the vehicle's engine management system to slow down the vehicle's speed to ensure that the vehicle arrives on time at the next green phase of the road intersection traffic light, as illustrated in Fig. 2.21. This will help to avoid unnecessary accelerating and braking actions, which waste energy, and reduce CO_2 emissions and pollution. Moreover, it would be possible for vehicles in the near future to adjust their speed and distance via intercar communication through the Internet of Things (Moeller et al. 2013).

There is already some statics technology available for speed optimization through intercommunication with traffic light systems in a certain distance from a traffic light road intersection with the impact of different advisory speeds according to the time to the next signal change at the road intersection.

2.9 Exercises

1. Explain what is meant by the term cost-effectiveness.
2. Give an example of cost-effectiveness in transportation.
3. Explain what is meant by the term cost-benefit analysis.
4. Give an example of a cost-benefit approach in transportation.
5. Explain what is meant by the term lifecycle cost analysis.
6. Give an example of a lifecycle cost analysis in transportation.
7. Explain what is meant by the term trip generation.
8. Give an example of trip generation in transportation.
9. Explain what is meant by the term trip distribution.

10. Give an example of trip distribution in transportation.
11. Explain what is meant by the term mode split.
12. Give an example of mode split in transportation.
13. Explain what is meant by the term route assignment.
14. Give an example of route assignment in transportation.
15. Explain what is meant by the term level of service.
16. Give an example of a level of service in transportation.
17. Explain what is meant by economic transportation and land use models.
18. List and define the main characteristic statements.
19. To what specific approaches do traffic flow models refer?
20. Describe the two approaches for flow rate versus density and speed versus flow rate.
21. Explain what is meant by the term free-flow speed.
22. Describe the mathematical equation for free-flow speed in detail.
23. Explain what is meant by the term macroscopic traffic flow model.
24. Give an example of a macroscopic traffic flow model.
25. Explain what is meant by the term queuing model.
26. Describe how queuing analysis allows determining the time needed under congested conditions.
27. Explain what is meant by Little's formula.
28. Describe the mathematical background for Little's formula.
29. Explain what is meant by Kendall's notation.
30. Describe a case study example by using Kendall's notation.
31. Explain what is meant by the term FCFS or FIFO.
32. Give an example of FCFS and FIFO in transportation.
33. Explain what is meant by the term preemptive shortest activity first.
34. Give an example of preemptive shortest activity first in transportation.
35. Explain what is meant by the term Erlang distribution.
36. Describe the mathematical equation for the Erlang distribution.
37. Explain what is meant by the term traffic demand model.
38. Give an example of a traffic demand model.
39. Explain what is meant by the congestion network model.
40. Describe the results shown in Fig. 2.10 in your own words.
41. Explain what is meant by the term graph model.
42. Give an example of a multigraph in transportation.
43. Explain what is meant by the term bottleneck analysis.
44. Give an example of a bottleneck in transportation.
45. Explain what is meant by the road intersection.
46. Give an example of a four-arm road intersection.
47. What is meant by the term radio frequency identification?
48. Give an example of a radio frequency identification application in the transportation system sector.
49. What is meant by Internet of Things?
50. Give an example of an Internet of Things application in the transportation system sector.

References and Further Readings

Bando M, Hasebe K, Nakayama A, Shibata A, Sugiyama Y (1995) Dynamical model of traffic congestion and numerical simulation. Phys Rev E Stat Phys Plasmas Fluids Relat Interdiscip Topics 51(2):1035–1042

Bartholomew K, Ewing R (2009) Land use-transportation scenarios and future vehicle travel and land consumption: a meta-analysis. J Am Plann Assoc 75(1):13–27

Cascetta E (2009) Transportation systems analysis: models and applications. Springer, New York

Crommelin RW (1974) Employing intersection capacity utilization values to estimate overall level of service. Trans Res Board 44(10):11–14, Transportation Research Record, Washington, DC

CUTR (2007) Economics of travel demand management: comparative cost effectiveness and public investment, Center for Urban Transportation Research, at www.nctr.usf.edu/pdf/77704.pdf

D'Agostino RB, Stephens MA (1986) Eds.: Goodness-of-fit tests. Marcel Decker Publ.

Dagpunar J (1988) Principles of random variate generation. Clarendon, Oxford

DeGroot MH (1975) Probability and statistics. Addison Wesley Publ., Reading

Doboszcarek S, Forstall V (2013) Mathematical modeling by differential equations. http://www.norbertwiener.umd.edu/Education/m3cdocs/Presentation2.pdf

Dong X, Ben-Akiva M, Bowman J, Walker J (2006) Moving from trip-based to activity-based measures of accessibility. Transp Res A 40(2):163–180

Donoso P, Martinez F, Zegras C (2006) The Kyoto protocol and sustainable cities: potential use of clean-development mechanism in structuring cities for carbon-efficient transportation. Transp Res Rec 1983:158–166

Ellis D, Glover B, Norboge N (2012) Refining a methodology for determining the economic impacts of transportation improvements, University Transportation Center for Mobility at Texas A&M University. http://utcm.tamu.edu/publications/final_reports/Ellis_11-00-68.pdf

(ERC) http://transport.epfl.ch/modelling-controlling-traffic-congestion

Evans M, Hastings N, Peacock B (2000) Statistical distributions. Wiley Publ., New York

Fishman GS (1973) Statistical analysis for queuing simulations. Management Science Vol. 20:363–369

Fishman GS (2006) Monte carlo: concepts, algorithms, and applications. Springer Publ., New York

Flynn MR, Kasimov AR, Nave J-C, Rosales RR, Seibold B (2009) Self-sustained nonlinear waves in traffic flow. Phys Rev E 79(5):056113. doi:10.1103/PhysRevE.79.056113, DOI:10.1103/PhysRevE.79.056113#_blank

Gupta D, Singla S, Singla P, Singh S (2013) Minimization of elapsed time in $N \times 3$ flow shop scheduling problem, the processing time associated with probabilities including transportation time. Int J Innovat Eng Technol Special Issue ICAECE-2013, pp 38–42

Harris GA (2008) Bridging the data & information gap, project report no. AL-26-7262-01

Hogg RV, Craig AF (1995) Introduction to mathematical statistics. Prentice-Hall Publ., Englewood Cliffs

Husch D (2003) Intersection capacity utilization. Trafficware, Albany

Immers LH, Logghe S (2002) Traffic flow theory, course H 111. Catholic University Leuven

Jehle IA (2014) Simulation of a traffic light junction, student project work, TU Clausthal, Germany

Kerner BS, Rehborn H (1997) Experimental properties of phase transitions in traffic flow. Phys Rev Lett 79:4030–4033

Kleijnen JPC (1974) Statistical techniques in simulation. Marcel Decker Publ, New York

Kockelman KM (2001) Modeling traffic's flow-density relation: accommodation of multiple flow regimes and traveler types. Transportation 29(4):363–373

Lee H-T, Romer TF (2011) Automating the process of terminal area node-link model generation. J Guid Control Dyn 34(4):1228–1237

Lee R, Niemeier D, Parker T, Handy S (2012) Evaluation of operation and accuracy of available smart growth trip generation methodologies for use in California, Transportation Research Record. J Transport Res Board, Serial Issue Number 2307:120–131

Lima E, Chwif L, Baretto M (2008) Methodology for selecting the best suitable bottleneck detection method. In: Proceedings SCS 32nd conference on winter simulation. Society for Computer Simulation International, San Diego, pp 740–754

Lindsey CR, Verhoef ET (1999) Congestion modelling, Series Tinbergen Institute discussion papers, Number 99–091/3

Lindsey CR, van den Berg VAV, Verhoef ET (2012) Step tolling with bottleneck queuing congestion. J Urb Econ 72(1):46–59

Little JDC, Graves SC (2008) Little's law, Chapter 5. In: Chhajed D, Loewe TJ (eds) Building intuition: insight from basis operations management models and principles. Springer Publ., pp 81–100

Litman T (2006) What's it worth? Economic evaluation for transportation decision making, transportation association of Canada, (www.tac-atc.ca); at www.vtpi.org/worth.pdf

Litman T (2013) Congestion costing critique: critical evaluation of the urban mobility report. In: VTPI, (www.vtpi.org); at www.vtpi.org/UMR_critique.pdf

Moeller DPF, Haas R, Vakilzadian H (2013) Ubiquitous learning: teaching modeling and simulation (M&S) with technology. In: Vakilzadain H, Crosbie R, Huntsinger R, Cooper K (eds) Proceedings of the 2013 summer simulation multiconference GCMS 2013. Curran Publication, Red Hook, pp 125–132

Muench S (2004) Traffic Concepts, http://www.google.de/url?sa=t&rct=j&q=&esrc=s&source=web&cd=1&ved=0CDEQFjAA&url=http%3A%2F%2Fcourses.washington.edu%2Fcee320w%2Flectures%2FTraffic%2520Concepts.ppt&ei=EkpRU82uC4e1tAajiYGYCw&usg=AFQjCNFOFglgi-iDYswasEvo91YzMZ_OKg&bvm=bv.65058239,d.Yms

Scheurer J, Horan E, Bajwa S (2009) Benchmarking public transport and land use integration in Melbourne and Hamburg: hints for policy makers. Presentation at 23rd AESOP 2009 congress, liverpool

Small KA (1998) Project evaluation, Chapter 5. In: Gómez-Ibáñez JA, Tye W, Winston C (eds) Transportation policy and economics: a handbook in honor of John R. Meyer. www.brookings.edu, www.uctc.net/papers/379.pdf

Stahl WA (2002) Statistical data analysis (in German). Vieweg Publ. Wiesbaden

Stopher PR, Greaves SP (2007) Household travel surveys: where are we going? Transp Res A 41(5):367–381

Sundarapandian V (2009) Queuing theory, Chapter 7. In: Probability, statistics and queuing theory. PHI Learning, New Delhi

Tan AC, Bowden RO (2004) The Virtual Transport System (VITS)—final report. Monograph, Transportation Research Board, Washington, DC

TDM Encyclopedia (2013) Transport model improvements: improving methods for evaluating the effects and value of transportation system changes; see http://www.vtpi.org/tdm/tdm125.htm

TDM Encyclopedia (2014) Victoria Transport Policy Institute, Victoria

Transportation Research Board, see http://onlinepubs.trb.org/onlinepubs/sr/sr288.pdf

TRB (2007) Metropolitan Travel Forecasting: Current Practice and Future Direction, Special Report 288, Transportation Research Board (www.trb.org); at http://onlinepubs.trb.org/onlinepubs/sr/sr288.pdf

TRB (2012) The effect of smart growth policies on travel demand, capacity project C16, Strategic Highway Research Program (SHRP 2), (www.trb.org); http://onlinepubs.trb.org/onlinepubs/shrp2/SHRP2prepubC16.pdf

USDOT (2003) US Department of Transportation, Economic Analysis Primer, Office of Asset Management, FHWA, USDOT, 2003. www.fhwa.dot.gov/infrastructure/asstmgmt/primer.pdf

Virginiadot (2014) http://www.virginiadot.org/projects/resources/vtm/what_is_travel_demand_modeling.pdf

Wang Y, Zhao Q, Zheng D (2005) Bottlenecks in production networks: an overview. J Syst Sci Syst Eng 14:18ff

Yücesan E, Schruben LW (1998) Complexity of simulation models: a graph theoretic approach. Journal of Comput 10:94–108

Links

(URL 1) http://pages.stern.nyu.edu/~adamodar/New_Home_Page/StatFile/statdistns.htm
(URL 2) http://www.tc.faa.gov/acb300/aasw.asp
(URL 3) http://www.tc.faa.gov/acb300/more_simmod.asp
(URL 4) http://ww1.jeppesen.com/support/technical_support_details.jsp?prodNameTxt2=Total
     %20Airspace%20and%20Airport%20Modeler%20%28TAAM%29

Traffic Assignments to Transportation Networks

3

This chapter begins with a brief overview of traffic assignment in transportation systems. Section 3.1 introduces the assignment problem in transportation as the distribution of traffic in a network considering the demand between locations and the transport supply of the network. Four trip assignment models relevant to transportation are presented and characterized. Section 3.2 covers traffic assignment to uncongested networks based on the assumption that cost does not depend on traffic flow. Section 3.3 introduces the topic of traffic assignment and congested models based on assumptions from traffic flow modeling, e.g., each vehicle is traveling at the legal velocity, v, and each vehicle driver is following the preceding vehicle at a legal safe velocity. Section 3.4 covers the important topic of equilibrium assignment which can be expressed by the so-called fixed-point models where origin to destination (O-D) demands are fixed, representing systems of nonlinear equations or variational inequalities. Equilibrium models are also used to predict traffic patterns in transportation networks that are subject to congestion phenomena. Section 3.5 presents the topic of multiclass assignment, which is based on the assumption that travel demand can be allocated as a number of distinct classes which share behavioral characteristics. In Sect. 3.6, dynamic traffic assignment is introduced which allows the simultaneous determination of a traveler's choice of departure time and path. With this approach, phenomenon such as peak spreading in response to congestion dynamics or time-varying tolls can be directly analyzed. In Sect. 3.7, transportation network synthesis is introduced which focuses on the modification of a transportation road network to fit a required demand. Section 3.8 covers a case study involving a diverging diamond interchange (DDI), an interchange in which the two directions of traffic on a nonfreeway road cross to the opposite side on both sides of a freeway overpass. The DDI requires traffic on the freeway overpass (or underpass) to briefly drive on the opposite side of the road. Section 3.9 contains comprehensive questions from the transportation system area. A final section includes references and suggestions for further reading.

© Springer-Verlag London 2014   109
D.P.F. Möller, *Introduction to Transportation Analysis, Modeling and Simulation*,
Simulation Foundations, Methods and Applications,
DOI 10.1007/978-1-4471-5637-6_3

3.1   Introduction

Transportation networks are used extensively and are often congested to varying
degrees, especially those in urban and metropolitan areas. Congestion in a transportation network is, in many cases, the result of either an increase in traffic demand
or a decrease in traffic supply. In either case, congestion reduces the efficiency
of the transportation network and increases the travel time of vehicles in these
networks. Therefore, the traffic demand and supply between locations must either
be known or estimated to achieve better traffic planning.

Assuming that information can be accessed on the aggregated traffic demands
between the origin and destination or within specific zones in a transportation
network, then traffic congestion can be determined by using the total travel time
of the system as an outcome of traffic assignment models. The assignment problem
in transportation can be defined as the distribution of traffic in a network considering the demand between locations and the transport supply of the network. Traffic
assignment models aim to determine the number of trips on different links, or road
sections, of the network given the travel demand between different pairs of nodes
(zones) in the transportation network. Hence, the traffic assignment models mathematically describe the route choice process using the sequential demand analysis
procedure. Furthermore, these models assume that travel time on the link will be the
major factor considered by the trip maker when choosing a route. However, the
models differ in their assumptions regarding the variation in link travel times based
on link volume or link traffic flow.

Traffic assignment models simulate the interaction of demand and supply and
their impact on a transportation network based on a set of constraints, typically
related to transport capacity, time, and cost (Rodrigue 2013). Therefore, the purchase of tickets to travel a certain distance with a chosen transportation mode at a
specific date and time and at a specific cost can be introduced as an example for
traffic assignment. This example allows the traveler to decide whether to travel a
direct path between origin O and destination D at a higher cost, e.g., via an airplane,
or to select a less costly mode of travel, e.g., a train, which may not necessarily be a
direct path. If a lot of travelers select the same path every day, this may result in
a problem with traveler assignments to a transportation mode and subsequent
assignment to the transportation mode supply through the respective transportation
companies. This may result in a less than optimal match of supply with regard to
actual demands.

As illustrated by the preceding example, the objective of traffic assignment
models is to identify the possible trips on different links in a network based
on the travel demand between different pairs of nodes. This assumes that cost
and travel time on the link are the most relevant factors which travelers consider
when choosing a route. Thus, traffic assignment models mimic the transportation
system itself, by observing the pattern of vehicular-based movements when travel
demand, represented by a travel matrix or matrices, is satisfied. However, there

are three basic categories of assignments which are often made in transportation studies:

- Existing trips to the existing network
- Future trips to the existing and committed network
- Future trips to the existing and committed network and proposed network

Thus, assignment methods are used to model the distribution of traffic in a network according to a set of constraints, notably related to transport capacity, time, and cost, which can be described as follows:

- Estimate the traffic volume on the links of the network and obtain aggregate network measures.
- Estimate the interzonal travel cost.
- Analyze travel pattern of each origin to destination (O-D) pair.
- Identify congested links and collect essential traffic data for the design of future junctions.

Therefore, traffic assigned on a network has to be in accordance to a sequence of links whereby every link has its own value and direction for which multiple conditions have to be fulfilled (Rodrigue 2013):

- There have to be nodes in the graph-oriented network model where traffic can be generated and attracted. These nodes are generally associated to centroids in an O-D matrix. The O-D matrix itself refers to the ratio of trips between origin and destination pairs and zones.
- The minimal ($l(a,b)$) and maximal ($k(a,b)$) capacities of every link has to be taken into account. k(a,b) is the transport supply on the link (a,b).
- Transport demand must be taken into account. The O-D matrix has equal inputs and outputs.
- The conservation of traffic at every node is not an origin or a destination.

At least four types of trip assignment models are relevant in transportation analysis, modeling, and simulation:

- *All-or-nothing assignment for uncongested networks*: focus on shortest path and assign all travels on the shortest path between an O-D pair.
- *User equilibrium*: follow the Wardrop principle assuming that every traveler is on the shortest path to his/her destination.
- *Stochastic travel assignment*: assume that travelers do not have perfect real-time information about cost, wherefore their route choice decisions have uncertain factors.
- *Dynamic assignment*: assume that early travel decisions have an impact on later traffic congestion situations. Therefore, it considers interactions over time.

3.2   Uncongested Network

Traffic assignment to uncongested networks is based on the assumption that cost does not depend on traffic flow. Therefore, traffic path flows and link flows are obtained from path choice probabilities that are themselves computed from flow-independent link performance attributes and costs (Cascetta 2009). The all-or-nothing assignment for uncongested networks is based on the following assumptions:

- Link costs are fixed.
- All drivers think alike.
- Every driver from A to B chooses the same route.
- All drivers are assigned to that route and none to other potential routes.

These assumptions are appropriate for uncongested networks with few alternative routes. The all-or-nothing assignment offers a desirable option—what drivers would do if all choices were available and if congestion was not an influence. Thus in this method, trips from any origin zone to any destination zone are allocated to a single, minimum cost path between them. But using only one path between every O-D pair is an unrealistic constraint if there is another path with equivalent or nearly equivalent travel cost. Moreover, traffic on links is assigned without consideration of whether or not there is adequate capacity or heavy congestion; travel time is a fixed input and does not vary depending on the congestion on a link. However, this approach may be reasonable in sparse and uncongested networks where there are few alternative routes and they have large differences in travel cost. This approach may also be used to identify the desired path, which is the path drivers would like to travel in the absence of congestion.

Uncongested assignment models are used for analyzing relatively uncongested traffic networks. They are used for several applications in the transportation system, such as public and private transport systems. The mathematical notation for traffic flow in uncongested networks was introduced in Sect. 2.2. Moreover, uncongested network assignment models are a key component of congested network assignment models, which are described in Sect. 3.3.

The aforementioned uncongested traffic assignment models can be defined as path flows expressed as a function of path costs and demand flow, which can be described as follows (Cascetta 2009):

$$h_{UN} = h_{UN} \frac{c_p}{d} = P(-c_P)d$$

where h_{UN} is the uncongested path flow; c_p is the path cost; d is the demand vector, whose components are the demand values for each O-D pair; and P is the vector of path choice probabilities.

3.3   Congested Network

An equivalent continuous-time, optimal control problem is formulated to predict the temporal evolution of traffic flow patterns on congested multiple *O-D* road networks, corresponding to a dynamic generalization of the Wardropian user equilibrium. Optimality conditions are derived using the Pontryagin minimum principle and given economic interpretations, which are generalizations of similar results previously reported for single-destination networks. Analyses of sufficient conditions for optimality and of singular controls are also given. Under the steady-state assumptions, the model is shown to be a proper dynamic extension of Beckmann's mathematical programming problem for a static user equilibrium traffic assignment.

Congestion modeling is based on assumptions from traffic flow modeling with the following boundary conditions:

- Each vehicle has the legal velocity v.
- Each vehicle driver follows the preceding vehicle with legal safe velocity.
- The notation of the traffic flow can be introduced as follows:

$$x_n^{''} = d_r[v(\Delta x_n) - x_n] \tag{3.1}$$

and

$$\Delta x_n = x_{n+1} - x_n \tag{3.2}$$

where n indicate number of each vehicle ($n = 1,2,3,\ldots,N$), N represent the total number of vehicles, d_r represent a constant describing drivers response which is assumed being independent of n, and x_n represent the coordinate of the nth vehicle. The two dots denote the second derivative with respect to time of x_n. Furthermore, (3.1) neglects the length of the vehicles for simplicity reasons and assumes that each driver has the same awareness. It is also assumed that legal velocity v of vehicle n depends on the following distance of the preceding vehicle number $n+1$ (Bando et al. 1995).

Two cases in traffic flow modeling are important for velocity v (see also Chap. 2):

- Distance between vehicles is short: velocity must be reduced to become slow enough to prevent crashing into a preceding vehicle.
- Distance between vehicles is wide: velocity of following vehicle can be higher, but should not exceed speed limit.

Therefore, velocity v becomes a function with the following properties:

- Monotonic increasing function f which is a function between ordered sets that preserve the given order which increase if x in $v(x)$ increase. A monotonic increasing function's derivative is always positive as shown in Fig. 3.1.
- $v(\Delta x_n)$ has an upper bound $v^{max} \equiv v(\Delta x_n \to \infty)$.

Fig. 3.1 Monotonic function

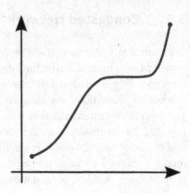

The following steady-state flow solution satisfies the dynamic traffic flow system notation:

$$x_n^{(0)} = b^* n + v_c^* t \tag{3.3}$$

with

$$b = \frac{l}{N} \tag{3.4}$$

where b represents a distant constant between successive vehicles with l as the length of the road the vehicles are moving on, N as the number of the nth vehicle, and v_c as the constant velocity or steady-state traffic flow.

These equations describe the steady-state traffic flow between vehicles without congestion, where vehicles are uniformly distributed with identical car spacing moving with the same constant velocity v (Bando et al. 1995). For a congested situation, b and v_c show different numbers, meaning that the number of vehicles moving on length l is so high that v_c could not hold the steady-state traffic flow condition any longer and v_c now holds the congested traffic flow condition. Thus an equivalent time optimal control problem can be formulated to predict the traffic flow patterns on congested multiple O-D networks.

Therefore, the management of severe congestion in complex urban networks can be achieved with regard to congested network assignment models that can replicate real traffic situations with long queues and spillbacks.

3.4   Equilibrium Assignment

Equilibrium assignment is generally expressed by the so-called fixed-point models, where O-D demands are fixed, representing systems of nonlinear equations or variational inequalities (Cascetta 2009). Equilibrium models are also used to predict traffic patterns in transportation networks that are subject to congestion phenomena. The idea of traffic equilibrium models goes back to the early 1920s when Knight gave a simple and intuitive postulate of traffic behavior under congested conditions (Knight 1924; Florian and Hearn 2008):

Suppose that between two points e.g. two highways, one of which is broad enough to accommodate without crowding the traffic which may use it, but is poorly graded and surfaced, while the other is a much better road, but narrow and quite limited in capacity. If a large number of trucks operate between the two termini and are free to choose either of the two routes, they will tend to distribute themselves between the roads in such proportions that the cost per unit of transportation, or effective returns per units of investment, will be the same for every truck on both routes. As more trucks use the narrower and better road, congestion develops, until a certain point it becomes equally profitable to use the broader but poorer graded and surfaced highway.

Therefore, a deterministic traffic equilibrium assignment model of route choice can be described as follows:

- Transportation network consists of nodes $n \in N$ which represent origins and destinations (O-D) of traffic.
- Transportation network consist of arcs $a \in A$ which represent the road network.
- The number of vehicles V at link a is V_a $(a \in A)$.
- Cost of traveling on a link is given by a cost function $c_a (Q)$ $(a \in A)$, where Q is the vector of link traffic flows over the entire network.
- Cost functions model of the time-dependent behavior (Florian and Hearn 2008).

Knight's traffic pattern at equilibrium is now known as Wardrop (or user) equilibrium (Dafermos and Sparrow 1969), and it is effectively thought of as a steady state evolving after a transient phase in which travelers successively adjust their route choices until a situation with stable route travel costs and route flows has been reached (Larsson and Patriksson 1999). In a seminal contribution, Wardrop (1952) stated two principles that formalize this notion of equilibrium and the alternative postulate of the minimization of the total travel costs. His first principle reads:

The journey times on all the routes actually used are equal, and less than those which would be experienced by a single vehicle on any unused route.

Wardrop's first principle of route choice, which is identical to the notion postulated by Knight (Knight 1924), became accepted as a sound and simple behavioral principle to describe the spreading of trips over alternate routes due to congested conditions (Florian 1999; Correa and Stier-Moses 2010).

Assigning traffic to paths and links requires a rule like the Wardrop equilibrium (Wardrop 1952). Hence, it is known that travelers have to find the shortest path from origin to destination; and network equilibrium occurs if no traveler can decrease his/her travel effort by shifting to a new path. These are user optimal conditions; no user gains from changing travel paths once the system is in equilibrium.

Let us consider a directed network graph, $G = (N, A)$, and a set, $C \subseteq N \times N$, of contingencies represented by O-D pairs. For each $k \in C$, a demand flow at a rate equal to q_{dk} must be routed from the corresponding origin to its destination. Therefore, the basic equilibrium assignment model assumes that demands are arbitrarily divisible, meaning that the routing decision of a single individual has only an infinitesimal impact on other users.

Let R_k be the set of routes in G connecting the corresponding O-D pairs with constraint $k \in C$, and let $R := \cup_{k \in C} R$.

Assume that a link flow introduced as a nonnegative vector, $f = (f_a)_{a \in A}$, describes the traffic rate in each link.

Furthermore, a nonnegative, nondecreasing, continuous link travel cost function c_a (•) with values in $\mathfrak{R}_{\geq 0} \cup \{\infty\}$ maps the flow q_a on arc a to the time needed to traverse a.

Let a route flow with a nonnegative vector, $q = (q_r)_{r \in R}$, meet the demand, i.e., $\sum_{r \in R_k} q_r = q_{dk}$ for $k \in C$. For a given route traffic flow, the corresponding link traffic flow can be computed as $q_a = \sum_{r \in R_k} q_r$ for each $a \in A$. For a traffic flow, q, the travel cost along a route, r, is $c_r(q) := \sum_{a \in r} c_a(q_a)$ (Correa and Stier-Moses 2010).

Interpreting Wardrop's first principle as requiring that all traffic flow travels along the shortest paths, a flow q is called a Wardrop equilibrium if, and only if, for all

$$k \in C, \text{ we have } c_r(q) = \min_{q=R_k} c_q(q).$$

Thus, the equilibrium assignment based on Wardrop's first principle states that no driver can unilaterally reduce the travel cost by shifting to another route.

Besides, the previously mentioned model assignment approach also measures the traffic flow on a road network and can be used as the maximum traffic load and traffic link cost minimization. The maximum traffic load approach is based on the number of vehicles in a traffic flow that a road network can support at a point in time. The maximum load (maxLD) in this approach can be expressed as a summation of the capacity of all links, which can be described as follows:

$$\max LD = \sum_a \sum_b k(a, b)$$

where $k(a, b)$ represents the traffic supply on the links (a, b) and its summation represents the maximal capacities of every link to be considered. The number of vehicles in the traffic flow supported by a road network while fulfilling a transport demand can be expressed as load (LD) which is the summation of the traffic flows of all links, described as follows:

$$LD = \sum_a \sum_b q(a, b)$$

where $q(a, b)$ is the traffic-flow-dependent supply on the links (a, b). If the traffic load of a road network reaches the maximum load capacity, a congestion state is reached.

Traffic flow maximization involves determining the maximum traffic demand that a road network or a section of a road network can support between its nodes. This can be expressed as follows:

$$\max : q(a,b), \forall (a,b)\_subject\_to\_q(a,b) \leq k(a,b)$$

which involves maximizing traffic flow for all links, where traffic flow on links must be equal or lower to the capacity of the link. The heuristic method is the easiest way to solve this equation for simple road networks.

Cost minimization involves determination of the minimum traffic flow costs considering a known demand. Traffic flow costs on a road network link can be expressed by $c[q(a,b)]$ and the minimization function by

$$\max : \sum_{a}\sum_{b} c[q(a,b)]\_subject\_to\_q(a,b) \leq k(a,b); or\_q(a,b) \geq l(a,b)$$

where $l(a,b)$.is the traffic supply on links $l(a,b)$, $k(a,b)$ represents the traffic supply on links (a,b), $q(a,b)$ is the traffic-flow-dependent supply on the links (a,b), and c is a cost functional.

The goal of this equation is to minimize the summation of traffic flow costs of each link subject to capacity constraints. Once again, the heuristic method is the easiest way to solve this equation for simple road networks. It should be noted that several types of costs are involved in the minimization procedure.

3.5   Multiclass Assignment

Let $G = (N, A)$ be a directed traffic road network defined by a set, N, of nodes and a set, A, of directed links. Each link, $a \in A$, has an associated traffic-flow-dependent travel time denoting the travel time per unit traffic flow or average travel time on each link. Travel demand at each link, $a \in A$, can be subdivided into classes corresponding to groups of users with different characteristics. Let λ_m ($\lambda_m \geq 0$) be the average value of time for users of class m, and let d_w^m be the demand for travel of class m between O-D pair with $w \in W$. Assume, for simplicity reasons, that d_w^m is given and the assignment problem is a fixed-demand, multiclass traffic network equilibrium problem (Yang and Huang 2004). Hence, multiclass assignment is supposed to be based on the assumption that travel demand can be allocated as a number of distinct classes which share behavioral characteristics of the relevant traffic demand models (see Chap. 2), including path choice, such as:

- Attributes
- Parameters
- Specification

In urban and/or metropolitan areas, classes may be identified on the basis of the following because different travel costs and different travel time values are associated with these characteristics (Cascetta 2009):

- Purpose of travel, such as business, private, etc.
- Socioeconomic category, such as residents, nonresidents, etc.
- Travel duration, such as minutes, hours, etc.

In intercity areas, classes can be defined by:

- Purpose of travel such as business, private, etc.
- Socioeconomic category such as residents, nonresidents, etc.
- Vehicle type such as aircraft, bus, private car, public transportation system, train, etc.

They can be defined as such because:

- Additive path costs in case of congestion
- Nonadditive path costs which include all specific path and/or class cost and are assumed being independent of congestion
- Path choice models
- Service charges such as tolls, etc.
- Travel times

These may be different, as shown in Fig. 3.2.

Therefore, the cost functions for different paths and vehicles used are different, but it is assumed that they all depend on the overall link flow.

Consistency between link and path costs for each O-D pair, OD, and each class, I, can be expressed in the following relationship, given in Cascetta (2009):

$$g_{OD,I}^{ADD} = \Delta_{OD,I}^{T} c^{I} \quad \forall OD \, \forall I$$

$$g_{OD,I} = g_{OD,I}^{ADD} + g_{OD,I}^{NA} = \Delta_{OD,I}^{T} c^{I} + g_{OD,I}^{NA} \quad \forall OD \, \forall I$$

where $g_{OD,I}^{ADD}$ is the additive path cost vector for O-D pair OD and class I, $g_{OD,I}^{NA}$ is the nonadditive path cost vector for O-D pair OD and class I, and $g_{OD,I}$ is the total path cost vector for O-D pair OD and class I.

Congestion can be modeled and simulated by assuming that cost c_a^{I} is a function of class flows on the same link, a, $a \in A$, and possibly on other links. As described in Cascetta (2009), we can consider cost functions that are nonseparable with regard to class flows as well as link flows. This effect is usually represented using cost functions in which the congested link performance attributes for each class depend on the total link flows:

$$c^{I} = c^{I}(f^{1}, \ldots, f^{I}) = c^{I}(f) = c^{I}\left(\sum_{I} f^{I}\right) \quad \forall I$$

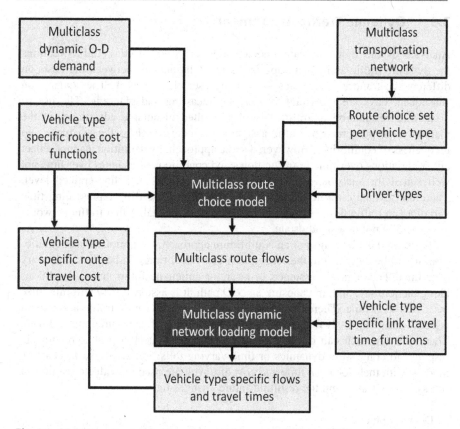

Fig. 3.2 Multiclass dynamic traffic assignment framework (Bliemer 2001)

Following the statements given in Bliemer (2001), the extension towards multiple vehicle types necessitates, among others, the following modifications:

- Defining multiclass specific input for each vehicle type and dedicated road infrastructure for a certain vehicle type (e.g., truck lanes).
- Adapt route choice model such that each vehicle type has its own route cost function and its own route choice sets.
- Adapt dynamic network loading model such that traffic flows are consistent with driving behavior of vehicle types. This means each vehicle type has its own link travel time function such that not all vehicle types move at the same speed through a link. Furthermore, vehicle types on the same link influence each other (i.e., a vehicle type can have a certain impact on link travel times of other vehicle types), possibly in an asymmetric fashion.

Details on these modifications can be found in Bliemer (2001).

3.6   Dynamic Traffic Assignment

Analysis of the dynamic path choice behavior of day-to-day traffic requires the dynamic traffic assignment approach. The relationships between the costs on different days and the attributes influencing user choices as well as updates on subsequent days are important for travel forecasting and planning. The use of a static representation of traffic flow has resulted in notable advantages in the mathematical properties of traffic assignment models, such as the existence and uniqueness of equilibrium. However, a static approach, by definition, cannot reflect either variations over time in traffic flows and conditions or changes over time not well suited for analysis, e.g., traffic congestion effects at a fine-grained level. Therefore, dynamic traffic assignment has been introduced by representing time variations in traffic flows and conditions, to reflect the reality that traffic networks are generally not in a steady state.

To retain the advantages of an equilibrium approach, the notion of user equilibrium has to be extended to the recognition that travel times on network links vary over time. Travelers are assumed to know or anticipate future travel conditions along the journey; and, in choosing an *O-D* path, it is assumed they will minimize the *O-D* travel time. Therefore, an important issue of dynamic traffic assignment is that it simultaneously determines a traveler's choice of departure time and path. Thus, this approach can directly analyze phenomena such as peak spreading in response to congestion dynamics or time-varying tolls. Hence, travel forecasting models, with their focus on time and cost of travel, are used to evaluate the impact of day-to-day traffic and the resulting future changes in:

- Demographics
- Land use
- Transportation facilities

All of these influence the performance of a region's transportation system. For this reason, traveler behavior is introduced into forecasting models as travel choices made by groups of homogeneous travelers in aggregated trip-based models or, in more advanced activity-based processes, to introduce travel choices made by individual travelers (Chiu et al. 2011). Travel time and cost measures can be based on static network analysis using time-invariant variables of interest. But it has become increasingly evident that these procedures are inadequate in explaining influences on travel choices and as measures used to evaluate impacts when deciding how to:

- Develop policies for managing transportation systems.
- Fund transportation system improvements.
- Measure environmental impacts related to system-wide travel.

Thus, dynamic network analysis models provide a more detailed means of representing the interaction between travel choices, traffic flows, and time and cost measures in a temporally coherent manner, such as further improvement upon the existing time-of-day static assignment approach (Chiu et al. 2011). Dynamic traffic assignment (DTA) models describe time-varying networks and demand interactions using a behavioral approach, and their results are used to evaluate meaningful measures related to individual travel time and cost as well as system-wide network measures for regional planning purposes.

In static models, inflow to a link is always equal to outflow, meaning that travel time simply increases as inflow and outflow increase. Furthermore, link volume increases indefinitely and exceeds the physical capacity of a link, represented by its volume-to-capacity ratio $\frac{V}{C} > 1$. Since the link volume does not conform to the traffic flow limit, which results from the physical characteristics of the roadway, the assigned link volume should be considered as demand instead of actual flow. Therefore, $\frac{V}{C} > 1$ means that demand exceeds capacity and, subsequently, congestion will occur. The drawback of using $\frac{V}{C}$ is that it does not directly correlate with any physical measure describing congestion, such as speed, density, or queue (Chiu et al. 2011).

Using the dynamic traffic assignment models, modeling of traffic flow dynamics explicitly ensures a direct link between travel time and congestion. In case the link outflow is lower than the link inflow, the link density increases (congestion) and the speed decreases, what is known as the fundamental speed-density relationship; therefore, link travel time will increase.

If the link outflow is reduced and is, thereby, potentially less than the inflow, various reasons can be discussed, such as:

- Merging two lanes into one at a freeway on-ramp effectively reduces capacity of each of the two merging lanes.
- Weaving lane change maneuvers that cross over each other also reduce link capacity.
- On arterial streets, traffic signals reduce outflow capacity of links.
- On freeways and arterial streets, significant oversaturation for one exiting movement from a link can result in reduced flow rates on other exiting movements, due to a local choke-off effect.
- In dynamic models, each link can be defined by their own fundamental diagram, if desired which is sometimes thought as dynamic analogy static, but this analogy is loose as the two mathematical relationships actually perform very different functions in the context of their respective models. In static models this actually represent a congested condition, while in a dynamic model, the fundamental diagram describes how congestion at exit node (reduced link outflow) is propagated upstream though the link, until it spills back onto next upstream links.
- This phenomenon raises the question of congestion spillback, which is not represented in static models. At the moment link inflow becomes equal to outflow, congestion continues to spread upstream into whichever upstream links are feeding traffic into congested link. Outflows of these links are reduced,

and process repeats as described. Queue spillback process also describes how a long queue (congested traffic) can be represented over a sequence of links in a dynamic traffic model (Chiu et al. 2011).

In traffic flow modeling, the goal of traffic assignment is to determine the road network traffic flows and conditions that result from mutual interactions among path choices that travelers make in traversing their O-D, and congestion that results from their travel over the road network.

To achieve satisfying results, several assumptions need to be made, particularly regarding how traveler path choice behavior is modeled and how traffic flows and conditions are represented.

In practice, a common behavioral assumption is that travelers choose an available path that has the least travel time between their O and D, reflecting the idea that travel is rarely a goal in and of itself but involves time, cost, or disutility, things that travelers would prefer to avoid (Chiu et al. 2011).

As a solution, a set of time-varying link and path volumes and travel times that satisfy the dynamic user equilibrium condition for a given road network and time-varying O-D demand pattern is a nontrivial problem, because each traveler's best path choice depends on the congestion levels throughout the trip, which in turn depends on path choices and progress through the road network of other travelers who depart earlier, at the same time, or later, as follows:

- Vehicles departing at different times are assigned with different paths.
- Vehicles departing at the same departure time between the same O-D pair but taking different paths should have the same travel time.
- Travel time cannot be realized at departure, but only at the end of the trip.

This interdependence means that the solution must be found through an iterative process, starting from an initial set of path choices and improving them. The improvement process can continue indefinitely, and in realistic-sized road networks, finding the exact equilibrium is challenging. Thus, the most common method of finding equilibrium in dynamic traffic assignments is to apply the following algorithmic components in an iterative sequence, until a defined stopping criterion is reached (Chiu et al. 2011):

- Network loading: given a set of path choices, i.e., paths and path flows, what are the resulting path travel times?
- Path set update: given the current paths' travel times, what are the new shortest paths (per O-D pair and departure-time interval)?
- Path assignment adjustment: given the updated path sets, how vehicles (or flows) should be assigned to paths to better approximate the dynamic user equilibrium?

This step results in the following general dynamic traffic-assigned algorithm (Fig. 3.3).

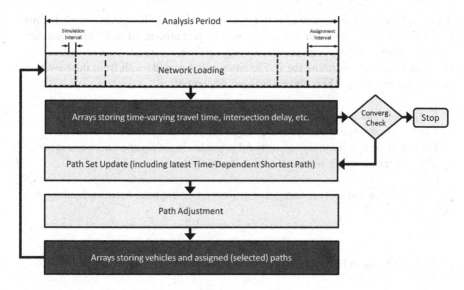

Fig. 3.3 General DTA algorithmic procedure (Chiu et al. 2011)

The path set update involves analyzing the results of the road network loading. Based on congestion patterns and travel times identified in the network loading step, paths with the lowest experienced travel between every *O-D* pair, for each departure-time period, called the assignment interval, are found by a time-dependent shortest path (TDSP) algorithm.

Sharing a similar overall model structure, most dynamic traffic assignment models differ from each other in how they implement these components. In path evaluation, the effect of time-varying link path flows and travel times resulting from vehicles following a given set of path choices are determined through a road network loading process. A variety of analytical and simulation-based road network loading approaches exists. Analytical models typically use exit functions to predict how traffic propagates in the road network, while most simulation-based approaches use some type of mesoscopic simulation approach that represents changes in traffic flow.

The path assignment adjustment follows logically from the path setup date. If travelers shift their path choices towards the least travel time path and away from longer paths, the assignment can be brought closer to equilibrium. Some care is required since a major complication in finding an equilibrium solution is the interdependence between different travelers' path choices and travel times. If all travelers shift to the shortest paths, those paths would become highly congested and would no longer be the shortest. Therefore, only some travelers' route choices should be adjusted in order to avoid overcorrecting. Generally this step involves finding which paths in the set need to be increased with assignment flow vehicles and which need to be decreased and by how much. Underperforming routes, e.g., long travel time, are decreased with flow. It is also noteworthy that at this step, not all vehicles will select or be assigned to a new path. Adjustments are made only to what is necessary in order to achieve equal travel among all paths in current set.

Performing the path assignment adjustment, the algorithm returns to the path evaluation step to determine the traffic pattern that would result from the new path choices (path flows). These three steps are sequential:

- Output of road network loading provides the input for path set update.
- Output of path set update provides input for path assignment adjustment.
- Output of path assignment adjustment provides input for road network loading. These three steps are repeated until a stopping criterion is met.

The stopping criterion is computed at the end of the road network loading step (Chiu et al. 2011).

3.7 Transportation Network Synthesis

Transportation network synthesis is required to modify a transportation road network to fit it with a required demand. But choosing a good alternative solution is not a trivial task due to the:

- Number of interdependent variables which can be rather high
- Interaction between public and private usage for which often no exact numbers exist
- Planning which often is to imprecise and/or to vague to be expressed in mathematical terms

In general, transportation networks can generally be grouped into:

- Industrial zones
- Urban motorways
- Residential zones

Transportation networks consist of links and nodes which can be expressed mathematically as follows:

$$G = (A, N)$$

where $a \in A$ is the set of links (with cardinality n_A), $n \in N\ N$ the set of nodes (with cardinality n_N), and G is the maximal road network. The links in A are introduced by the length of the road segments $l_{rs} \forall\ a \in A$.

The travel demand can be identified through traffic census and supervision and/or by trip distribution models (Wilson 1967). Therefore, the transportation network synthesis requires constraints with regard to the composition of needs to be fulfilled.

Assume that trip matrix $T_M = [T_{(zi,zj)}]$ $(z_i, z_j) \in N^2$ is given for individuals who change from one zone to another zone in the early morning and the late afternoon (Dubois et al. 1979).

An optimal subset of links in a maximal set for a transportation network synthesis problem is given as follows (Dubois et al. 1979):

$$\text{Minimize } T = \sum_{(zi,zj)} T_{(zi,zj)} t_{(zi,zj)}(X)$$

where T is total travel time

$$T = \sum_{(zi,zj)} T_{(zi,zj)} t_{(zi,zj)}$$

with $t_{(zi,zj)}$ as travel time between zones i and j, including trip waiting times and investment costs for (using) links A with C_T as an investment constraint

$$\sum_{(zi,zj) \in A} C_{(zi,zj)} x_{(zi,zj)} \leq C_T$$

$$x_{(zi,zj)} = 0 \quad \text{or} \quad 1 \forall (zi, zj) \in A$$

where $x_{(zi,zj)} = 1$ only if $(zi,zj) \in A$; X is vector of $x_{(zi,zj)}, (zi,zj) \in A$; and X is a binary representation of a road network $G = (A, N)$.

The transportation network synthesis with a single source, where cost functions of flows along the road network edges are concave functions, is presented in Lozovanu and Solomon (1995). This problem is a generalization of the problems of optimal forest finding in oriented graphs with weight and has a direct application in research and solving problems of the allocation of points of production and transportation planning in transportation networks. For this problem, solving a combinatorial algorithm based on the analysis of optimal flows in the transport network with a single source and a generation of admissible trees with a given source which corresponds to admissible flows in the network is proposed in Lozovanu and Solomon (1995).

3.8 Case Study: Diverging Diamond Interchange

A diverging diamond interchange (DDI) is an interchange in which the two directions of traffic on the nonfreeway road cross to the opposite side on both sides of the freeway overpass. The DDI requires traffic on the freeway overpass (or underpass) to briefly drive on the opposite side of the road.

The DDI eliminates the left-hand turns for the roadway passing over or under the freeway. Consequently, vehicles do not have to cross in front of oncoming cars, reducing conflict points and increasing safety; and the elimination of left-hand turns

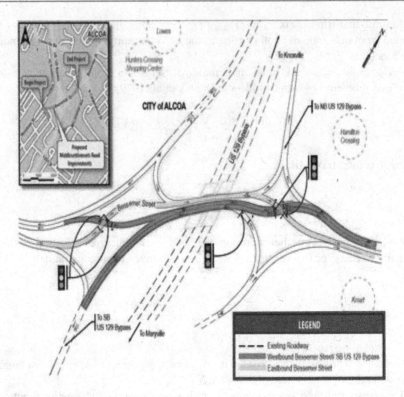

Fig. 3.4 Road network with traffic signals

allows for the removal of the left-turn phase from the traffic signal, which increases the amount of time that can be allocated to through movement and increases the capacity of the roadway, as shown in Fig. 3.4.

The DDI uses traffic lights to briefly shift vehicles to the left side of the road. The vehicles traveling from the freeway to the arterial street can then directly turn left on the access ramp, as if passing between two one-way roads. The sketch in Fig. 3.4 shows the DDI currently under construction at Alcoa, TN, USA. The diverging diamond interchange allows for two-phase operation at all signaled intersections in the interchange. This is a significant improvement in safety, since no left turns must clear opposing traffic; and all traffic is controlled by traffic signals.

The Federal Highway Administration has modeled the Springfield, Missouri, DDI using the highway driving simulator to evaluate the human factor aspects of the proposed design (Inman, et al. 2010). Figure 3.5 is one of the screens from the simulator showing the approach to the crossover on the west side of the interchange. In the screen are the arrows in the nearside signage heads, wrong-way arrows, and a glare screen that is intended to mask headlights of opposing traffic at the crossover.

Figure 3.6 shows the crossover back to the right side of the roadway on the east side of the interchange. This screen shows the use of signage to guide drivers through the intersection. The interchange also includes regulatory signage,

Fig. 3.5 Highway driving
simulator screen showing the
approach to the crossover

Fig. 3.6 Highway driving
simulator screen of the
crossover

including lane restriction, left and right turn restriction, keep right, do not enter, and
wrong-way signs.

The Federal Highway Administration (FHWA) has released a technical brief
covering four intersection designs and two interchange designs that offer substantial advantages of conventional at-grade intersections and grade-separated diamond
interchanges (Hughes and Jagannathan 2010). This report provides information on
each alternative design covering geometric design features, operational and safety
issues, access management, cost, construction sequencing, environmental benefits,
and applicability.

The Missouri Department of Transportation has constructed a simulation of
driving through a diverging diamond interchange at Springfield.

A process model (1999) was developed from the DDI given in Fig. 3.7. The
interstate traffic travels East (E) and West (W). The overpass traffic travels North
(N) and South (S). There are a total of 12 traffic routes through the DDI. The routes
are from S to N, S to W, S to E, N to S, N to E, N to W, W to N, W to S, W

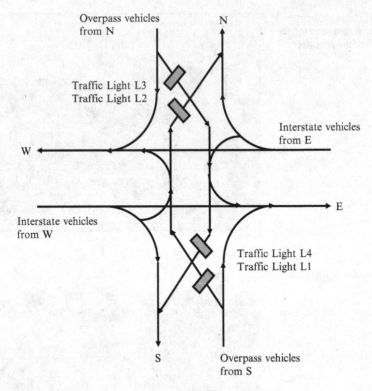

Fig. 3.7 Process model diverging diamond interchange

to W, E to S, E to N, and E to E. The W to E and E to W interstate through traffic is
not modeled since this traffic does not enter the DDI.

The process model has the following submodels:

- Traffic entering from N and E
- Traffic entering from S and W
- Traffic lights L1 and L4
- Traffic lights L2 and L3
- Parameter initialization

Data are passed between the submodels using global variables. These global
variables are also used to control movement of vehicles or process model entities.
The process model has 35 activity blocks, 22 global variables, and 12 entity types.
Also, 12 label blocks are used to display the global variables during the simulation.

The input data for Baseline Run 1 of the experimental design are given in
Table 3.1. The $T\ (a,b,c)$ is a triangular distribution with parameters a, b, and c.
Parameter a is the smallest value, b is the most likely value, and c is the largest
value.

Table 3.1 Input data for Baseline Run 1

Data input	Value	
Time between arrivals of vehicles		
From W	15 s	
From E	15 s	
From N	10 s	
From S	10 s	
Vehicle routing		
From W to W	5 %	
From W to S	50 %	
From E to E	5 %	
From E to N	50 %	
From S to E	25 %	
From S to W	25 %	
From N to E	25 %	
From N to W	25 %	
Traffic lights green minimum time	90 s	
Traffic lights change after minimum time	30 s	
Or traffic lights change after minimum cars waiting	10 cars	
Activity	*Time*	*Capacity*
Vehicles enter from S or N	T(4,5,6) s	10
% vehicles from S exit E and from N exit W	0	1
Vehicles from S move to L1 or from N move to L3	T(9,10,11) s	20
Vehicles from S at L1 or from N at L3	0	1
Vehicles from S move through L1 or from N move through L3	T(5,6,7) s	2
% vehicles from S turn W or from N turn E	0	1
Vehicles from S move to L2 or from N move to L4	T(9,10,11) s	20
Vehicles from S at L2 or from N at L4	0	1
Vehicles from S move through L2 or from N through L4	T(5,6,7) s	2
Vehicles from S exit going N or from N exit going S	T(4,5,6) s	10
Vehicles enter from W or E	T(4,5,6) s	10
% vehicles from W turn S or from E turn N	0	1
Vehicles from W move to L2 or from E move to L4	T(9,10,11) s	20
% vehicles from W turn W or from E turn E	0	1

The model assumes two lanes of traffic in each direction. The capacity at each process model activity varies between 1 and 20. For example, if the capacity is 20, then the maximum capacity is 10 vehicles per lane.

The capacity at activity Vehicles_from_S_move_through_L1_ is two vehicles or one vehicle per lane. Therefore, only two vehicles can go through a green light at a time. If there are more than two cars, a queue will form.

The mean vehicle times through the DDI, assuming no delays, are:

- N to S or S to N 5 + 10 + 6 + 10 + 6 + 5 = 42 s
- S to W or N to E 5 + 10 + 6 + 10 = 31 s

Table 3.2 Results for Baseline Run 1

Vehicle route	Average time through DDI (min)	Average delay (min)
From S to N	1.31	0.61
From S to E	0.08	0.00
From S to W	0.97	0.46
From N to S	1.27	0.57
From N to W	0.08	0.00
From N to E	0.96	0.45
From W to S	0.08	0.00
From W to N	0.83	0.40
From W to W	0.25	0.00
From E to N	0.08	0.00
From E to S	0.81	0.38
From E to E	0.24	0.00
Location of delays		
Vehicles from S waiting at L1		0.45
Vehicles from S waiting at L2		0.24
Vehicles from N waiting at L3		0.45
Vehicles from N waiting at L4		0.20

- W to W or E to E $5 + 10 = 15$ s
- W to S or E to N   5 s

These times are computed by adding all the corresponding process model activity block times in Table 3.1.

The Baseline Run 1 results are given in Table 3.2. Overpass vehicles going from S to N had an average wait of 37 s and from N to S an average wait of 34 s. Vehicles going from S to W had an average wait of 28 s and from N to E an average wait of 28 s. Also, interstate vehicles going from W to N had an average wait of 24 s and from E to S an average wait of 23 s.

All vehicle delays occurred at the traffic lights. The average delay at Light 1 was 27 s; at Light 2, 14 s; at Light 3, 27 s; and at Light 4, 12 s.

The following factors that affect traffic through a DDI were included in the experimental design in Table 3.3:

- Traffic volume (Runs 2–7)
- Length of time traffic light is green (Runs 8–10)
- Traffic signal that turns green after a minimum time or after a minimum number of vehicles are waiting (Runs 11–14)

Runs 2–14 are variations of the Baseline Run 1. The variation for each run, or the experimental design, is given in Table 3.3.

Each process model was run four hours to warm up or to reach steady state and an additional 40 h to collect data.

Table 3.3 Experimental design

Run	Description
Run 1 (Baseline)	See Table 3.1
Run 2	Time between arrivals (TBA) E or W = 18 s, N or S = 12 s
Run 3	TBA E or W = 20 s, N or S = 14 s
Run 4	TBA E or W = 12 s, N or S = 8 s
Run 5	TBA E or W = 10 s, N or S = 6 s
Run 6	TBA E or W = 9 s, N or S = 5 s
Run 7	TBA E or W = 8 s, N or S = 4 s
Run 8	Traffic light green 120 s
Run 9	Traffic light green 60 s
Run 10	Traffic light green 30 s
Run 11	Light changes when min time waiting = 30 s or min cars waiting = 15
Run 12	Light changes when min time waiting = 40 s or min cars waiting = 15
Run 13	Light changes when min time waiting = 30 s or min cars waiting = 5
Run 14	Light changes when min time waiting = 20 s or min cars waiting = 5

Model verification consists of determining if the model is correctly represented in the simulation code. Model validation consists of determining if the model is an accurate representation of the real-world system.

The process model has label blocks that display data from the global variables during the simulation and that are often used in the model verification. By slowing down the simulation, it is possible to observe these values as entities move through the simulation (e.g., if it was possible to only have traffic enter the model from one direction at a time). Then, by limiting the number of entities to one, the movement of the entity could be readily followed on the screen.

Model validation was not possible since data were not readily available. However, a team of individuals familiar with the operations of interstate interchanges was assembled to visually observe the operations of the DDI during the simulation. The process model was stopped throughout the simulation and values observed, such as queues, delays, and vehicle routings. Many of these values were displayed in the label blocks.

3.8.1 Model Results: Traffic Volume

The results of increasing traffic volume (Runs 2–7) are given in Table 3.4. Figure 3.6 is a plot of the average vehicle delays through the DDI for three vehicle routes: S to N, S to W, and E to S. The average vehicle delays for N to S, N to E, and W to N were similar and, therefore, not plotted (see Table 3.4).

The system became unstable with time between arrivals from N or S of 4 s and from E or W of 8 s. That is, the arrival rates exceed the service rates. Therefore, since the system is unstable, the delays and vehicle times in the system will continue to increase.

Table 3.4 Model results for Runs 2–7

Input	Run 7	Run 6	Run 5	Run 4	Run 1	Run 2	Run 3
TBA E or W	8 s	9 s	10 s	12 s	15 s	18 s	20 s
TBA N or S	4 s	5 s	6 s	8 s	10 s	12 s	14 s
Results							
Vehicles through DDI	2,460	2,245	1,918	1,524	1,198	1,000	875
Vehicle route	*Average time through DDI (min)*						
S to N	224.0	2.1	1.43	1.29	1.31	1.23	1.25
S to E	0.08	0.08	0.08	0.08	0.08	0.08	0.08
S to W	112.0	1.17	1.11	1.01	0.97	0.95	0.93
N to S	230.0	2.06	1.57	1.40	1.27	1.26	1.23
N to W	0.08	0.08	0.08	0.08	0.08	0.08	0.08
N to E	116.0	1.17	1.08	1.00	0.96	0.95	0.94
W to S	0.08	0.08	0.08	0.08	0.08	0.08	0.08
W to N	125.0	1.42	0.90	0.84	0.83	0.80	0.82
W to W	0.24	0.24	0.24	0.25	0.25	0.25	0.25
E to N	0.08	0.08	0.08	0.08	0.08	0.08	0.08
E to S	129.0	1.27	0.96	0.89	0.81	0.80	0.82
E to E	0.25	0.24	0.24	0.24	0.24	0.24	0.25
Location of delays	*Average delay at traffic light (min)*						
S at L1	2.01	0.64	0.41	0.48	0.43	0.37	0.40
S at L2	2.13	0.76	0.24	0.21	0.24	0.19	0.22
N at L3	2.24	0.65	0.56	0.48	0.45	0.43	0.41
N at L4	2.26	0.67	0.36	0.29	0.20	0.20	0.21

For S to N traffic, the delay increased from 39 s for Run 3 with 875 vehicles per hour to 85 s for Run 6 with 2,245 vehicles per hour. For Run 7, the delay for S to N vehicles was 224 min with 2,460 vehicles per hour. The theoretical value was 2,700 vehicles per hour; therefore, the difference was the vehicles in queues.

3.8.2 Model Results: Length of Time Traffic Light Green

The results of varying the length of time the traffic lights were green (Runs 8–10) are given in Table 3.5.

Figure 3.9 is a plot of the average vehicle delays through the DDI for three vehicle routes S to N, S to W, and E to S.

The delays reduced considerably by increasing the length of time the traffic lights were green. For example, for S to N traffic, the delay dropped from 95 s with a 120-s green light (Run 8) to 22 s with a 30-s green light (Run 10).

Table 3.5 Model results for Runs 8–10

Input	Run 10	Run 9	Run 1	Run 8
Green traffic light length	30 s	60 s	90 s	120 s
Results				
Vehicle route	*Average time through DDI (min)*			
S to N	1.07	1.05	1.31	2.28
S to E	0.08	0.08	0.08	0.08
S to W	0.66	0.81	0.97	1.16
N to S	1.00	1.21	1.27	2.30
N to W	0.08	0.08	0.08	0.08
N to E	0.66	0.81	0.96	1.11
W to S	0.08	0.08	0.08	0.08
W to N	0.60	0.65	0.83	1.17
E to N	0.08	0.08	0.08	0.08
E to S	0.58	0.77	0.81	1.12
E to E	0.24	0.25	0.24	0.25
Location of delays	*Average delay at traffic light (min)*			
S at L1	0.13	0.29	0.45	0.62
S at L2	0.19	0.10	0.24	0.61
N at L3	0.14	0.28	0.45	0.60
N at L4	0.15	0.26	0.20	0.83

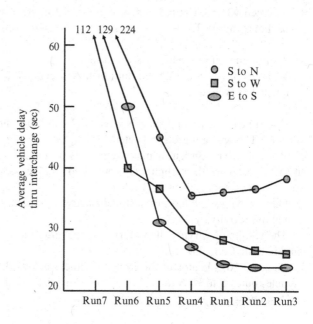

Fig. 3.8 Average vehicle delay through DDI as function of traffic volume

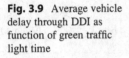

Fig. 3.9 Average vehicle delay through DDI as function of green traffic light time

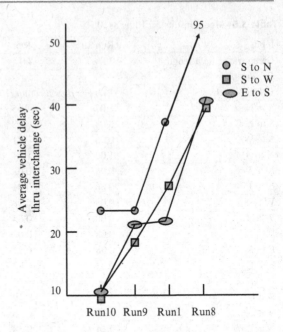

3.8.3 Model Results: Rule for Traffic Light

All traffic lights have the identical rules of operation, and the lights operate in pairs: L1 is green when L4 is red and vice versa; L2 is green when L3 is red and vice versa. For example, L1 will turn green when one of the following conditions is met:

- A minimum of ten vehicles are waiting.
- Vehicles have waited a minimum of thirty seconds (at least one vehicle in the queue has waited 30 s).

Once one of these conditions is satisfied, L1 will turn green (provided L4 is red) for 90 s. The above logic is repeated for L4.

When L1 turns green, a maximum of two vehicles can go through the light at a time. The throughput is controlled by the capacity of the activity Cars_from_S_move_through_Light1.

Other vehicles will have to wait in the activity queue until one of the vehicles has exited the activity.

The results of varying the criteria for changing the traffic lights (Runs 11–14) are given in Table 3.6.

Figure 3.10 is a plot of the average vehicle delays through the DDI for three vehicle routes S to N, S to W, and E to S. Surprisingly the average vehicle delays did not change.

Table 3.6 Model results for Runs 11–14

Input	Run 14	Run 13	Run 1	Run 11	Run 12
Light changes after time	20 s	30 s	30 s	30 s	40 s
Or light changes after cars waiting	5	5	10	15	15
Results					
Vehicle route	*Average time through DDI (min)*				
S to N	1.31	1.31	1.31	1.31	1.31
S to E	0.09	0.08	0.08	0.08	0.08
S to W	0.97	0.97	0.97	0.97	0.97
N to S	1.27	1.27	1.27	1.27	1.27
N to W	0.08	0.08	0.08	0.08	0.08
N to E	0.96	0.96	0.96	0.96	0.96
W to S	0.08	0.08	0.08	0.08	0.08
W to N	0.83	0.83	0.83	0.83	0.83
W to W	0.25	0.25	0.25	0.25	0.25
E to N	0.08	0.08	0.08	0.08	0.08
E to S	0.81	0.81	0.81	0.81	0.81
E to E	0.24	0.24	0.24	0.24	0.24
Location of delays	*Average delay at traffic light (min)*				
S at L1	0.45	0.45	0.45	0.45	0.45
S at L2	0.24	0.24	0.24	0.24	0.24
N at L3	0.45	0.45	0.45	0.45	0.45
N at L4	0.20	0.20	0.20	0.20	0.20

Fig. 3.10 Average vehicle delay through DDI as function of traffic light rule

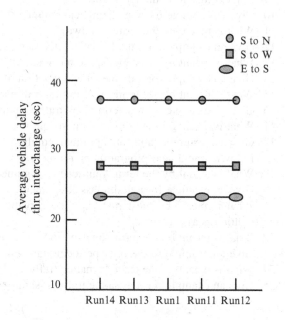

The following conclusions are based on varying selected parameters of the Baseline Run 1:

- The DDI was rapidly modeled using ProcessModel.
- The locations of vehicle delays were only at the traffic lights. For the Baseline Run 1, the average vehicle delays were between 12 and 27 s.
- A decrease in the length of the green traffic lights decreased the average vehicle delays.
- An increase in the vehicle traffic through the DDI increased the average vehicle delay.
- The criteria for changing the light to green had no impact on the average vehicle delay. The results may change with a difference traffic light criteria.

3.9   Exercises

1. What is meant by the term "traffic assignment models"?
2. Describe the structure of the traffic assignment models.
3. What is meant by the term "traffic regime"?
4. Give examples for the traffic regime.
5. What is meant by the term "uncongested traffic"?
6. Give an example for an uncongested traffic.
7. What is meant by the term "congested traffic"?
8. Give an example for a congested traffic.
9. What is meant by the term "flow rate"?
10. Give an example for a mathematical notation flow rate.
11. What is meant by the term "headway"?
12. Give an example for a mathematical notation headway.
13. What is meant by the term "space mean speed"?
14. Give an example for a mathematical notation of space mean speed.
15. What is meant by the term "equilibrium traffic assignment"?
16. Give an example for a deterministic traffic equilibrium assignment model.
17. What is meant by the term "optimum equilibrium"?
18. Give an example how an optimum equilibrium can be found solving the following nonlinear programming problem.
19. What is meant by the term "multiclass assignment"?
20. Give an example for a multiclass assignment.
21. What is meant by the terms "flow conservation" and "flow propagation" in multiclass assignment?
22. What is meant by the term "inter-period assignments"?
23. Give an example for an inter-period assignment model.
24. What is meant by the term "dynamic traffic assignments"?
25. Give an example for a dynamic traffic assignment model.

References and Further Readings

Bando M, Hasebe K, Nakayama A, Shibata A, Sugiyama Y (1995) Dynamic model of traffic congestion and numerical simulation. Phys Rev E 51(2):1035–1042

Bliemer MCJ (2001) Analytical dynamic traffic assignment with interacting user-classes: theoretical advances and applications using a variational inequality approach. PhD thesis, Delft University of Technology, The Netherlands

Bliemer MCJ, Castenmiller RJ, Bovy PHL (2002) Analytical multiclass dynamic traffic assignment using a dynamic network loading procedure. In: Proceedings of the 9th meeting EURO Working Group on Transportation. Tayler & Francis Publication, pp 473–477

Cascetta E (2009) Transportation systems analysis: models and application. Springer Science + Business Media, LLV, New York

Chiu YC, Bottom J, Mahut M, Paz A, Balakrishna R, Waller T, Hicks J (2011) Dynamic Traffic Assignment, A Primer for the Transportation Network Modeling Committee, Transportation Research Circular, Number E-C153, June 2011

Chlewicki G (2003) New interchange and intersection designs: the synchronized split-phasing intersection and the diverging diamond interchange. In: Proceedings of the 2nd urban street symposium, Anaheim

Correa ER, Stier-Moses NE (2010) Wardrop equilibria. In: Cochran JJ (ed) Encyclopedia of operations research and management science. Wiley, Hoboken

Dafermos SC, Sparrow FT (1969) The traffic assignment problem for a general network. J Res US Nat Bur Stand 73B:91–118

Dubois D, Bel G, Llibre M (1979) A set of methods in transportation network synthesis and analysis. J Opl Res Soc 30(9):797–808

Florian M (1999) Untangling traffic congestion: application of network equilibrium models in transportation planning. OR/MS Today 26(2):52–57

Florian M, Hearn DW (2008) Traffic assignment: equilibrium models. In: Optimization and its applications, vol 17. Springer Publ., pp 571–592

Hughes W, Jagannathan R (2010) Double crossover diamond interchange. TECHBRIEF FHWA-HRT-09-054, U.S. Department of Transportation, Federal Highway Administration, Washington, DC, FHWA contact: J. Bared, 202-493-3314

Inman V, Williams J, Cartwright R, Wallick B, Chou P, Baumgartner M (2010) Drivers' evaluation of the diverging diamond interchange. TECHBRIEF FHWA-HRT-07-048, U.S. Department of Transportation, Federal Highway Administration, Washington, DC. FHWA contact: J. Bared, 202-493-3314

Knight FH (1924) Some fallacies in the interpretation of social cost. Q J Econ 38:582–606

Larsson T, Patriksson M (1999) Side constrained traffic equilibrium models—analysis, computation and applications. Transport Res 33B:233–264

Lozovanu D, Solomon D (1995) The problem of the synthesis of a transport network with a single source and the algorithm for its solution. Comput Sci J Moldova 3(2(8)):161–167

ProcessModel (1999) Users Manual, ProcessModel Corporation, Provo, UT

Rodrigus J-P (2013) The geography of transportation systems. Taylor & Francis, Routledge

Steinmetz K (2011) How it works, traffic gem, diverging-diamond interchanges can save time and lives. Time Magazine, pp. 54–55, 7 Feb 2011

Wardrop JG (1952) Some theoretical aspects of road traffic research. In: Proceedings of the institute of civil engineers, Part II, vol 1, ICE Virtual Library, Thomas Telford Limited, pp 325–378

Wilson AG (1967) A statistical theory of spatial distribution models. Transport Res 1:253–269

Yang H, Huang H-J (2004) The multi-class multi-criteria traffic network equilibrium and systems optimum problem. Transport Res Part B 38:1–15

Integration Framework and Empirical Evaluation

4

This chapter begins (Sect. 4.1) with a brief introduction to computer simulation integration platforms and their use in the transportation systems sector. Section 4.2 gives an overview of framework architectures and introduces the reader to the high-level architecture (HLA) framework and rules that provide a specification for the common technical architecture for modeling and simulation. The primary goal is to facilitate interoperability among simulations, to promote reuse of simulations and their components, and to provide a case study for a land-based transportation simulation developing the respective simulator framework architecture. Thereafter, Sect. 4.3 introduces ontology-based modeling and its integration into transportation. Section 4.4 covers the important topic of workflow-based application integration in transportation, while Sect. 4.5 describes a marine terminal-traffic network simulation and its empirical evaluation. Section 4.6 contains an airport operation simulation and its empirical evaluation, while Sect. 4.7 describes the case study for a highway ramp control simulation and its empirical evaluation. Section 4.8 describes a case study for vehicle tracking using the Internet of Things paradigm. Section 4.9 contains comprehensive questions from the integration framework area, and a final section contains references and further readings.

4.1 Introduction

Transportation analysis, modeling, and simulation are expediently introduced into mathematical modeling of transportation systems through the application of simulation software to support planning, designing, and operating transportation systems. This requires a common framework approach for the several modalities existing in transportation supply models to study their manifold aspects. Thus, Chapter 4 introduces the modes of air-, sea-, and land-based simulation of transportation systems and their evaluation as well as aspects of sustainability and green visions in the transportation system sector. Furthermore, the impact of the paradigm of the

© Springer-Verlag London 2014
D.P.F. Möller, *Introduction to Transportation Analysis, Modeling and Simulation*,
Simulation Foundations, Methods and Applications,
DOI 10.1007/978-1-4471-5637-6_4

Internet of Things will also be discussed in Chapter 4 to address unprecedented economic opportunities in the transportation system sector, such as vehicle tracking.

With regard to the complexity and the risks which may appear while planning, designing, and operating transportation systems, computer simulation has been identified as a key technology for overcoming these difficulties. Computer simulation increasingly appears to be a self-reliant discipline where theoretical solutions developed for highly complex systems have conventionally been confirmed by experiments and, subsequently, applied to practice. Hence, computer simulation provides access to true systems, possessing few limitations with regard to system operational conditions and parameters, and has, consequently, also been credited with saving time, labor, and cost.

The importance of computer simulation has been recognized worldwide. The National Science Foundation report, Simulation-Based Engineering Science (NSF 2006), the update, International Assessment of Research and Development in Simulation-Based Engineering and Science (WTEC 2009), and the more recent published position paper of the Germann Scientific Council "Significance and development of simulation in science" (URL 1). Conclude that computer simulation is the key enabling technology of the scientific and engineering world of the twenty-first century. Thus, computer simulation methods are referred to as the third pillar in terms of grounding science with theory and experiments which finally lead to the idea of integration platforms. Because research and development needs are diverse, covering applications, programming models and tools, data analysis and visualization tools, and middleware, integration platforms are composed of several collaborating computing components that interact through embedded communication facilities. Therefore, these systems require advanced integration of abstractions and techniques which have been developed over the past few years in the diverging areas of the transportation systems sector.

One of the widely used solutions developed by the US Modeling and Simulation Coordination Office (M&S CO) is the high-level architecture framework (HLA 2002). The HLA framework and rules are the capstone document that specifies the common technical architecture for modeling and simulation with the primary goals of facilitating interoperability among simulations and promoting reuse of simulations and their components. Thus, HLA is defined by its components and the rules that outline the responsibilities of HLA federates and federations to ensure a consistent implementation. This enables broader integration of different simulation tools using different models of computation.

A key benefit of HLA is that its distributed discrete-event model of computation allows full flexibility to individual subsystems in using any internal solver and model of computation. Moreover, this flexibility permits multirate simulations. However, the HLA standard lacks some key facilities for developing integrated distributed heterogeneous simulations. For example, the HLA standard does not formalize methods for developing interactions and objects used by HLA federates; and it does not provide facilities for easily moving simulations from one computational node to another. Consequently, HLA-based simulations also require a significant amount of tedious and error-prone hand-developed integration code. The HLA is a standardized framework for distributed computer simulation systems.

Communication between different federates is managed via the runtime infrastructure (RTI) layer. The RTI provides a set of services such as time management, data distribution, message passing, and ownership management. Other components of the HLA standard are the object model template (OMT) and the federate interface specification (FIS) (Neema et al. 2014). In general, HLA is defined by three major elements (Moeller and Popescu 2000):

- *HLA Rules or Federation Rules (FR)*: ensure proper runtime interaction of simulations (or federates) in a federation, describing the simulation and federation responsibilities
- *Interface Specification (IS)*: defines the interfaces between federates and the RTI services and provides the means for federates to exchange data
- *Object Model Template*: describes the data federates exchange, providing a common method for recording information and establishing the format of key models:
  - Federation object model (FOM)
  - Simulation object model (SOM)
  - Management object model (MOM)

At the highest level, HLA consists of a set of HLA rules which must be obeyed if a federate or federation is to be regarded as HLA compliant. The execution of a simulation is built directly into the HLA standard.

A transportation systems model relies upon these services during runtime. In this way, a transportation system model can be manipulated in accordance with the scope of the simulation study, i.e., by changing the model structure, parameters, inputs, and outputs to accurately match the real-world system behavior. In order to ensure that the computer simulation generates a sufficiently accurate solution, which means that the computer outputs are in close vicinity to the corresponding features of the transportation system being simulated, it is necessary to have a thorough understanding of the real-world object modeled and the numerical algorithms of the simulation tools.

Let a model of a land-based transportation system be based on assumptions such as:

- Traffic flow is described as a sequence of situations.
- Within each situation, the vehicle driver has a plan of action for different situations.
- Each vehicle driver decides on a plan of action from his/her point of view.

With these situation-action-model assumptions, the traffic-specific situation can be analyzed for decision support of the driver-vehicle-element for each time segment. With this in mind, the following interactions of the discrete-event, land-based transportation, simulation system framework will include:

- Driver-vehicle-element holds the actual lane, adapting to the actual traffic situation.

- Driver-vehicle-element decides to change lanes in the case of a slow vehicle in front or to fall into line in case of a branch off.
- Driver-vehicle-element has to adapt to a right-of-way situation.

For this purpose, the distributed transportation simulation framework contains a federation and several federate which could run on a distributed computer network. Distribution of the computational load onto different computers yields an effective simulation framework for big road networks because one single computer could not process the burden.

4.2   Overview of the Framework Architecture

From Sect. 4.1, it is evident that the integration framework architecture should be an open platform. Open platform architectures describe software systems based on the so-called open standards, published and fully documented external application programming interfaces (APIs). An API specifies how the software components should interact with each other. Thus, the API allows the software to be used in ways other than originally intended without requiring modification of the source code. Using these interfaces, a third party could integrate with the platform adding new functionalities. Hence, an open platform can consist of software components or modules that are either commercial or open source or both. Using an open platform, a developer could add features or functionalities within the platform that have not been thought of or completed before. Therefore, an open platform allows the developer to change existing functionality, as the specifications are publicly available open standards.

4.2.1   SOA

A major focus is on using web services to make functional building blocks accessible over standard Internet protocols that are independent of platforms and programming languages. Therefore, using an open, service-oriented architecture platform (SOA) allows anyone to access and interact with the building blocks of such an open, service-oriented architecture platform. Thus, SOA enables the development of applications that are built by combining loosely coupled and interoperable services which interoperate based on a formal definition that is independent of the underlying platform and programming language. Hence, SOA makes it easy for computers connected over a network to cooperate. Every computer can run an arbitrary number of services, and each service is built in a way that ensures that it can exchange information with any other service in the network without human interaction and without the need to make changes to the underlying program itself (URL 2). The SOA interface definition hides the implementation of the language-specific service. Therefore, SOA-based systems can therefore work independently of development technologies and platforms such as Java, .NET, etc., meaning that services written in

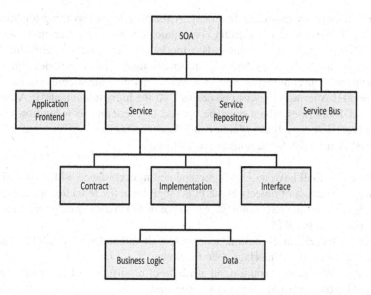

Fig. 4.1 Elements of a SOA

C# running on .NET platforms and services written in Java running on Java EE platforms, for example. Both can be allocated by a common composite application or client. Applications running on either platform can also consume services running on the other as web services that facilitate reuse. Thus, SOA can support integration and consolidation activities within complex systems, but an SOA does not specify or provide a methodology or framework for documenting capabilities or services. The elements of a SOA are shown in Fig. 4.1 (Krafzig et al. 2005).

Based on the SOA elements, introduced in Fig. 4.1, the overall SOA architecture can be viewed as architecture with five horizontal layers (URL 2):

1. Consumer interface layer: graphical user interface (GUI) for end users or application accessing the application/service interface.
2. Business process layer: choreographed services representing use cases in terms of applications.
3. Services: consolidated together for whole enterprise in service inventory.
4. Service components: components used to build the services, like functional and technical libraries, technological interfaces, etc.
5. Operational systems: this layer contains the data models, enterprise data repository, technological platforms, etc.

4.2.2   HLA

In addition to an SOA, the IEEE Standard 1516 HLA is a framework architecture for simulation reuse and interoperability, linking simulations and interfaces to systems, collectively known as federates. A set of federates working together is a federation. Federates do not directly communicate with each other but through an RTI. An HLA approach separates data model and architecture semantics from functions or methods of exchanging information. An HLA includes three core specifications as introduced in Sect. 4.1. With regard to these core specifications, a common HLA terminology can be derived. At the highest level, an HLA consists of a set of HLA rules which must be obeyed if a federate or federation is to be regarded as HLA compliant (Moeller and Popescu 2000).

The HLA rules for federations are as follows:

- Federations shall have an FOM, documented in accordance with the OMT.
- All representation of objects in the FOM shall be in the federate, not in the RTI.
- During a federation execution, all exchange of FOM data among federates shall take place via the RTI.
- During a federation execution, federates shall interact with the RTI in accordance with the HLA interface specification.
- During a federation execution, an attribute of an instance of an object shall be owned by only one federate at any given time.

The HLA rules for federates are as follows:

- Federates shall have an SOM, documented in accordance with the OMT.
- Federates shall be able to update and/or reflect any attributes of objects in their SOM and send and/or receive SOM interactions externally, as specified in their SOM.
- Federates shall be able to transfer and/or accept ownership of attributes dynamically during a federation execution, as specified in their SOM.
- Federates shall be able to vary the conditions under which they provide updates of attributes of objects, as specified in their SOM.
- Federates shall be able to manage local time in a way which will allow them to coordinate data exchange with other members of a federation.

The interface specification identifies how federates interact with a federation and, ultimately, with one another.

The RTI consists of:

- Software that provides common services to simulation systems
- Implementations of an HLA interface specification
- An architectural foundation encouraging portability and interoperability

RTI services:

- Separate simulation and communication.
- Improve on older standards.
- Facilitate construction and destruction of federations.

- Support object declaration and management between federates.
- Assist with federation time management.
- Provide efficient communications to logical groups of federates.

The interface specification management areas belong to:

- Federation management
- Declaration management
- Object management
- Ownership management
- Data distribution management
- Time management

Reusability and interoperability require that objects and interactions managed by a federate, and visible outside the federate, be specified in detail and with a common format. The OMT provides a standard for documenting HLA object model information as follows:

- Provides a common framework for HLA object model documentation
- Fosters interoperability and reuse of simulations and their components

The following information is required:

- Object class structure table
- Object interaction table
- Attribute/parameter table
- FOM/SOM lexicon

OMT extensions represent optional information:

- Component structure table
- Associations table
- Object model metadata

OMT defines the FOM, SOM, and the MOM.

FOM
- Consists of one per federation
- Introduces all shared information, e.g., objects, interactions
- Contemplates interfederate issues, e.g., data encoding schemes

SOM
- Consists of one per federates
- Describes salient characteristics of a federate
- Presents objects and interactions that can be used externally
- Focuses on the federator's internal operation

MOM
- Uses a universal definition
- Identifies objects and interactions used to manage federations

As introduced above, the HLA rules for federations and federates include ten rules of interaction and responsibilities. The federate interface specification includes the services and interfaces required of the RTI and the callback function which federates are required to provide. The OMT specification includes the means to specify data exchange capabilities of federates (SOM) and data to be exchanged during federation execution (FOM).

Federates shall have an HLA SOM documented in accordance with the HLA OMT. The suggested steps for model building within the HLA framework architecture are as follows:

1. Identify essential objects.
2. Identify attributes used to describe the objects identified in Step 1.
3. Build class hierarchy based on common attribute groupings.
4. Classify each object and prepare an object class structure table.
5. Repeat Steps 1–4 for interactions. Identify interactions and associated parameters, build hierarchy, classify interactions, and prepare the interaction class table.
6. Prepare the initial attribute and parameter tables.
7. While constructing data types, lexicons, and routing space tables, iterate with earlier tables, especially the attribute and parameter tables. Verify that the potential attributes or parameters have not been overlooked and/or modify existing ones as necessary.

Based on the HLA framework architecture, one can develop a transportation simulator for computer simulation of the complexity and heterogeneity of land-based transportation. The model building process of a land-based transportation system entails the utilization of three types of information sources:

- Goals and purposes of modeling, e.g., boundaries, components of relevance, and level of details
- A priori knowledge of the traffic system to be modeled
- Experimental data consisting of measurements of the system inputs and outputs

For an HLA land-based transportation simulator, we need to:

- Simulate:
  - Single transport vehicles
  - Transport vehicle bundles
  - Individual lanes
  - Any kind of road network

- Realize a distributed interactive simulator in order to fit big road network situations.
- Develop a simple but flexible interface to adapt the control strategies, e.g., traffic lights.

To build a realistic land-based transportation simulator, a library of simulation models is needed which consider different:

- Types of traffic situations
- Types of road conditions
- Types of car following behavior
- Types of velocity profiles
- Interaction profiles of traffic participants
- Types of lane changing
- Types of dangerous traffic situations
- Types of interactions with pedestrians

Model building of land-based transportation situations will be based on assumptions such as:

- Traffic flow described as a sequence of situations.
- Within each situation, the driver has a plan of action which could interact with different ensuing situations.
- Each driver decides from a plan of action based on his/her point of view.

With these situation-action-model assumptions, the specific traffic situation can be analyzed for each time segment for decision support of the so-called driver-vehicle-element. With this in mind, the following interactions of the discrete-event, land-based transportation simulator could be realized:

- Driver-vehicle-element holds the lane, adapting to the actual traffic situation.
- Driver-vehicle-element decides to change lanes in the case of a slow vehicle in front or fall into line in case of a branch off.
- Driver-vehicle-element has to adapt to a right-of-way situation.

The distributed, land-based transportation simulation system contains a federation and several federate which could run on a distributed computer network. Distribution of the computational load onto different computers yields an effective simulation of big road networks as one single computer could not process the burden.

The components of the system are (see Fig. 4.2):

- RTIexec: a global process that manages the creation and destruction of federation execution
- FedExec: one running process per executing federation that manages the federation, allows federates to join, and resigns from the federation and facilitates data exchange between federates

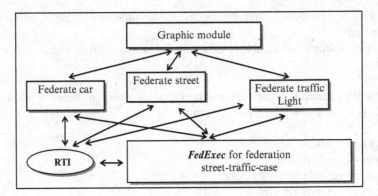

Fig. 4.2 Schematic diagram of the HLA traffic simulation system framework

- Federates (Simulation Modules): perform the computation of the different types of traffic flows
- Graphic Module: displays the actual intrinsic traffic dynamics of the involved driver-vehicle-elements within the road network.

By using the land-based transportation simulator framework architecture, the road network has to be modeled. For easy modeling of road networks, an HLA-specific notation has been introduced, which deals with:

- Clear description of road networks
- Separation of a global road network into local road networks as part of the simulation modules

The specific traffic components investigated with the HLA-RTI land-based transportation simulator consist of cars, streets, possible interrupt requests for traffic lights or traffic signs, and so on. Streets consisting of lanes for each direction, such as west to east, east to west, north to south, and south to north, are realized with blocks of equal length. For this case study, the streets are assumed to have one lane per direction. The cars are driving—as a simplification of the simulation system realized in this case study—with the same speed. Hence, traffic flow can be described as a sequence of blocks, as shown in Fig. 4.3.

The initialization of the RTI contains the following steps:

- Instantiating the objects for the RTI ambassador and federate ambassador
- Creating federation execution with specified name
- Joining the federate to the federation
- Setting the initial time management parameter

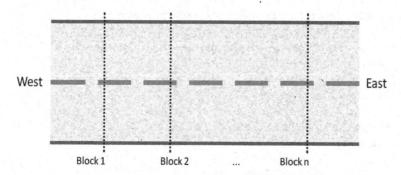

Fig. 4.3 Schematic diagram of federation street with traffic flow west/east and vice versa

```
//-----------------------------------
// Create RTI objects
//
// The federate communicates to the RTI through the RTIambassador object
   and the RTI communicates back to the federate through the
   FederateAmbas-sador object.
//-----------------------------------
RTI::RTIambassador rtiAmb;
// libRTI provided
TrFederateAmbassador fedAmb;
// User-defined
// Named value placeholder for the federates handle
RTI::FederateHandle federateId;
//-----------------------------------
// Create federation execution.
// The RTI_CONFIG environment variable must be set in the shell's
   environment to the directory that contains the RTI.rid file and the
   Traffic.fed
//-----------------------------------
try
   {
//-----------------------------------
// A successful createFederation
// Execution will cause the fedex process to be executed on this machine.
// A "Traffic.fed" file must exist in the current directory. This file
   specifies the FOM object, interaction class structures, default/ini-
   tial transport and ordering information for object attributes and
   interaction classes
//-----------------------------------
   cout << "FED_TR: CREATING FEDERATION EXECUTION" << endl;
   rtiAmb.createFederationExecution( fedExecName, "Traffic.fed" );
   cout << "FED_TR: SUCCESSFUL CREATE FEDERATION EXECUTION" << endl;
   }
```

```
catch (RTI:: FederationExecutionAlreadyExists& e )
    {
    cerr << "FED_TR: Note: Federation execution already exists. " << &e <<
       endl;
    }
catch ( RTI::Exception& e )
    {
    cerr<<"FED_TR:ERROR:"<<&e<<endl;
    return -1;
    }
RTI::Boolean Joined=RTI::RTI_FALSE;
int numTries = 0;
//--------------------------------
// Join federation execution
// Here we loop around the joinFederationExecution call until we try to
   many times or the Join is successful.
//--------------------------------
while(!Joined && (numTries++ < 20))
{
//--------------------------------
try
{
cout<<"FED_TR: JOINING FEDERATION EXECUTION: " << exeName << endl;
federateId =  rtiAmb.joinFederationExecution (myStreet->GetName(),
fedE ecName, &fedAmb);
    Joined = RTI::RTI_TRUE;
}
catch (RTI:: FederateAlreadyExecutionMember& e)
{
cerr<<"FED_TR:ERROR:"<<
       myStreet->GetName()<<"already    exists    in    the    Federation
       Execution"<<fedExecName<<". "<<endl;
cerr << &e << endl;
return -1;
}
catch (RTI::FederationExecutionDoesNotExist&)
{
cerr<<"FED_TR:ERROR:"<<fedExecName<<"Federation Excution"<<
does not exists"<<endl;
rtiAmb.tick(2.0, 2.0);
}
catch ( RTI::Exception& e )
{
cerr<<"FED_TR:ERROR:"<<&e<<endl;
```

```
return -1;
}
} // end of while
cout<< "FED_TR:JOINED SUCCESSFULLY: "<<exeName<< ":Federate
Handle=" << federateId << endl;
```

Each federate has to:

- Define what data are to be published for each update or event.
- Declare which updates and interactions (events) it is interested in receiving by subscribing to those attributes/messages.
- Specify if it is interested in controlling unnecessary message traffic.

The simulation consists of:

- Calculating state and updating to RTI
- Asking for a time advance
- RTI waiting

When a federate has completed a simulation, it deletes the objects it has created (streets, lane, intersection, cars, etc.), resigns from the federation execution, and tries to destroy the federation.

Assume that the federation consists of more federates responsible for the traffic flow. Therefore, the following object classes can be used: *Street, Lane, Block, Car, TrafficLight, Intersection, EntryInIntersection.* Now, object classes can be described as follows:

```
//---------------------------------
class Street contains:
name of the street
number of lane per direction (1 lane in this case study)
number of blocks per lane direction *char[2];
        (direction (0) = "west"direction (1) = "east")
lane for direction[0]
lane for direction[1]
class Lane contains:
direction
array of Block
class Block contains:
array of TrafficLight
array of TrafficSign
array of Car
class Car contains:
pointer to Street
pointer to Lane
```

```
pointer to Block
max velocity
current velocit
class TrafficLight characterised by:
ID
state (colour)
entry in street
entry in block
class Intersection contains:
array of EntryInIntersection
class EntryInIntersection characterised by:
street
lane
numberOfBlock
//------------------------------------
```

The HLA land-based transportation simulator version developed can be expanded by more features such as:

- Streets with more than one lane in each direction
- Cars with different velocity profiles (in this case, it has to be taken into account that a driver-vehicle-element can decide to overtake another driver-vehicle-element driving with a slower velocity)
- More complex intersections, traffic signs, etc.

In the case of several streets, the *class Street* must contain a supplemental array with pointers to each street (a static variable), in order to access them and perform the communication.

```
static StreetPtr        ms_StreetExtent [MAX_STREETS + 1];
```

In addition, instead of the two attributes of type *Lane* (lane for direction [0] and lane for direction [1]), the class has two arrays of lanes, each of the length "number of lanes per direction."

4.3   Ontology-Based Modeling and Integration in Transportation

Integrating the opportunities offered by the Internet for the transportation system sector requires integrating the heterogeneous data sources essential for planning and controlling in the transportation system sector into a semantic web approach. In this case, the semantics of the heterogeneous data sources in the transportation system sector are captured by their ontologies representing the terms and relationships. Therefore, ontologies become an important method of building sharable and reusable knowledge repositories and supporting their interaction (Zhai et al. 2007). Hence, ontology can be defined as an abstract representation of real-world objects of the system under investigation which means that the ontology constitutes a domain-specific model defining the essential domain concepts, their properties, and the relationships between them, represented as a knowledge base. Thus, we can summarize that an ontology (O) organizes domain knowledge in terms of concepts (C), properties (P), and relations (R). In other words, we can say that an ontology (O) is a triplet of the form

$$O = (C, P, R)$$

where C is a set of concepts essential for the domain, P is a set of concept properties essential for the domain, and R is a set of binary semantic relations defined between concepts in O which can be expressed as follows:

$$R = \{r \perp r \subseteq C \times C \times R_I\}$$

where R_t is the set of relation types. A set of basic relations is defined as $R_b = \{\approx, \uparrow, \nabla\}$ with the following interpretations (Zhai et al. 2007):

- For any two ontological concepts, $c_i, c_j \in C$, $\approx$ denotes the equivalent relation, meaning $c_i \approx c_j$. If two concepts, c_i and c_j, are declared equivalent in ontology, then instances of concept c_i can also be inferred as instances of c_j and vice versa.
- $\uparrow$ is the generalization notation. In cases where the ontology specifies $c_i \uparrow c_j$, then c_j inhibits all property descriptors associated with c_i; and these need not be repeated for c_j while specifying the ontology.
- $c_i \nabla c_j$ means c_i has part c_j. If a concept in ontology is specified as an aggregation of other concepts, it can be expressed by using ∇.

Figure 4.4 contains an example of multimodal transportation alternatives in tourist information domain ontology. The figure depicts the mechanisms for mapping and transformation by enabling co-use of different domain models as well as co-use of models of different levels of detail. This allows using individual information services as part of traveler information services, generating correlative information from multidata sources to satisfy the individual traveler's needs.

The ontological concepts in Fig. 4.4 use the equivalence relation meaning "synonym," the generalization notation meaning "*is-a*," and the "has part" meaning the whole part. The related-to semantic is used for generic associations between components (Zhai et al. 2007). The resource description framework scheme, which is a data model supporting mechanisms for representing the metadata of schemas, and the extended markup language are today's standard for establishing semantic interoperability on the Web.

The ontological concepts in Fig. 4.4 use the equivalence relation meaning "*synonym*," the generalization notation meaning "*is-a*," and the "has part" meaning the *whole part*. The *related-to* semantic is used for generic associations between components (Zhai et al. 2007). The resource description framework scheme, which

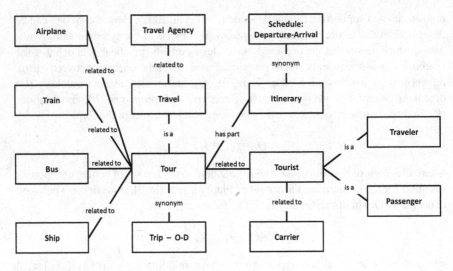

Fig. 4.4 Tourist information domain ontology

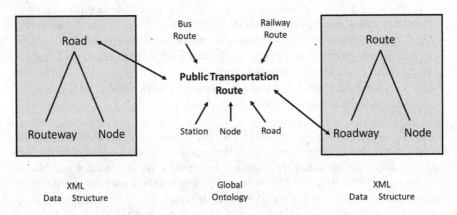

Fig. 4.5 Global ontology and XML documents (Zhai et al. 2007)

is a data model supporting mechanisms for representing the metadata of schemas, and the extended markup language are today's standard for establishing semantic interoperability on the Web.

The example in Fig. 4.5 shows the elements "Road" and "Routeway" of two XML data structures which are semantically integrated into the "Public Transportation Route" concept of the global ontology.

4.4 Workflow-Based Application Integration in Transportation

Workflow-driven analysis allows all kinds of transportation processes to be managed more effectively and costs to be reduced. Thus, workflow-based integration in transportation focuses on optimal modeling and implementation of operational transportation processes within a given infrastructure, including the required traffic management. In 1996, the Workflow Management Coalition published a glossary of terms related to workflow. It defines workflow as: *"The automation of a business process, in whole or part, during which documents, forwarded information or tasks are passed from one participant to another for action, according to a set of procedural rules."* Another useful definition is the one for Workflow Management Systems that: *"A system that defines creates and manages the execution of workflows through the use of software, running on one or more workflow engines, which is able to interpret the process definition, interact with workflow participants and, where required, invoke the use of IT tools and applications."* Workflow Systems consists of the following main components (Espinosa and Pulido 2002):

- *Workflow engine*: responsible for creating, assigning, controlling the tasks, and deciding, in each moment, the next action to be performed. In fact, it is the core for executing processes as all of the information generated during runtime must be stored.
- *Database module*: responsible for the persistence of all workflow items, cases, tasks, users, and data related to processes.
- *Process designer*: develops the processes (quite often graphically) that will be interpreted by the workflow engine.
- *Clients*: provides access to workflow objects, usually software modules and applications that allow users in the workflow to interact with the workflow engine.

Example

We describe the workflow for a transportation case study illustrating how a customer can access an online service to buy a railway ticket via the Internet. For this purpose, an online train ticket portal is dedicated to presenting departure and arrival times and connection options in order to deliver an e-ticket sold through the Internet. The process starts when the customer accesses the portal to request a ticket. After determining optimum departure and arrival times with as few train changes as possible, the customer submits an e-ticket order. Once this is submitted, the train ticket portal machine will check the customer credit card data to determine if it is correct and that the credit card has not expired. If correct and not expired, two activities are carried out:

1. Ticket and the itinerary are printed.
2. Bill is prepared and printed after the ticket and the itinerary have been printed.

When both activities are complete, the train portal ticket machine is ready for a new customer request. The workflow architectural principle behind this example is a finite state machine, which can be expressed in a Unified Modeling Language (UML) diagram specifying the process described.

4.5   Marine Terminal Operation Simulation and Its Empirical Evaluation

Shipping is the most cost-effective modality of providing transport over large distances. This is why more than 90 % of global trade is carried out by sea. But the recent well-documented economic fluctuations have had a huge impact on the seaborne trade volume. Thus, during the last economic crisis, seaborne trade got back on track in 2010 with a volume of goods loaded of 8.4 billion tons. Moreover, maritime traffic has increased in parallel as the main rules of logistics are to deliver cargo to the right location at the right time. Therefore, gateway seaports have to respond to the challenges of growing maritime traffic volumes which necessitates serious planning and development, e.g., of container yard capacity, multimodal transportation, and logistics capacity, etc., in relation to practical constraints. In addition, the policies and regulations of the following sectors must be considered:

- Environmental issues
- Infrastructure
- Investment
- Land use
- Logistics
- Port size
- Intermodal and multimodal transport
- Transport and trade facilities
- Etc.

Moreover, the environmental impact of heavy road transport to and from gateway seaports and the lack of sustainable technical solutions for reducing the carbon footprint in transportation clearly require alternatives and bridging solutions to overcome today's problems. One such possibility is to transfer shipments to more sustainable modalities of transport and/or rectify the allocation of maritime container load units at the respective nodes. Nodes represent components in the supply chains and, together with edges, an intermodal cluster formation in logistics. Thus, ports have evolved from simply intermodal locations to logistical clusters, as illustrated in Fig. 4.6.

The medium gray-colored node in Fig. 4.6 represents a gateway seaport (GSP) focusing on transshipment between maritime and inland transport systems. The light gray-colored and the dark gray-colored nodes represent two different types of destination ports, part of a multidestination port concept. DP1 characterizes a complex large

Fig. 4.6 Simple maritime
supply chain transportation
network model showing
nodes and edges

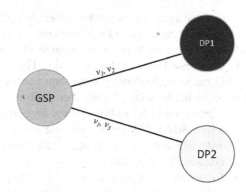

freight distribution terminal, while DP2 represents a destination port load center. The nodes indicate the capacity of inbound and/or outbound maritime container load units. The edges, the lines in between two nodes, e.g., GSP and DP2, indicate the intermodal transportation weighting factors. Transportation capacity from GSP to both destination ports is indicated by v_1, v_2 indicates the truck capacity from GSP to DP1, and v_3 characterizes barge capacity from GSP to a DP2 satellite terminal.

The mode and edge representation shown in Fig. 4.6 originate from mathematical graph theory, whereby a graph is an ordered pair $G = (V,E)$ where:

- V is the vertex set whose elements are vertices, or nodes, of the graph. This set is often denoted $V(G)$ or just V.
- E is the edge set whose elements are the edges, or connections, between vertices, of the graph. This set is often denoted $E(G)$ or just E. Individual edges are ordered pairs (u, v) where u and v are vertices in V.

Two graphs G and H are considered to be equal when

$$V(G) = V(H) \text{ and } E(G) = E(H).$$

The order of a graph is the number of vertices in it, usually denoted $|V|$ or sometimes n. The size of a graph is the number of edges in it, denoted $|E|$ or sometimes m. If $n = 0$ or $m = 0$, the graph is called empty or null. If $n = 1$, the graph is considered trivial.

This graph theoretical approach can be used for planning port development, e.g., building new ports and/or new destination ports and/or destination port facilities represented by a node for which high-capacity transport modalities must be available for the majority of transportation, represented by the nodes edge. As a result of the constraints of the node and the edge, a modality shift, e.g., from trucks to trains, can occur.

The port of Hamburg operates four container terminals:

- Container Terminal Altenwerder (CTA): one of the most modern container handling facilities in the world. Container handling is almost entirely automated because autonomous guided driverless vehicles (AGVs) transport the containers between the quay and the container yards. Fifteen container gantry cranes are operated at the four berths for large vessels at CTA.

- Container Terminal Burchardkai (CTB): the largest sea freight handling facility in the port of Hamburg. More than 5,000 ships per year are loaded and unloaded at 10 berths with 27 container gantries. It is assumed that the capacity will increase up to 5.2 million twenty-foot equivalent units (TEUs) (which means it will nearly double) in the coming years whereby Twin-Forty container cranes will help achieve this goal by enabling loading or unloading of two 40-foot containers in one move.
- *Container Terminal Tollerort (CTT)*: provides four berths with eight container gantries enabling CTT to handle post-Panamax size ships. The terminal has its own container rail station with 720 m of track, and 3 new Transtainer® cranes enable the terminal to handle block trains quickly without shunting. The Transtainer cranes, especially the rubber-tired Transtainer® crane (RTTC), is the key element invented by PACECO® for handling containers in a terminal. The RTTC is a rigid gantry crane mounted on large rubber tires which travels in a straight line on a horizontal path, just the same as if it were running on rails.
- Container Terminal Eurogate: connects directly to interstate highway A7. Six large-ship berths with 21 container cranes (of which there are 19 Post-Panamax types) and more than 140 van carriers ensure rapid handling. Handling 2.7 million TEU in 2008, the EUROGATE Container Terminal in Hamburg is the second largest terminal of the EUROGATE Group in Germany.

Containers arriving at and departing from the seaport terminals require intermodal transfers. The most frequently used are vessel to truck and vice versa, closely followed by vessel to train and vice versa. Both of these require the respective infrastructure resources. Planning these resources can be achieved based on scenario analysis, including the evaluation of the impact on future intermodal transportation system needs; the results of the analysis can be obtained from simulation. Therefore, a traffic network model has to be developed that supports intermodal traffic. Such a traffic network simulation model is based on a number of nodes and edges describing the overall traffic network lane capacity and performance with regard to travel time, load, and more. A general model of a traffic network composed of nodes and edges is shown in Fig. 4.6.

This graph theoretical approach can be used for planning transportation traffic network infrastructure, e.g., building new traffic roads and/or new port facilities modeled through nodes for which high-capacity transport modalities must be available for the majority of transportation chains represented by the nodes edges. As a result of the constraints of nodes and edges, a modality shift from trucks to trains and vice versa can occur. Vehicles in the transportation traffic network can be modeled individually by attributes, including current location, load, speed, and final destination. Vehicles stochastically appear at any node because the interarrival time is assumed to be exponentially distributed and traverse fixed routes, e.g., a sequence of road links represented by the nodes and edges, to reach the final destination.

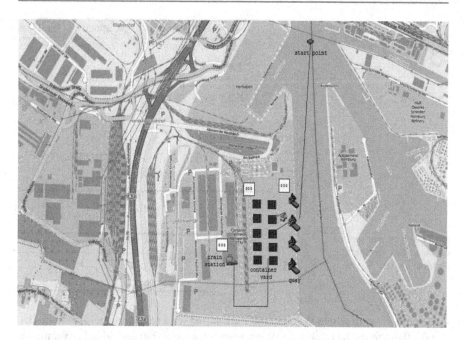

Fig. 4.7 Container Terminal Altenwerder (CTA) model based on Open Street Map (2014)

For the international student team project in Sect. 4.5.1 (Moeller et al. 2013), the CTA was chosen. The overall simulation model is shown in Fig. 4.7.

4.5.1 Marine Terminal Operation Simulation Model

The terminal operation is quite complex; thus, we developed a case study model which is quite a bit simpler, considering only the container ship and the crane between the quay and the container yard as resources. Trucks which bring containers from the container yard to the *train_station* are considered only as a network and are not shown in the model. For the terminal model, we used four locations which are *start_point*, *quay*, *container_yard*, and *train_station*. Details about the locations are provided below, and the icons used are shown in Fig. 4.8.

- *start_point*: where the ship arrives; we put the capacity to infinity, so that there is no limit for the arriving containers.
- *quay*: where the container ships anchor. According to the real description of the CTA, the quay has four berths. We used the capacity of the quay which means that a maximum of 200 containers will arrive at the quay at a time. A counter was added in order to check how many containers are there.

Fig. 4.8 Icons and specs of
locations in the terminal
model

Icon	Name	Capacity	Units
	quay	200	1
	container_yard	500	1
	start_point	INFINITE	1
	train_station	INFINITE	1

Fig. 4.9 Icon and spec of
container and speed in the
terminal model

Icon	Name	Speed (fpm)
	con-tainer	15

- *container_yard*: where the containers are stored for a short time before the trucks take them to the next location. We set the capacity at 200 and added a counter to check how many containers are stored in the container yard.
- *train_station*: where the trains are loaded with the containers. We also added a counter in order to count.

A container has only one entity as shown in Fig. 4.9.

Additional resources are ship and fork. The details are provided below and in Fig. 4.10.

- *ship*: container ship that brings the containers from the *start_point* to the quay
- fork: brings the container from the quay to the *container_yard*
- *SHIP_PATH*: path along which the ship arrives from the entry of the harbor to the *quay*; the ship is the resource which takes it to the containers.
- *CONTAINER_PATH*: path along which the containers are moved from the resource fork to the container yard.
- *TRAIN_PATH*: path along which the containers are taken from trucks and moved to the train station (Table 4.1).

Containers arrive at the *starting_point* with an assumed exponential distribution for the quantity of 100 (Table 4.2).

Containers can be moved from the ship when the ship reaches the berth, as shown in Table 4.3.

When the ship reaches the berth containers, it can be unloaded after 20 min because it has to anchor first and then start the unloading process, meaning that the unloaded containers are moved to the container yard by fork, as shown in Table 4.4

Fig. 4.10 Icons and specs of ship and fork in the terminal model

Icon	Name	Units	Specifications
	ship	100	SHIP_PATH, N1
	fork	15	CONTAINER_PATH, N1

Table 4.1 Specs of ship-, container-, and train-path

Name	Paths	Interfaces
SHIP_PATH	1	2
CONTAINER_PATH	1	2
TRAIN_PATH	1	2

Table 4.2 Specs of containers

Entity	Location	Qty each	First time	Occurrences	Frequency
Container	start_point	100	0	INF	e(120)

Table 4.3 Process and routing of containers

Process			Routing			
Entity	Location	Operation	Output	Destination	Rule	Move logic
Container	*Start_point*		Container	Quay	FIRST 1	MOVE WITH ship THEN FREE

Table 4.4 Process and routing of container movements

Process			Routing			
Entity	Location	Operation	Output	Destination	Rule	Move logic
Container	Quay	WAIT 20 min.	Container	*container_yard*	FIRST 1	MOVE WITH fork THEN FREE

As soon as the containers reach the *container_yard*, it is assumed that they wait for an exponentially distributed time of 240 min for the trucks to pick them up for transfer to *the train_station*, as shown in Table 4.5.

As soon as the containers arrive at the *train_station,* they have to wait there for 60 min because it is assumed that the train is loaded with containers and will leave thereafter, which costs the operation time 60 min (Table 4.6).

The simulation runs for 200 h, which is considered a long enough time period to discern the processes for *entities*, *resources*, and *locations*. The settings used are shown in Fig. 4.11.

Table 4.5 Process and routing of container from the *container_yard* to the *train_station*

Process			Routing			
Entity	Location	Operation	Output	Destination	Rule	Move logic
Container	container_yard	WAIT e(240)	Container	train_station	FIRST 1	MOVE ON TRAIN_PATH

Table 4.6 Process and routing of container at the *train_station* and departure of the train

Process			Routing			
Entity	Location	Operation	Output	Destination	Rule	Move logic
Container	train_station	WAIT 60 min.	Container	EXIT	FIRST 1	MOVE ON TRAIN_PATH

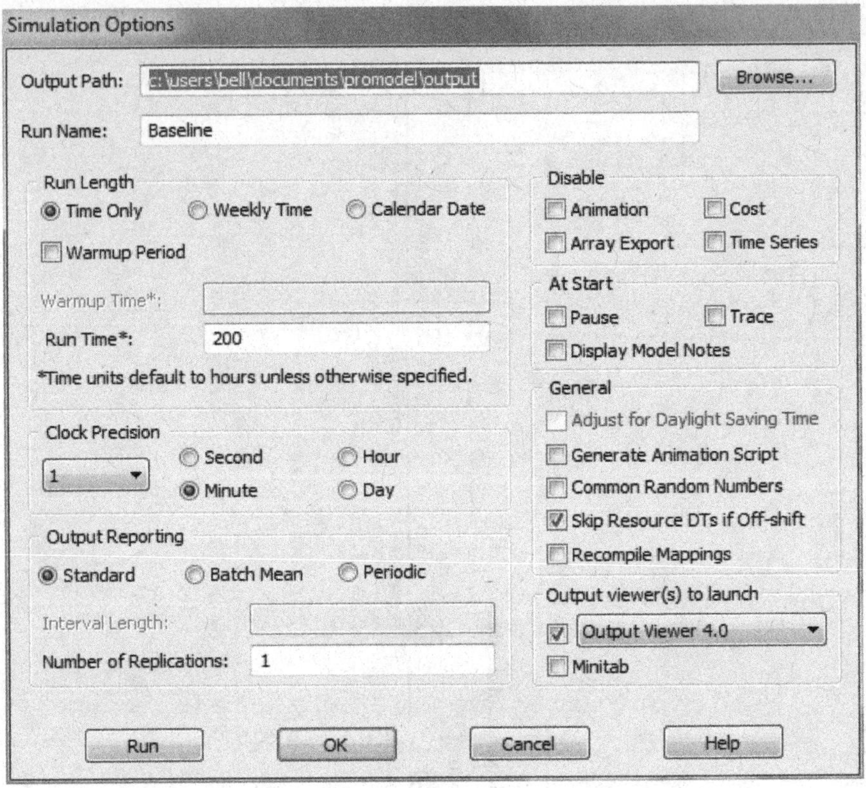

Fig. 4.11 Simulation options

After running a simulation, ProModel provides an Output Viewer window which shows a collection of different charts and reports. Thus, it is possible to examine the collected charts by selecting one of them. For this purpose, the charts shown in Fig. 4.12 are available.

The most significant result in this maritime container simulation case study is the time plot showing the container quantities at the quay, the container yard, and the train station over the simulation runtime. The output chart is shown in Fig. 4.13.

This graph shows the most important result—the *container_yard* is never totally filled up with containers because truck arrivals are on time. The reason is that the assumed random behavior (exponentially distributed) causes the *container_yard* to have fewer units. This is a very important result for the harbor operation because if the *container_yard* got full, there would not be enough space to store the unloaded containers, which would result in the ship having to wait longer in the berth to unload.

Considering the total count of containers handled on different days, the entity count chart is shown in Fig. 4.14.

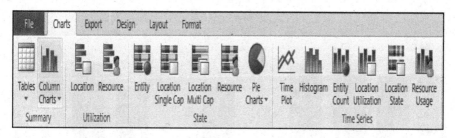

Fig. 4.12 Output viewer option in ProModel

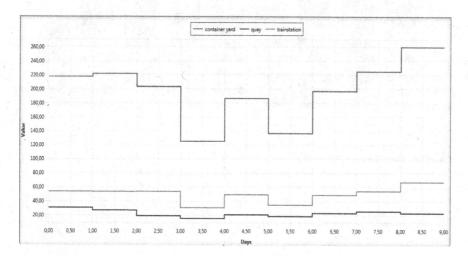

Fig. 4.13 Container quantities over time at container_yard (*upper graph*), train_station (*middle graph*), and quay

From Fig. 4.14, it can be seen that the maximum number of containers that can be handled per day is 1,300, based on the assumptions made for our scenario.

4.6 Airport Operation Simulation and Its Empirical Evaluation

Aviation has shown tremendous development because since the 1970s, commercial aviation has doubled about every 15 years, as shown in Fig. 4.15, based on an investigation by Airbus (GreenAir 2010). This trend proved to be very stable; and even if the numbers declined during years of crisis, the aviation industry recovered and kept the pace. Also, the US National Airspace System (NAS) is expected to grow around 2.4 % per year over the next 20 years and accommodate around 1.6 times today's traffic level by 2028 (GreenAir 2010; JPDO 2007; Ky and Miaillier 2006; Arbuckle et al. 2006). The anticipated growth in air traffic is expected to bring additional concerns to an already congested system (JPDO 2004; Thanh Le 2006).

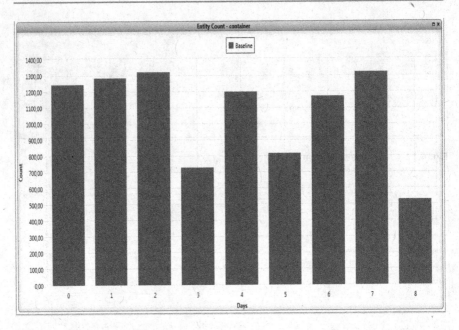

Fig. 4.14 Container quantities over time

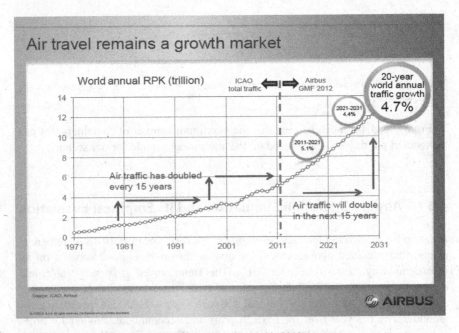

Fig. 4.15 Past and forecasted air traffic growth (GreenAir 2010)

Along with increasing infrastructure requirements, the gap in available airport capacities can be identified as shortages in today's air transportation system because it is reported that major airports always run at full capacity (JPDO 2007). This requires that these airports optimize their many operations to allow them to operate at full capacity during operational peaks, which can be introduced as a transportation system operating in close vicinity of an already congested level (JPDO 2004; Thanh Le 2006). To overcome this shortage requires developing new operational concepts and/or enhancing capacity. But there will be constraints where airports are unable to expand their capacity to adapt to the increasing demand. This can be due to the lack of space for new runways, ramps, terminal buildings, and so on. Thus, these airports are likely to experience air traffic shortages.

Airports, airlines, and air traffic service providers strategically try to optimize their operations to run their businesses at full capacity during peak activity periods, based on the assumption of providing favorable conditions for their customers. But sometimes airline flights are overscheduled regardless of available airport resources, meaning that departure and arrival schedules are created which assume that available arrival and departure rates are not limited for any reason other than nominal runway capacity and safety constraints. In addition to tight planning, the stochastic nature of surface operations implies that there are enough aircraft taxiing out to ensure a constant, near maximum runway service rate (Burgain 2010; Carr et al. 2002). Therefore, airports are very sensitive to unexpected events, such as bad weather conditions, which can immediately disrupt throughput resulting in congestion and a huge number of passengers waiting for their new departure schedules. To cope with congestion, air traffic service providers, airports, and airlines have several options:

- Reduce congestion by:
  - Increasing capacity by building more runways
  - Reducing aircraft separation using new technologies, such as runway incursion alerting systems (Schönefeld and Möller 2012)
  - Forcing restrictions on departure planning through new, innovative methods such as value-based departure sequencing (Brelie 2014)
- Optimize operations by:
  - Centralizing operations around airport authorities to ease information sharing, optimize throughput, and lower inefficiencies (Moeller 2013)
  - Using collaborative decision making to improve operations while respecting the competitive environment (Burgain 2010)

Thus, technological awareness and management is a priority for airports' short-, mid-, and long-term business for airport ramp management, introduced as modular, reliable, secure, and stable workflows, where humans proactively facilitate collaboration between airport authorities, airlines, ground handling, and air traffic control to minimize operational failures. Figure 4.16 depicts an aircraft at its parking position with the corresponding ground handling facilities for aircraft turnaround.

Each turnaround process is tied to a limited period of time. Some processes depend on each other, while others may not. Thus, for example, passengers cannot

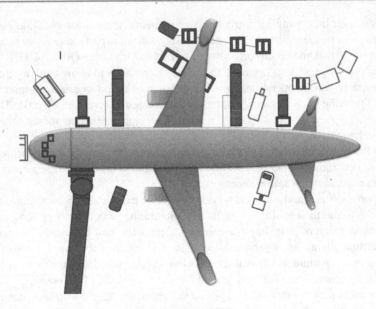

Fig. 4.16 Aircraft during ground handling (Airbus 2011)

board or deboard while the aircraft is refueled, for safety reasons. Moreover, maintenance of sanitation and fresh water supplies does not take place in parallel for hygiene reasons when passengers are boarding or deboarding. Therefore, these two processes must be completed consecutively. Thus, timing of turnaround services and facilities depends on the different ground handling processes. In Fig. 4.17, the sequence of individual ground handling processes are identified.

Figure 4.17 gives an overview of the timing and potential concurrency of the turnaround process at the airport ramp as well as the processes that can never run concurrent. Processes such as taxi-in, pushback, taxi-out, and runway may not be executed in parallel to other processes. Parallel work can only take place when the aircraft is in its parking position, the engines are switched off, and the break blocks have been placed.

The other processes are altogether time interdependent. For example, the ground power supply is independent of all other dispatch ground services and runs all of the time while the aircraft is in its parking position. Moreover, ground handling staff must complete their tasks within the predetermined time. For example, baggage handling staff has to unload the luggage of first-class passengers 7 min after the engines have been switched off, which is agreed to by contract with the airlines (Farschtschi et al. 2011). But optimizing individual processes locally in turnaround is not enough. They are interdependent for several reasons, i.e., because they are carried out on the same object and at the same place. Therefore, the processes being optimized have to be adapted globally since they use ground handling resource facilities to avoid having ground handling service providers block each other. The following examples of optimization approaches focus only on the local improvement of the turnaround

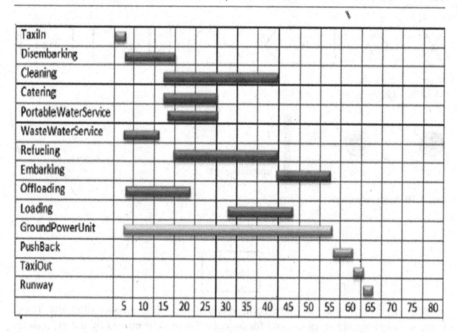

Fig. 4.17 Timing of turnaround processes (Moussaoui 2013)

process to make it easy to understand while introducing UML activity diagrams. A detailed report of the issues involved with a more global optimization of the turn-around process and associated solution approaches can be found in Farschtschi et al. (2011).

4.6.1   UML Activity Diagrams

UML is a general purpose modeling language designated to provide a standard way of visualizing the design of a system. In 2000, UML was accepted by the International Organization for Standardization (ISO). UML describes the order in which actions are carried out. All actions taken together describe a process. Case study diagrams show interrelationships between objects. Activity diagrams represent processes. The elements of UML are shown in Fig. 4.18.

UML activity diagrams consist of nodes and edges. Certain events occur on the node. Edges connect nodes. Tokens are spread out over the entire activity diagram. A UML activity diagram begins with a start node and ends with an end node.

action

A rectangle with rounded corners represents individual actions in the entire activity diagram. Within the element, a short and precise description of the action is written.

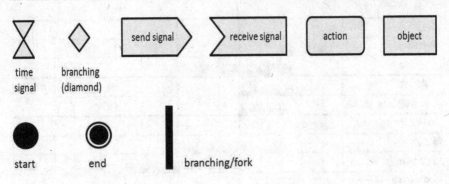

time signal   branching (diamond)   send signal   receive signal   action   object

start   end   branching/fork

Fig. 4.18 UML elements

Rectangle objects are another component of the UML activity diagram. They serve as intermediate storage units for objects. The data is moved by the preceding action to the following action, so data is passed from one action to the next.

branching
(diamond)

The diamond denotes a division or branch of the path (e.g., the edges) or even a merger of two paths. The incoming token will continue but only on one of the outgoing lines. The selection criteria can be determined in advance and then recorded to the corresponding edges.

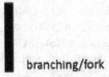

branching/fork

The black bar is a branch known as a bifurcation in the road. The difference is that the token on a line not only continues on that line but on all lines by copying. Conversely, bringing an incoming line and an outgoing line together, the first incoming token must wait for the other tokens for conjunction; and then the unified token can go ahead. To allow other processes to access the elements, a signal transmitter and signal receiver are used. When a signal is sent, then all processes associated with it will not execute.

time
signal

The time event sign means that an action is time control triggered. The time signal stops the signal flow of the diagram for a certain time. Once the time has expired, the diagram or the token continues.

In Fig. 4.19, the UML activity diagram-based turnaround model is shown. The model begins with the aircraft landing and ends with the aircraft departing. After landing, the plane has to wait for the assignment of the parking position: direct to the gate, on the terminal ramp, or a position farther away from the terminal on the ramp. After the assignment, the aircraft taxies to this position. The engines are switched off, and the brake blocks are applied. Now the (terminal) ramp-based ground handling processes can begin, as shown in Fig. 4.19, using the parallel signal transmitter blocks.

The diamond in Fig. 4.20 splits into two outgoing paths. The upper path is chosen if the parking position of the aircraft is between No. 4 and No. 40 and the lower path is selected, if the assigned parking position is not between No. 4 and No. 40.

In general, the UML activity diagram allows the creation of models that represent an abstracted effigy of the real world to analyze the reality by reducing important attributes that are restricted to certain periods of time. The model shown in Fig. 4.21 covers the turnaround process at Hamburg Airport, which takes place on the airport airside. From this model, the actual sequence of each process is clearly seen. The time-dependent processes and concurrencies depicted in this version of the model are close to reality. But the final model differs from this version because of the ground power supply and its dependence on processing time.

Processing time for using the ground power supply corresponds with the sum of processing times for all other clearance processes. Since modeling is based on the assumption of a distribution function, the sum of processing times can only be determined at the end of the processes because the times currently used by MATLAB are not known in advance. This would be totally different if we assume that constant times are passed through the present model. In that case, the processing times are known in advance and could be taken out of the workspace by a MATLAB function and totaled. Thereafter, the result has to pass by one corresponding Simulink block into the model.

However, for a cost analysis of ground handling tasks, the fact that ground power supply takes place in parallel to the remaining clearance processes has to be taken into account. Thus, for cost analysis, the ground handling cost model has to cover all other trials because the entire clearance time must be calculated as part of the cost analysis of the ground power supply process.

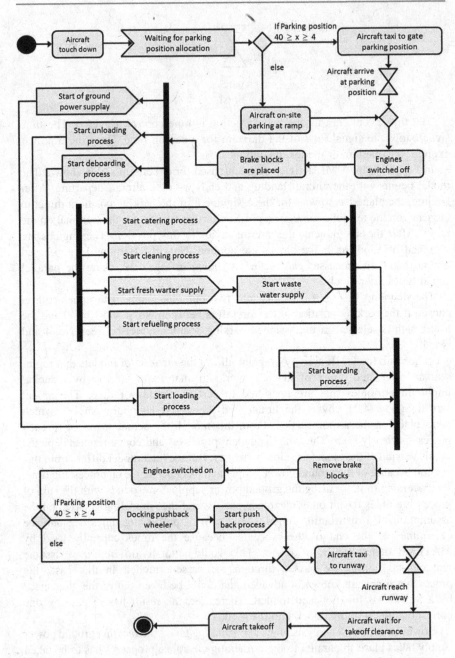

Fig. 4.19 Ramp turnaround process represented as UML activity diagram (Moussaoui 2013)

Fig. 4.20 Branching out of the turnaround UML activity diagram of Fig. 4.19

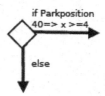

4.6.2   SimEvents Model

Using SimEvents for the aforementioned ground handling process of aircraft, turnaround blocks are introduced such as blocks *Start Timer* (before the replicate block) and *Read Timer* (after the entity combiner block), the time required for the entire clearance is analyzed. If all categories of ground handling services are completed and the time provided for this purpose has been identified, the cost for ground power supply is calculated. In Fig. 4.22, the SimEvents part of the ground power supply is shown.

Based on the SimEvents model in Fig. 4.22, the ground power supply cost is calculated shortly before pushback. Thus, the cost of the ground power supply (CGPS) can be calculated as follows:

$$CGPS = t * x * k_{GP}$$

where t is the overall time for ground handling activities excluding pushback, taxi-out, and runway use; x is the rate charged for ground power supply usage based on a 20-min usage time base; and k_{GP} is a weighting factor. Assuming the cost for a default 20-min ground power supply usage is $x = 25$ US\$, the overall time for aircraft ground handling activities is $t = 40$ min, and the weighting factor is $k = 0.05$, then the ground power supply usage cost is 50 US\$.

To reduce the clearance time at airports, some turnaround ground handling processes are done concurrently. For example, after the brake blocks have been applied on the aircraft, deboarding and unloading luggage are done concurrently as well as ground power supply.

From Fig. 4.21, it can be seen that at least three processes must be performed on an entity at the same time. Thus, each entity arriving in the model must be replicated and reassembled at the end of the concurrent processes. Hence, the replicate block copies the incoming entities for each output port. The number of output ports is then embedded manually. The time required for the processes is assumed as a distribution function, realized by the *event-based random number block pass*. Providing a precise period of time for the individual processes would be unrealistic because at a real airport, the duration of each ground handling process could become extended by external influences and/or also shortened. Moreover, the time allocated for each process cannot always be precisely maintained because MATLAB only processes decimal numbers which requires the time to be transferred as a decimal time number.

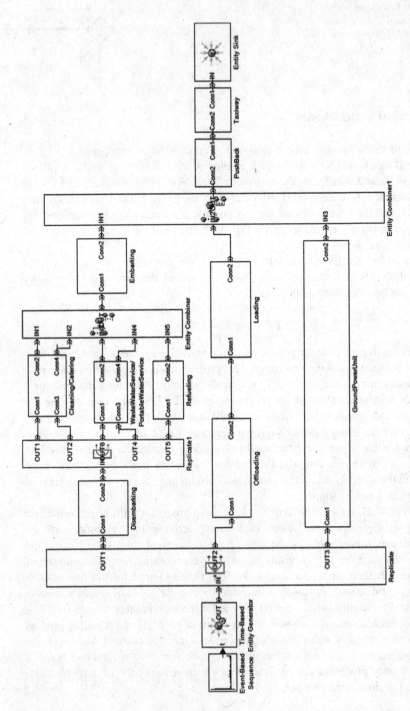

Fig. 4.21 Proof-of-concept turnaround model in SimEvents

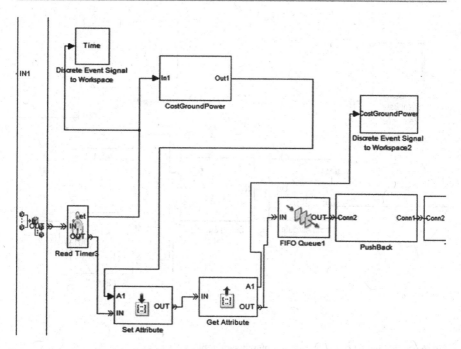

Fig. 4.22 SimEvents part of the ground power supply

Since SimEvents does not have all of the desired functionalities required for the airport operation case study, specific tasks associated with the airport ramp operational function have to be realized using MATLAB in conjunction with a functional programming language. Based on the following MATLAB code, the possible parking positions can be separated into two columns for the table "FlightPark-position." One is the terminal position, and the second is the ramp position, as shown in the following MATLAB code string for parking positions:

```
for i = 1: (length(FlightParkposition))
x = FlightParkposition(i,1);
if x >=4 && x <= 40;
FlightParkposition1(i,1) = 1;
else
FlightParkposition1(i,1) = 2;
end
end
```

The separated parking positions are passed through the Simulink block as attributes using the attributes block set. In Fig. 4.23, a submodel of the turnaround output switch is shown.

The output switch block sends the incoming entities on different paths. Based on the attributes used, it is decided which aircraft will be moved to output Out1 or

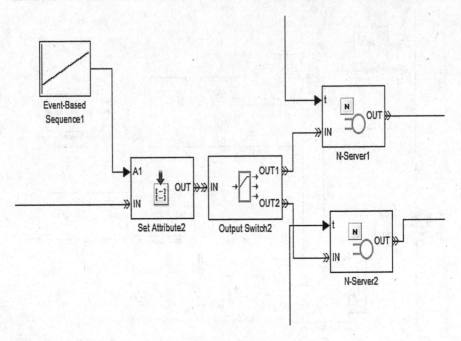

Fig. 4.23 Submodel of the turnaround output switch block

Out2. The output switch block makes it possible to differentiate between incoming entities based on the various processes in the model.

The aircraft which the model passes along to OUT1 have their parking positions directly at the gate. At this position, passengers leave the plane through the passenger bridge which connects the aircraft with the terminal so that the passengers and flight staff can directly deboard. A cost analysis for deboarding of passengers through the passenger bridge depends on the weight of the aircraft and the duration of the deboarding process. Per ton of weight and per 15 min usage of the passenger bridge, the average cost is assumed to be 0.80 US$ which results in the following cost equation for deboarding via the terminal bridge CDTB:

$$CDTB = t * w * k_{TB}$$

where $t =$ the number of minutes required for the process of deboarding the aircraft, $w =$ the weight of the aircraft, and k_{TB} is a weighting factor. The Simulink cost calculation model for passenger deboarding the aircraft via a terminal bridge makes use of the deboarding time, meaning the time this process actually takes, because costs are calculated on a 15-min time window for terminal bridge usage. The weight of the aircraft is transferred from the data structure in tons and multiplied by the factor k_{TB}. The factor refers to the 15-min time window in relation to 0.80 US$ average cost which results in $k_{TB} = 0.054$.

Incoming aircraft which pass through the model to OUT2 have their parking position on the ramp apron slightly farther away from the terminal. To allow passenger deboarding, a height-adjustable passenger stair and buses are required to transport the passengers to the airport terminal. The deboarding cost for the ramp depends on the number of passenger stairs and the number of buses needed. Therefore, costs for passenger deboarding on the ramp (CDBR) include the costs for using one or two passenger stairs and one or more buses. Assuming a bus can transport 50 passengers, the CDBR can be calculated using the following notation:

$$CDBR = k_{DB} * t * k_{PB} * PS * k_B * PB * k_{DBR}$$

where $t =$ the number of minutes required for the process of deboarding the aircraft on the ramp, based on a 20-min time window for calculation expressed by factor k_{DB}, the cost for a passenger stair $= PS$, $k_{PB} =$ factor indicating the number of passenger stairs, the cost for a passenger bus $= PB$, and $k_{DBR} =$ factor indicating the number of passenger buses.

The previously described airport operation simulation models described the terminal-related processes of passengers and luggage flow with the corresponding ramp operation with a focus on the aircraft turnaround. The aim of such a simulation is to obtain valid data for the capacities, process time, and bottlenecks of clearance processes. Moreover, these simulation models can be used for process optimization. This objective meets a Level 1 simulation with the constraint of correct input data as well as an adequate process model based on model building. Model building can be done based on UML activity diagrams which are suitable for formal mapping of these processes into model-based, event-driven process chains.

The implementation of the airport operation models is easily done within the SimEvents framework. SimEvents extends Simulink by a discrete event-driven simulation engine and a graphical components library. The analysis and optimization of latency between components, data throughput, packet loss, and other performance indicators is a major part of SimEvents. Transportation systems can be modeled using the component library correctly and comprehensibly. There are predefined blocks, such as queues, servers, and switches, which can be assembled, using drag and drop, into a model file for a simulation model. The block features extensive options so that tracing process interruptions, prioritization, and other operations of the simulation system are freely configurable. A generator block, for example, can serve as the starting point of a simulation model. SimEvents creates optional entities, events, or signals. Processing entities are characteristic of SimEvents. Events and signals are inherited Simulink functions and enable the (almost) unlimited combination of components of the various libraries. Thus, SimEvents is suitable for the simulation of event-driven transportation processes with mobile and stationary objects while identifying resource requirements, shortages, and optimization potentials.

4.6.3   A-SMGCS

Besides the terminal- and turnaround-related airport operational aspects, other issues to be handled are:

- Safety with growing air traffic volume
- Utilization of airport capacities
- Efficiency of air transport

These require a focus on approaching and departing as well as taxiing traffic on runways at airports which can be achieved by integrating A-SMGCS into airport operation. At the first level, aircraft in the movement area and transponder-equipped vehicles in the maneuvering area are identified via a Human-Machine Interface (HMI). The second level provides an alerting function to air traffic controllers (ATCOs) in case of intrusion of aircraft or vehicles in a predefined, protected area around the runway system. The identification is realized by using primary and secondary radar functions, data fusion, and representation on an HMI to the air traffic controller. A-SMGCS Levels 1 and 2 are the baseline for further enhanced surface management functions currently under development in the SESAR program, such as routing and guidance (URL 3). Currently, A-SMGCS usually consists of the following equipment:

- Secondary Surveillance Radar (SSR) Mode S Multilateration System to localize and positively identify aircraft and vehicles equipped with squitter units.
- Surface Movement Radars (SMRs) to localize all aircraft and vehicles. Vehicles without cooperative equipment can be localized but not identified.
- Airport Surveillance Radar (ASR) interface to ensure seamless tracking of approaching and departing aircraft.
- Interface to flight data bases to correlate A-SMGCS data with additional flight information in the identification process.
- Sensor Data Fusion (SDF) to merge all data from all sensors into one data set for each individual aircraft or vehicle.
- Runway incursion detection and handling logic.
- Displays of traffic situation for several controller working positions.

Based on A-SMGCS, airport ramp control can be achieved, which requires data including the duration of:

- Departure airlines waiting on the ramp
- Taxi-out operations
- Taxi-in operations

Also required:

- Number and duration of delays
- Number of arrivals and departures

With regard to this data, the essential information for ramp control is available and can be analyzed and optimized through simulation. Such an analysis can be the duration that each departure aircraft waited on the ramp by recording the time elapsed from when an aircraft first occupied the ramp until the aircraft made its first taxi movement. Based on this analysis, it is possible to simulate the mean duration of the departing aircraft's wait on the ramp before beginning to taxi. Another important analysis is the duration of the taxi-out time for a departing aircraft by recording the time elapsed from the first taxi movement until the aircraft has left the runway, meaning takeoff. For a given time window, the mean duration of all taxi-out operations can be simulated. The taxi-in operations for arriving aircraft can be analyzed and simulated as part of airport ramp control. For this purpose, the duration of taxi-in operations is analyzed based on the time elapsed from when an arriving aircraft touches down on the runway until the aircraft reaches its final parking position. Managing taxi-in and taxi-out traffic at ramps is a major concern in optimizing the turnaround process through simulation.

Economic benefits for airports and, therefore, better competitive positions can be realized through vehicle management on-ramps to maintain an optimized turnaround process. Today, ground handlers often suffer under economic pressure due to liberalization of ground handling services at airports. Additional competition from traditional and low-cost airlines results in more efficient utilization of aircraft with regard to shorter turnaround times. Optimized turnaround times can only be achieved when doing the same work in less time with better resource utilization, which is not easy to achieve due to constraints which have to be fulfilled, as shown in Table 4.7 (Moeller 2013).

Therefore, airport ramp control maintains the airport's airside operation, which is primarily tasked with transportation of passengers from remote aircraft parking positions to the passport stations and/or luggage claim and opposite transportation of passengers and luggage to the aircraft for departure.

From Table 4.7, it can be seen that optimizing the ground handling process requires a turnaround process workflow which is less sensitive to disturbances such as delayed plane arrivals, bad weather conditions, etc. This makes it necessary for ground handling services to embed the past, current, and future involvement of

Table 4.7 Different cases for scenario analysis

Best case scenario analysis	Worst case scenario analysis	Real case scenario analysis
Resources for ground handling are available; no shortage appears	*Resources* for ground handling not available in the required amount or at the worst only one resource is available but several are needed; shortages appear	*Resources* available for ground handling are well balanced. Cost of solution in between best case and worst case
Result: Costly solution; basically resources available cannot be used in an optimal way because more resources are available than necessary	*Result*: Cheap solution; available ground handling resources are not adequate	*Result*: Obtained solution for the real case analysis is suboptimal

individual ground handling vehicles in the turnaround process for decision making in the management of an apron system. Ground services need status information and predicted schedules for the turnaround processes for planning resource usage and utilization based on the vehicle fleet. For deployment purposes, vehicle ground services need the appropriate means for sending assignments to individual vehicle drivers and obtaining acknowledgements from them. A large number of vehicles for ground handling need to be equipped with communication tools requiring inexpensive onboard units (OBU) including possible ad hoc installations for vehicles rarely accessing the airport. Obvious options for such OBUs are smart devices, such as PDAs, smartphones, and tablet PCs, commercially available with GPS for positioning, and communication equipment such as WLAN and GSM. This also includes safe operation of aircraft taxiing in and taxiing out. Therefore, besides managing vehicles on aprons, managing and optimizing taxiing traffic on runways is much more technology and innovation driven.

4.6.4 Runway Incursion

Runway incursions are events where two or more vehicles use the same runway, resulting in a conflicting situation. Therefore, alerting of runway incursions is one of the key airport technology management issues for safe and secure taxiing and takeoff at airports. Runway incursions are occurrences at an aerodrome that involve the unapproved presence of an aircraft, a ground vehicle, or a person in the protected area designated for the landing and takeoff of aircraft. The growing air traffic volume has kept avoiding runway incursions on the National Transportation Safety Board's (NTSB) most wanted list of safety improvements for over a decade (Schönefeld and Möller 2012). In the past, runway incursions have led to accidents with significant loss of life. The worst runway incursion accident was at Tenerife, Canary Islands, Spain, in 1997, where two Boeing 747 collided. Recent incidents (Reuss 2009; NTSB 2010) indicate that runway incursions are still a problem. Although the number of runway incursions that result in an accident is small, the number of runway incursions has not significantly declined over the last decade. Statistics and results from simulation studies strongly indicate that the number of runway incursions increases much more rapidly than the traffic volume. Depending on the airport topography, an increase of 20 % in traffic volume may result in a 140 % increase in potential runway incursions (ALPAI 2007).

Runway incursion prevention technology is based on protective measures against causes that lead to a runway incursion and providing alerts during a runway incursion. For example, safety logic could prevent air traffic control (ATC) from assigning more than one aircraft to the same runway, thus providing protection against this type of operational error. The primary input to the system is given by information from various sensors and from traffic information service networks. This information is usually fused by multisensor data fusion, integrating background information, such as maps and movement models, into tracks describing the movements at the airport. This description is evaluated; and ATC commands, such as route information, are integrated to assess the traffic situation, to predict conflicts, and to detect runway incursions. Information about the traffic situation is given to Air Traffic Control (ATC), pilots, and vehicles via a Human-Machine Interface (HMI), e.g., an Electronic Flight Bag (EFB), if available, and signals at the airport or via radiotelephony (RTF) from ATC. The fact that the communication architecture of the distributed components belongs to the technology used is important because a communication infrastructure to support high speed data transfers to/from sensors and signals distributed across the airport is not always available. For example, the operation of intelligent signals on serial circuits requires the use of power line communication technology with sophisticated algorithms to ensure real-time constraint compliance (Schönefeld and Möller 2012). Runway incursion prediction and detection algorithms are used to detect runway incursions in their early stages or to predict them before they happen. Often, dedicated areas/volumes on the airport surface and on the final approach paths are used to determine specific variables needed for the detection of a runway conflict.

Simulation studies for runway incursion scenario analysis are of importance to study the surveillance performance of such type of advanced airport technology observation systems to provide an independent comparison of the different technologies that are used for a runway incursion protection and alerting system (RIPAS). Such a simulation study can examine, e.g., two vehicle movements at a taxiway/runway intersection with a number of different setups. The required data fusion algorithm is performed with a particle filter, to estimate the position, heading, and speed of the traffic. In the first traffic scenario, it has been assumed that a vehicle approaches the hold line from the taxiway, decelerates (app. 1.2 m/s^2) approximately 35 m before the hold line until it is approximately 7 m from the hold line, and comes to a complete stop approximately 3 m in front of the hold line. It stays there for approximately 30 s and then accelerates (approx. 2.5 m/s^2) to continue its way on the runway.

In the second scenario, a vehicle approaches the hold line from the taxiway, crosses the hold line, and continues its way on the runway.

These two scenarios are typical examples of the interfering traffic movement during a runway incursion from the taxiway and the non-inferring traffic during normal operations.

The primary finding of the simulation study was that, in the second scenario, all of the configurations follow the movement of the vehicle quite well. However, in the first scenario, baseline surveillance (BServ) configurations usually follow the vehicle during deceleration and acceleration not as well as during continuous movements. This scenario results in an overshoot when the vehicle stops at the time window 20–30 s (see Fig. 4.24). This overshoot makes it difficult to distinguish between a vehicle stopping directly in front of the hold line and one that crosses the hold line for a few seconds immediately after the vehicle stops or crosses the hold line. The BSurv configuration also lags behind the position of the vehicle once the vehicle begins to accelerate and crosses the hold line at the time window of 50–60 s (see Fig. 4.24). During the hold of the vehicle, the estimated distance to the hold line fluctuates a few meters around the true distance to the hold

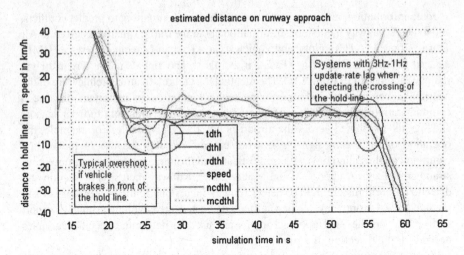

Fig. 4.24 Representative example of surveillance performance results from simulation for a specific airport traffic scenario. Line dthl plots the true distance of the vehicle to the hold line, and line dthl plots the estimated distance to the hold line (EDTHL) with Baseline Surveillance (BSurv) and a cooperative target. Line ncdthl plots EDTHL for BSurv and a noncooperative target. Typical for these approaches are the overshoot when the vehicle brakes and the lag when the vehicle accelerates again. Line rdthl plots the EDTHL for Extended Surveillance (ESurv) and a cooperative target, and the line rncdthl plots the EDTHL for ESurv and a noncooperative target

line. These findings apply specifically to the noncooperative target setup. The Extended surveillance (EServ) configuration, in contrast, has no overshoot and immediately follows the accelerating vehicle.

Since the RIPAS evaluation studies are not directly comparable because of different test setups, environmental conditions, and other factors, the alerting performance has been estimated by simulations, which allows for a comparison of different test setups. To characterize the alerting performance, the following evaluation criteria are introduced (Schönefeld and Möller 2012):

1. *Time to detect conflict t_{dc}*: the time that passed from when the runway conflict occurred until the system reliably detects the incident. Generally, an increase in t_{dc} can make up for a lack of accuracy in the surveillance with respect to the probability of missing an alert p_{ma} due to insufficient space for a safe separation of vehicles.
2. *Total time to alert t_{tta}*: the time until the alert has arrived at the involved parties.
3. *Probability of missed alert p_{ma}*: the probability of missing a runway incursion alert strongly depends on the airport topography, especially the distances between hold lines on runways and the accuracy of the surveillance equipment.
4. *Probability of false alert p_{fa}*: the probability of creating an alert without a conflicting situation. This probability strongly depends on the airport topography, especially the distances between hold lines and runways and the accuracy of the surveillance equipment.

For most systems, the performance in each of these criteria depends strongly on the surveillance/tracking performance of the system and is different for the various runway incursion scenarios as indicated by many authors (Cassel et al. 2002; Cassel 2005; Ir 2002). Therefore, the alerting performance of the systems was determined based on the results of the simulation.

4.7   Highway Operation Simulation and Its Empirical Evaluation

The rapid growth of traffic flow and the increase in urbanization has resulted in higher traffic demand, which has in turn resulted in a growing saturation of the road network. This has led to the need for investments to increase road network capacity and accessibility, a demand difficult to meet considering the financial, physical, and ecological situations in most major urban areas. Therefore, the road network traffic system is a vital component of the infrastructure which has a substantial impact on the development of a country's economy and socioeconomic standard. Hence, the performance of the road network traffic system largely depends on the relationships between transportation demands and supplies, i.e., capacity and accessibility. But traffic performance is also heavily dependent on demand variability and the occurrence of traffic incidents, restricting the availability of road network space. Therefore, congestion has become a common phenomenon, especially during peak hours in major urban areas. The negative impact of congestion on transportation sensitizes transportation authorities to the need for finding solutions that ensure more efficient, safe, and reliable traffic flows which requires road network, traffic operation simulation.

In Sect. 4.2, the traffic situation of land-based transportation, consisting of cars, streets, possible interruptions for traffic lights or traffic signs, lane changes, and so on, was introduced as part of an HLA simulation framework. With such a simulation framework, lane changing at on-ramps can be simulated. But this requires a couple of prerequisites which have to be fulfilled. The first prerequisite for modeling lane changing at on-ramps is that the model vehicle planning a lane change has to update the model settings for the desired lane change, which requires identifying if there will be any improvement in the traffic conditions for the model vehicle as a result of the lane change. The next essential issue is to check if it is possible to change lanes, which requires verifying if there is a sufficient gap for the planned lane change. This requires a more accurate representation of the model vehicle's behavior in the lane change decision process, which can be considered by different zones of a road section, as shown in Fig. 4.3, each one corresponding to a different lane changing intention. The distance up to the end of the road section characterizes these intentions.

Models for lane changing can be classified as "mandatory" and "discretionary" lane changes (Gipps 1986; Kesting et al. 2007; Laval and Daganzo 2006). Mandatory lane changes are necessary to keep to the route, such as merging at a freeway on-ramp. Discretionary lane changes are performed to improve driving conditions, to overtake a slower driving vehicle to increase speed, for example.

The framework model developed for lane changing by Gipps is still in use, based on the abovementioned three decision constraints to change lanes:

- Is it possible to change lanes?
- Is it necessary to change lanes?
- Is it desirable to change lanes?

Based on Gipps framework model, the following factors and their effects are considered to be the most important for a model (Kolen 2013):

- *Physically possible and safe to change lanes*: vehicle driver will not change lanes if there is any unacceptable risk of a collision.
- *Location of permanent obstructions*: vehicle drivers familiar with a road try to avoid being trapped behind known obstructions by selecting lanes that will give them free passage.
- *Presence of transit lanes*: vehicle drivers entitled to use a transit lane move into other lanes only when necessary to pass obstructions or slow vehicles.
- *Vehicle driver's intended turning movement*: readiness of a vehicle driver to change lanes affected by the distance from the intended turn and the direction of that turn.
- *Presence of heavy vehicles*: vehicle drivers try to avoid being trapped behind heavier vehicles because of their lower accelerations.
- *Speed*: affects a vehicle driver's decision to change lanes if the traffic in the present lane or the target lane is more likely to limit his/her speed in the short term.

Based on these constraints, the model assumes that a lane change only takes place when a gap of sufficient size is available in the target lane. With regard to the zone concept, shown in Fig. 4.3, a properly chosen zone algorithm can allow the application of the model in urban driving situations and on freeways.

As mentioned previously, the road network is divided into lengths of metering zones. The upstream boundary of a zone is required to be a free traffic flow area and the downstream is a critical bottleneck where demand-capacity ratio is usually high. A zone may have an arbitrary number of entrance and exit ramps. However, not all of the entrance ramps in a zone are metered ramps. Thus, the zone algorithm tries to balance the volume of the traffic entering and leaving the zone. The following equation can be used to calculate the ramp metering rate (Zhang et al. 2001):

$$UMV + SVMR + VMR + FRV = ERV + DBC + SZ$$

where UMV is the upstream mainline volume, $SVMR$ is the sum of volumes from nonmetered ramps, VMR is the sum of the volume from metered ramps, FRV is the sum of metered freeway to freeway ramp volumes, ERV is the sum of exit ramp volume, *DBC* is the downstream bottleneck capacity, and *SZ* is the space available in the zone. Let $S = 0$, and then the maximum volume that can enter the road network system through the metered ramp becomes.

$$VMR + FRV = (ERV + DBC)_i(UMV + SVMR)$$

The metering rate for each metered ramp is obtained based on the ramp factor and ($VMR + FRV$):

$$R_r = f_r(VMR + FRV)$$

where *Rr* is the metering rate of ramp *r* and *fr* is the ramp factor that defines the share the current ramp should take to balance the zone traffic demand-supply. Implementing the zone algorithm, an API is described in Zhang et al. (2001) and shown as follows:

```
typedef struct ramp_zone RAMP_ZONE;
struct ramp_zone {
      char *node;
      char *loop;
      float targetOcc;
      float Regulator; // for local ALINEA
      int NumOfLanes; // # of lanes for ramp
      float queueLength; // the length of the critical queue
      int measuredVol;
      float measuredOcc;
      float oldRampRate;
      float newRampRate;
      int index; /* internal identify */
      float *zoneWeight; /* point to zone weight */
      float help; /* total number of vehicle this ramp should reduce */
};
```

The simulation results obtained for the ramp zones are analyzed with regard to the congestion phenomena. Thus, the general pattern of the congestion is analyzed with regard to the situation when and where the congestion begins and how it propagates and dissipates. This analysis is of importance because it is essential to understanding the general pattern of the congestion in the study site for further analysis because it could uncover some hidden errors in the simulation or reveal site-specific traffic characteristics and limitations of the simulation.

This investigation can be executed, analyzing the simulation results obtained for the arrival time/travel time diagrams. Another option is a sensitivity analysis, which can be obtained through a number of simulation runs with different parameter sets. In this case, we try to find a parameter set that seems to give the best performance of the ramp metering algorithm. The effectiveness of the ramp control algorithms can vary according to the parameter values and a comparison of the performance of different ramp metering algorithms and is valid only when these algorithms are equally well calibrated (Zhang et al. 2001).

4.7.1 Swarm Behavior

Besides the aforementioned road network analysis, swarm behavior can also be used to study the collective vehicle behavior at on-ramp situations, etc., because swarm behavior is the aggregation of similar objects (vehicles, drivers), generally cruising in the same direction. Thus, a swarm relies on a large number of objects, so-called agents. The power of a swarm is derived from large numbers of direct and/or indirect agent interactions, hence the need for a relatively large number of agents, which happens in the transportation domain with regard to the huge number of objects (trucks, cars, buses, etc.). It is through the interactions of the agents that their individual behaviors are magnified, resulting in a global level behavior for the entire swarm. Thus, the swarm behavior concept makes use of the ability to search a problem's solution space in a way that is similar to the foraging search by a colony of social insects (ants). But the development of an artificial swarm system does not entail the complete imitation of natural systems, but it explores them in search of ideas for modeling. Therefore, transportation-engineering problems can be introduced being of combinatorial nature which can be described by algorithms that combine the existing results in the area of swarm intelligence with approximate reasoning in solving transportation systems problems like a stochastic vehicle routing problem and schedule synchronization in public transportation.

Let the stochastic vehicle routing problem be assigned as follows:

- Single depot and n nodes to be serviced.
- Homogenous seats (vehicle capacity, denoted by C).
- Demand at each node is randomized (probability density function is known).
- Vehicles from depot D serve a number of nodes, and on completion of service, they return to the depot.

Increasing the number of nodes along vehicles routes decreases the availability capacity of the vehicles. After completing the service at one node, it is possible to calculate whether or not the vehicle is able to serve the next node assuming the demand at the nodes is deterministic. This follows the constraint that the vehicles have a remaining capacity after serving their first nodes, which means being able to serve the next nodes. In case a vehicle has an insufficient capacity after serving the first node with regard to the uncertainty in actual required capacity and assumed capacity, the vehicle may return to the depot, unloading what it has picked up and returning to the node it could not serve due to less capacity and continuous service along the predefined path. With the objective, using the vehicles capacity at its maximum may result in planning routes with shorter distances. However, this also increases the number of cases in which vehicles arrive at a node and are unable to service it. Smaller utilization of vehicle capacity, on the other hand, will result in longer routes and less additional distance covered.

Let us assume that the random demand values at all nodes are known. Let us also assume that the optimization problem of finding the best set of vehicle routes can be solved using a particular optimization technique or heuristic algorithm, and then the solution can be achieved by the following approach:

- Using the behavior of the biological bee system which allows solving the vehicle routing problem as a traveling salesman problem and generating the most frequently unfeasible solution to the original problem.
- Traveling along the created route, it is possible to decide when to finish one vehicle's route and when to start with the next vehicle's route (Lucic 2002).

As foregoing mentioned, swarm behavior was first investigated in biology and has been inspired by the foraging behavior of ants.

- Ants find the shortest path to a food source from the nest.
- Ants deposit pheromones along a traveled path which is used by other ants to follow the trail.
- Swarm behavior shows adaptability, robustness, and redundancy.

Thus, swarm behavior becomes an attractor in transportation systems too because it can be introduced based on only three simple constraints derived from observed biological behavior:

1. Avoid collision with neighboring objects.
2. Match the velocity of neighboring objects.
3. Stay near the neighboring objects.

Given a group of agents considered simple relative to the observer, swarm behavior can be defined as a group of agents, whose collective interactions magnify the effects of individual agent behaviors, resulting in the manifestation of swarm level behaviors beyond the capability of a small subgroup of agents (Kovacina 2006).

Within the swarm behavior paradigm, the probability of agent interaction increases with agents' similarity. Thus, the use of swarm algorithms in the transportation system sector results in innovative analytical and modeling and simulation solutions, whereby the swarm algorithm is built on objects represented by symbolic strings and arranged on a 2D grid. Developing a swarm algorithm for transportation analysis requires:

- Random selection of an object.
- Random selection of a neighbor of the object.
- Probability of interaction = (No. shared string values)/(string length).

- If interacts, then copy value on one feature of neighbor's string to the object's string.

This approach can be formalized as a metaheuristic approach with regard to:

- Artificial objects build solutions to an optimization problem and exchange information on their quality vis-à-vis real objects.
- A combinatorial optimization problem reduced to a construction graph.
- Objects build partial solutions in each iteration and deposit information on each vertex.

This results in the following code string:

```
Optimization Metaheuristic Algorithm
Set parameters, initialize ramp trails
while termination condition not met do
    ConstructObjectSolutions
    ApplyLocalSearch
    UpdateRampInformation
end while
```

There are a number of existing swarm simulation packages that are currently available. Perhaps the best known is SWARM developed at the Santa Fe Institute, which provides an extensive library of software components designed for multiagent simulations of complex systems. Besides SWARM, Repast (Recursive Porous Agent Simulation Toolkit) Simphony is an agent-based simulation toolkit. Repast Simphony permits the systematic study of complex system behaviors through controlled and replicable computational experiments. Repast Simphony has an extensive set of libraries relating to topics such as networking, statistics, and modeling. Another feature of Repast Simphony is the built-in adaptive tools, such as a genetic algorithm package, that allow a simulation to tune itself. This tool has been used for Ph.D. thesis research supervised by the author (Farschtschi 2014).

A code string for Repast Simphony for an agent arriving at a location checking if the resource is needed is available (Sweda 2014). The agent begins using the resource right away; otherwise, it enters the queue. The next agent arrival is also scheduled. When an agent finishes using the resource, it departs out of the system. If there are any agents waiting in the queue, the one at the head leaves the queue and begins using the resource.

```
// Initialize system before each replication
public void initialize () {
repCounter ++;
arrivals = 0;
departures = 0;
for ( queue q: qList )
```

```
q. clear ();
for ( resource r: rList )
r. reset ();
for ( Stat s: sList )
s. initialize ();
double firstArrival = schedule . getTickCount ()+ arrivalRNG .
nextDouble ();
nextArrival = schedule . schedule ( ScheduleParameters . createOneTime (
firstArrival , 1), this , " arrive ");
nextClear = schedule . schedule ( ScheduleParameters . createOneTime (
schedule . getTickCount ()+warmup , ScheduleParameters . LAST_PRIORITY ),
this , " clearStats ");
nextEnd = schedule . schedule ( ScheduleParameters . createOneTime (
schedule . getTickCount ()+ endTime , ScheduleParameters .
LAST_PRIORITY ),
this , "end ");
```

4.8 Vehicle Tracking Based on the Internet of Things Paradigm

We are on the brink of a new ubiquitous computing and communication era, one that will radically transform our corporate, community, and personal spheres. Over a decade ago, Mark Weiser developed a seminal vision of future technological ubiquity—one in which the increasing "availability" of processing power would be accompanied by its decreasing "visibility." As he observed, "the most profound technologies are those that disappear...they weave themselves into the fabric of everyday life until they are indistinguishable from it." Early forms of ubiquitous information and communication networks are evident in the widespread use of mobile devices. Today the Internet of Things connects anything with everything everywhere. Thus, the Internet of Things spans across many industries; and transportation is no exception. With increased communication and data collection abilities, such as global positioning system (GPS), cloud computing, machine-to-machine (M2M), and cell phone triangulation, more and more data is available. This data provides information on travel time, origin destination, vehicle volumes, and traffic movements. It can be applied to adaptive signal control, also utilizing vehicle to vehicle (V2V) and vehicle to infrastructure (V2I) to transmit and use traffic data and engineering and construction projects that rely on traffic data collection. With the pervasion of smartphones and devices, travel time, speed data, and origin destination, information can be more readily available. Using cell phone triangulation, users can collect traffic flow information through the transmission of cell phone signal information to the mobile phone network. The information is then applied to solutions that can be utilized by road users, such as red light warnings, automatic tolling, and routing and navigation information (URL 4).

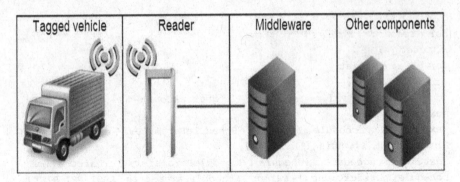

Fig. 4.25 Basic concept for vehicle tracking

Combining the ubiquity of access with mobile devices with data from traffic servers, which communicate with controllers on the side of the road, will, in time, accomplish a number of things, such as:

- Autonomous guided vehicles
- Vehicle tracking
- Smart mobility

In order to perform comprehensive vehicle tracking as part of the Internet of Things in the transportation application, tracking of tagged load units as a part of logistics and supply chain management and tagged vehicle tracking on road systems are used. This can be illustrated in the simplified Fig. 4.25, showing the common radio frequency identification (RFID) system structure and its interaction with other transportation system components.

The first issue is tracking of load units in logistics and supply chains. In this area, the items are usually tracked at several stages as they pass through the transportation process. There are several ways to use RFID systems on road networks. One application is the automatic payment collection on toll roads which makes it possible to overcome severe problems, such as traffic jams at toll points, and to reduce labor costs (Xiao 2008). The system consists of RFID tags usually fixed on windshields, bumpers, or license plate of moving vehicles and RFID readers located at toll stations. To ensure the system works effectively, each tag is associated with a corresponding payment account. As soon as the tagged vehicle enters the toll area and passes the reader, its tag is requested to provide information for identification. Once the information is read by the reader, the payment is charged to the linked account. Other examples of using RFID in road systems are managing parking lots and tank stations (Pala and Inan 2007; Mathis 2012).

The main goal of an RFID system is to provide stakeholders with valuable, complete, and reliable information on time and in a convenient form. Creating a system composed of RFID tags and readers for vehicle tracking results in the production of a certain amount of data, in some cases quite a significant amount.

Gathering this data and storing it in standalone form does not make sense. It has to be presented to the user. This results in an obvious requirement, that an RFID system should be integrated with other enterprise information system components with the help of middleware to provide users with data for further business use (Deriyenko 2012).

Let us consider the case of toll collection. Obviously, there should be no significant delays in obtaining vehicle data in order to perform payments. The same can be said about logistics and supply chain management activities. Users should be able to access most up-to-date information about vehicle movements; otherwise, the whole system will lose its advantage. Furthermore, it is important to avoid readers' signal collision, as their reading ranges overlap.

Another situation that can be modeled is scanning a tag that is supposed to be located beyond a reader's reading range. Therefore, tags and reader positions should be controlled properly using technical indicators of both devices, also keeping in mind environmental specifics. Moreover, all tag identifications should be stored in the system. For example, storing tag IDs in the system can help to detect a load pallet if one is missing or its tag is not readable; but, for obvious reasons, it is unfeasible for toll collection.

Corporate e-espionage is an important issue to consider (Weis et al. 2004) as information about tagged items and their movements can be of great value to business competitors. When discussing RFID use in road systems, the need for security of information about a tagged vehicle's location and, therefore, its driver must be contractually guaranteed. Loss of this information would constitute a serious breach of privacy, and an adequate level of security must be provided. Tag data access control techniques can be implemented, as well as cryptographic reader-to-tag authentication protocols, time bounding to defeat replay attacks, etc. (Weis et al. 2004; Ting 2006). However, the solutions mentioned are merely examples and do not represent the whole range of possible countermeasures. They can be modified based on developments in security technology, specific business requirements, available resources, potential return on investment (ROI) of the whole system, and potential losses from attacks. Currently, there are numerous providers of RFID solutions suitable for vehicle tracking being on the market. Some examples are Savant middleware by Auto-ID, solutions by SK RFID, or Ramp Holdings (Deriyenko 2012; Auto-ID 2003; SK 2012; Ramp 2012).

4.9   Exercises

1. Explain what is meant by the term "HLA federate."
2. Give an example of an HLA federate.
3. Explain what is meant by the term "HLA runtime infrastructure."
4. Describe the HLS RTI shown in Fig. 4.2 in your own words.
5. Explain what is meant by the term "SOA."
6. Describe the SOA elements shown in Fig. 4.1 in your own words.
7. Explain what is meant by the term "ontology-based modeling in transportation."

8. Describe the mathematical notation in your own words.
9. Explain what is meant by the term "XML."
10. Give an example of an XML data structure
11. Explain what is meant by the term workflow.
12. Give an example of a customer accessing a service station.
13. Explain what is meant by the term "vertex."
14. Give an example of a vertex in a graph structure.
15. Explain what is meant by the term "edge."
16. Give an example of an edge in a graph representation.
17. Explain what is meant by "flights are overscheduled."
18. List and define the main reasons that flights are overscheduled.
19. Explain what is meant by the term "turnaround."
20. List and define the main characteristics responsible for an aircraft turnaround.
21. Explain what is meant by the term "LEAP technology" in ProModel.
22. Describe the LEAP approach in ProModel in detail.
23. Explain what is meant by the term "UML."
24. Describe the most important UML elements.
25. Explain what is meant by the term "A-SMGCS."
26. Describe the most important components of A-SMBCS.
27. Explain what is meant by the term "lane change constraints."
28. Describe the three main constraints for lane change.
29. Explain what is meant by the term "swarm behavior."
30. Describe the three main constraints for swarm behavior.
31. Explain what is meant by the term "Internet of Things."
32. Describe the main constraints for the Internet of Things.
33. Explain what is meant by the term "vehicle tracking."
34. Describe the main components required for vehicle tracking.

References and Further Readings

Airbus SAS (2011) Airplane characteristics for airport planning- A320
Air Line Pilots Association International (2007) White paper-runway incursions – a call for action, A.L.P.A. International, Hrsg. Air Line Pilots Association International, Washington, DC
Arbuckle D, Rhodes CD, Andrews M, Roberts D, Hallowell S, Baker D (2006) (Auto-ID 20003) Auto-ID Savant Specification 1.0, Version of 1 Sept 2003
Auto-ID (2003) Auto-ID Savant Specification 1.0, Version of 1 Sept 2003
Banks J, Carson II JS, Nelson BL, Nicol DM (2001) Discrete-event system simulation. Prentice Hall
Biles WE (1996) Discrete-event systems. In: Kheir NA (ed) Systems modeling and computer simulation. Marcel Dekker, Inc., New York, pp 219–277
Brelie J von der (2014) PhD thesis under preparation at TU Clausthal
Burgain P (2010) On the control of airport departure operations. PhD thesis, Georgia Institute of Technology
Burleson C, Howell J, Anderegg A (2006) U.S. vision for 2025 air transportation. ATCA J Air Traffic Control 48(11):15–21
Carr F, Evans A, Clarke J, Feron E (2002) Modeling and control of airport queuing dynamics under severe flow restrictions. In: Proceedings of the 141 American Control Conference, IEEE

Cassel R (2005) Development of the Runway Incursion Advisory and Alerting System (RIAAS), NASA, 2005, 9. EUROCONTROL, Annual Safety Report 2010

Cassel R, Evers C, Esche J, Sleep B (2002) NASA Runway Incursion Prevention System (RIPS) Dallas–Fort Worth demonstration performance analysis

Deriyenko T (2012) RFID application in vehicle tracking, project work for lecture internet of things, TU Clausthal

Espinosa JA, Pulido AS (2002) IB (Integrated Business): a workflow based integration approach. In: Proceedings of the 35th IEEE Hawaii international conference on system sciences, pp 1–6

Farschtschi Y (2014) Conceptual design of agent-based simulation to investigate formation flight in civil aviation on the basis of the biological behavior of the swarm. PhD thesis (in German) submitted at TU Clausthal

Farschtschi Y, Moeller DPF, Widemann M (2011) Global optimization of interdependent turn-around processes at airports. In: Proceedings of grand challenges in M&S, Den Haag, pp 167–172

Gipps PG (1986) A model for the structure of lane-changing decisions. Transp Res Part B 20B (5):403–414

GreenAir Communications (2010) Commercial air traffic annual growth to continue at 4.7 percent over next 20 years, forecasts airbus. http://www.greenaironline.com/news.php?viewStory=596

HLA (2002) IEEE standard for modeling and simulation (M&S) high-level architecture (HLA) – framework and rules. http://www.ieeexplore.ieee.org/servlet/opac?punumber=7179

Ir DFG (2002) Runway safety monitor algorithm for runway incursion detection and alerting

Joint Planning and Development Office (2004) Next generation air transportation system integrated plan, technical report, Washington, DC

Joint Planning and Development Office (2007) Concept of operations for the next generation air transportation system

Kesting A, Treiber M, Helbing D (2007) General lane-changing model MOBIL for car-following models. Transport Res Rec J Transport Res Board 1999:86–94. doi:10.3141/1999-10

Kolen, H-P (2013) Modeling merging behaviour on freeway on ramps, Master thesis at TU Delft

Kovacina MA (2006) Swarm algorithm: simulation and generation. Master thesis at Case Western Reserve University

Krafzig D, Banke K, Slama D (2005) Enterprise SOA. Prentice Hall

Ky P, Miaillier B (2006) SESAR: towards the new generation of air traffic management systems in Europe. ATCA J Air Traffic Control 48(11):79–88

Laval JA, Daganzo CF (2006) Lane-changing in traffic streams. Transport Res Part B 40:251–264. doi:10.1016/j.trb.2005.04.003

Lucic P (2002) Modeling transportation problems using concepts of swarm intelligence and soft computing. PhD thesis Virginia Polytechnic Institute and State University, USA

Mathis R (2012) Neste oil launches automated vehicle identification at fueling stations. Available: http://secureidnews.com/news-item/neste-oil-launches-automatedvehicle-identification-at-fueling-stations/

Moeller DPF (2013) Airport technology management, Chapter 9, pp 105–119. In: Akhilesh KA (ed) Emerging dimensions of technology management. Springer

Moeller DPF, Wagner F, Jehle IA, Fermanelli V, Gao X (2013) Modeling and simulation workbench for international student team projects in transportation and logistics. In: Vakilzadain H, Crosbie R, Huntsinger R, Cooper K (eds) Proceedings of the 2013 summer simulation multiconference GCMS 2013. Curran Publication, Red Hook, pp 30–37

Moeller DPF, Popescu H (2000) HLA simulator for land based transportation. In: Waite WF (ed) Proceedings 2000 summer computer simulation conference. SCSI Publ., San Diego, pp 704–709

Moussaoui S (2013) Discrete event-based modeling and optimization approaches of the airport airside of Hamburg Airport using MATLAB Simulink SimEvents. Master thesis (in German) University of Hamburg

National Transportation Safety Board (NTSB) (2010) NTSB most wanted list – transportation safety improvements 2010–2011, NTSB (NTSB), Hrsg., NTSB

Neema H, Gohl J, Lattmann Z, Sztipanovits J, Karsani G, Neema S, Bapty T, Batteh J, Tummescheit H, Sureshkumar C (2014) Model-based integration platform for FMI co-simulation and heterogeneous simulations of cyber-physical systems. In: Proceedings of the 10th international Modelica conference. Modelica Association and Linkoping University Electronic Press, pp 235–245

NSF Blue Ribbon Panel (2006) Simulation-based engineering science, report, NSF, p 88ff, http://www.nsf.gov/pubs/reports/sbesfinalreport.pdf

Open Street Map (2014) http://www.openstreetmap.org/

Pala Z, Inan N (2007) Smart parking applications using RFID technology, in RFID Eurasia

ProModel (2011) http://www.promodel.com/solutionscafe/webinars/ProModel%202011%20Tutorial/PM2011Tutorial.html

Ramp (2012) http://www.ramp.com.au/rfidvehicletracking.html

Reuss J (2009) Investigation report EX006-1-2/04, German Federal Bureau of Aircraft Accidents Investigation

Schönefeld J, Möller DPF (2012) Runway incursion prevention systems: a review of review of runway incursion avoidance and alerting system approaches. Prog Aerosp Sci 51:31–49

SK (2012) http://www.skrfid.com/industrysolutions/electronic-toll-collection.html

Sweda T (2014) Discrete event simulation using repast java: a discrete event tutorial. https://code.google.com/p/repast-demos/wiki/ DiscreteEventSim

Thanh Le L (2006) Demand management at congested airports: how far are we from Utopia? PhD thesis, George Mason University, Fairfax

Ting Z (2006) A framework of networked RFID system supporting location tracking. In: 2nd IEEE/IFIP international conference in central Asia, Beijing

Weis SA, Sarma SE, Rivest RL, Engels DW (2004) Security and privacy aspects of low-cost radio frequency identification systems. In: Hutter D (ed) Security in pervasive computing, vol 2802, Lecture notes in computer science. Springer Publication, Cambridge, MA, pp 201–212

WTEC Panel (2009) International assessment of research and development in simulation-based engineering and science, report, World Technology Evaluation Center, Inc., p 426ff

Xiao Z (2008) The research and development of the highways electronic toll collection system. In: Knowledge discovery and data mining. Vol. 3, pp 359–362

Zhai J, Zhou Z, Shi Z, Shen L (2007) An integrated information platform for intelligent transportation systems based on ontology. In: Xu I, Tjoa A, Chaudhary S (eds) IFIP Vol. 254, research and practical issues of enterprise information systems. Springer, pp 787–796

Zhang M, Kim T, Nie X, Jin W, Chu L, Recker W (2001) Evaluation of on-ramp control algorithms, California PATH research report, UCB-ITS-PRR-2001-36

Links

(URL 1) http://www.wissenschaftsrat.de/download/archiv/4032-14.pdf

(URL 2) http://en.wikipedia.org/wiki/Service-oriented_architecture

(URL 3) https://www.eurocontrol.int/articles/advanced-surface-movement-guidance-and-control-systems-smgcs

(URL 4) http://miovision.com/blog/the-internet-of-things-and-transportation/

(URL 5) http://en.wikipedia.org/wiki/Architecture_framework

(URL 6) http://en.wikipedia.org/wiki/Ontology-based_data_integration

(URL 7) http://eolo.cps.unizar.es/docencia/MasterUPV/Articulos/Selles-Ontology_based%20Model%20Transformation.pdf

(URL 8) http://ceur-ws.org/Vol-766/paper08.pdf

Simulation Tools in Transportation

5

This chapter begins with a brief overview of simulation tools in transportation. Section 5.1 covers the model-building process and the vital importance of computer simulation for two different application domains: continuous and discrete systems. Section 5.2 introduces continuous systems simulation tools including block-oriented simulation tools (Sect. 5.2.1) and equation-oriented simulation tools (Sect. 5.2.2). Thereafter, Sect. 5.3 introduces discrete-event simulation tools by describing a sample of the many available simulation software packages, focusing on the ones used for the case study examples in this book. Section 5.4 covers the important topic of object-oriented simulation, while Sect. 5.5 introduces online simulation. Section 5.6 describes a ProModel-based case study for a maritime transportation analysis. Section 5.7 contains comprehensive questions from the simulation software area, and a final section includes references and suggestions for further reading.

5.1 Introduction

Computer simulation refers to methods for studying a wide variety of models of real-world systems by numerical evaluation using simulation software designed to mimic the system's operations and/or characteristics, over time. Thus, computer simulation entails the implementation of a model developed on a computer. In this way, a transportation system model can be manipulated in accordance with the scope of the simulation study, i.e., by changing the model structure, parameters, inputs, and outputs to accurately match real-world system behavior. In order to ensure that the computer simulation generates a sufficiently accurate solution, meaning that the computer outputs are in close vicinity to the corresponding features of the transportation system being simulated, it is necessary to have a thorough understanding of the real-world object modeled and the numerical algorithms of the simulation tools.

© Springer-Verlag London 2014
D.P.F. Möller, *Introduction to Transportation Analysis, Modeling and Simulation*,
Simulation Foundations, Methods and Applications,
DOI 10.1007/978-1-4471-5637-6_5

The use of modeling and simulation for solving problems that overlap the disciplines of science and engineering has improved cooperation among these disciplines and lowered previously rigid barriers between them. The most important step in modeling and simulation, as applied to a particular transportation system, is the translation of a real-world transportation system into the mathematical language, which is universal while independent due to the application and domain.

Based on phenomenological and physical principles, relevant to describing a particular system, the equations that characterize the system's behavior are carried out in a number of ways, which requires the use of simulation like:

- Continuous simulation, which concerns modeling over time of a system by a representation in which the state variable change continuously with respect to time.
- Discrete-event simulation, which concerns modeling of a system as it evolves over time by a representation in which the state variables change instantaneously at separate points in time at which an event occurs.
- Discrete-continuous simulation, which concerns modeling of a system by a representation with an interaction between discretely changing and continuously changing state variables as follows (Law 2007):
  - A discrete event may cause a discrete change in the value of a continuous state variable.
  - A discrete event may cause the relationship governing a continuous state variable to change at a particular time.
  - A continuous state variable achieving a threshold value may cause a discrete event to occur or to be scheduled.
- Monte Carlo simulation, which is a scheme employing random numbers, that is, U(0,1) random variates, which is used for solving certain stochastic or deterministic problems.
- Spreadsheet simulation can be done in spreadsheets such as Excel if the problem of interest is not too complex. But spreadsheets have the following limitations (Law 2007):
  - Only simple data structures are available.
  - Complex algorithms are difficult to implement.
  - Data storage is limited.
  Spreadsheet simulations are widely used for performing risk analysis.

5.2 Classification of Simulation Systems

Historically simulation systems were classified to be of two major types, namely:

- Simulation languages: are general in nature, and model development was done by writing code. Furthermore, simulation languages provided a good possibility of modeling flexibility, but were often difficult to use.

- Application-oriented simulators: are oriented towards a particular application, and a model was developed by using graphics, dialog boxes, and pull-down menus. Furthermore, application-oriented simulators were sometimes easier to learn and to use, but often were not flexible enough to use for some application problems.

If the selected simulator and/or simulation language is not flexible enough or too difficult to use, then the simulation project can generate erogenous results or may not even be completed. However, in recent years, vendors of simulation languages have attempted to make their software easier to use by employing a graphical model-building approach. A typical scenario is to have a library of icons, so-called blocks, located on one side of the computer screen. The blocks are selected from the library with a mouse and are placed on the work area. The blocks are then connected to indicate the flow of entities through the system under test. Furthermore, double-clicking a block brings up a dialog box where details are added. SIDAS, PSI, and ModelMaker, to name a few of which, are of this type of block-oriented simulation tools. Each of these tools has salient features; therefore, regardless of the block-oriented simulation tool selected, several requirements for a simulation run must be fulfilled by the user, such as the description of the real-world system to be modeled and simulated. This involves specifying the type of equations in terms of differential equations, transfer functions, and state space models. Continuous-time systems representation by block diagrams and/or signal flow graphs is given with simplification and reduction techniques leading, for example, to decomposition of overall transfer functions. This is accomplished by means of structure statements or commands to build up the model of the real-world system.

Using the so-called block-oriented simulation tool, blocks have to be specified by parameters, inputs and outputs, the appropriate functions, initial conditions, arbitrary functions, runtime, time interval, and other control commands. These simulation tools are, for the most part, very user oriented and incorporate default conditions that enable a novice to obtain meaningful results immediately. Moreover, these simulation tools can solve linear and nonlinear differential equations of the nth order which describe the real-world system through the mathematical model.

In most cases, high-order differential equations are reduced to sets of first-order differential equations, which means that for block-oriented simulation tools, a decomposition into a block-oriented scheme of first-order blocks which can easily be solved using numeric integration methods is used, as shown in Fig. 5.1.

5.2.1 Block-Oriented Simulation Systems

As an example, the interactive block-oriented simulation system, PSI, will be introduced briefly. PSI can be used for studying the behavior of continuous and discrete systems. The notation used in PSI is similar to most other block-oriented simulation tools. Block-oriented simulation systems use differential equations

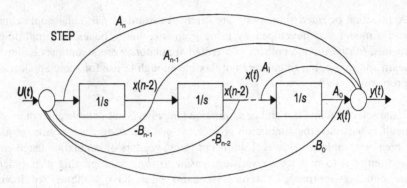

Fig. 5.1 Decomposition of a dynamic system of the nth order

which represent the simulation model, as shown for the second-order differential equation:

$$y''(t) = -y'(t) - y(t) + u(t), \tag{5.1}$$

which can be rewritten as a set of first-order integral equations, yielding

$$y'(t) = y'(0) + \int_0^t y''(\tau)d\tau \tag{5.2}$$

$$y(t) = y(0) + \int_0^t y'(\tau)d\tau, \tag{5.3}$$

and modeled as a block-oriented representation as shown in Fig. 5.2.

In Fig. 5.2, Y2DOT represents the second derivative, YDOT represents the first derivative, Y is the primitive function, and STEP represents a constant $U(t)$.

The second-order system, shown in Fig. 5.2, consists of specific facilities such as:

- Structure
- Parameters
- Numeric integration
- Output

The structure is given by defining the inputs for each block. When the inputs of all blocks are defined, the structure of the simulation model is known; and the block-oriented simulation software is able to calculate the behavior of the system. The idea behind the block-oriented simulation approach is that any simulation

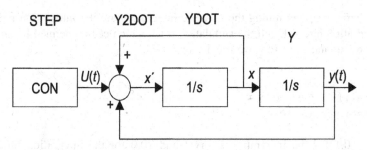

Fig. 5.2 Block diagram of a second-order system

model can be built up from basic block elements, such as integrators, gain-transfer functions, table lookup facilities, nonlinear function blocks, constants, etc. Each block is identified by its name, which determines the block and its output, as well as the block type and its inputs. In the example, shown in Fig. 5.2, we have:

```
STEP  = u(t)
Y     = y(t)
YDOT  = y'(t)
Y2DOT = y''(t)
```

Together with the simulation tool's specific control commands, the block structure can be defined. For PSI we use B as follows:

```
PSI·B
Configuration Specification
Block, Type, Input1, Input2, Input 3
B·STEP, CON              : STEP is a CON block
B·YDOT, INT, Y2DOT       : YDOT is an integrator with Y2DOT as input
B·Y, INT, YDOT           : Y is an integrator with YDOT as input
B·Y2DOT, SUM, STEP, Y, YDOT: Y2DOT is a summer to add all inputs
B
```

The parameters of each block can be defined using specific control commands, which depends on the simulation software used. For PSI, we use P; hence:

```
PSI·P
Parameters
Blocks, Par1, Par2, Par3
P·STEP, 1          : STEP has value 1
P·Y, 0, 1          : initial condition=0; input gain=1
P·YDOT, 0, 1       : initial condition=0; input gain=1
P·Y2DOT, 1, -1, -1 : gains of the corresponding inputs
P
```

The variables determining the numeric integration are the integration method, the integration interval, and the simulation time, which can be defined by specific control commands. For PSI, we use T and obtain:

```
PSI·T
Integration interval=0.1
Integration time=10.0
```

Using 0.1 for the integration interval and 10.0 for the integration time, the simulation run will be calculated for 10 time units with an integration interval of 0.1 time unit.

All blocks are calculated during the simulation run; however, only some of which can be shown on the screen. Suppose that $y(t)$ is the output variable of interest which may be shown on the screen. Some specific control commands are needed, which depend on the simulation software used. For PSI, we use O; hence:

```
PSI·O
Name of blocks to be shown=?
```

Now, the required transient behavior can be calculated, and the output variable (s) indicated is shown on the PC screen. The simulation run will start using a specific control command. For PSI we use R:

```
PSI·R
```

PSI allows the presentation of more than one variable, meaning more complex figures can be shown representing the respective system variables, including inputs as well as outputs.

As mentioned, simulation tools for continuous systems have salient features but are all based on three types of instructions:

- Model instructions, which generate the algebraic block-oriented structure of the mathematical equations of the real-world system
- Data instructions, which are used to assign the respective parameter values, initial conditions, etc.
- Control instructions, which determine the simulation run and select the output variables

The simulation run itself can be divided into four steps as follows:

1. Interactive implementation of the block diagram of the mathematical model: for this feature, the simulation system outlines a frame with a number of cross-points, at each of which a block can be inserted representing the respective mathematical functional element. To insert a block, e.g., an integrator, one

chooses a cross-point with the cursor and types "I" for insert, followed by the symbolic name of the block to be inserted, which is "INT" for the integrator block. When connecting or modifying blocks, a set of additional commands is available, such as change a block, delete a block, link a block to different blocks, and end the input modus. Moreover, special blocks are used to implement specific algorithms, transient behavior, etc.

2. Set of block parameters: for this feature, the simulation system offers initial conditions; gain factors of integrators; simulation parameters, such as the length of the simulation interval; the desired accuracy of the numerical integration; the numeric integration method; etc.

3. Numerical solution of the sets of differential equations: for this feature, the simulation system offers successful sorting algorithms which determine the calculation sequence in such a way that only these elements are processed in the loop, the values of which are updated in the respective sequence.

4. Presentation of simulation results.

Sorting algorithms handle the algebraic loop problem of a simulation run.

The simulation run has an algebraic loop problem; if it is not possible to calculate all blocks $b \in B$ in such a way that the output of block b_i is connected to the input of block b_j, then $i < j$. The value of algebraic blocks can be calculated if:

- Algebraic block is a constant block.
- Output of the algebraic block at t_{n+1} is known either by the initial values or the output values one step behind the actual integration step.
- Output value of the algebraic block has already been calculated in the actual time interval.

Sorting algorithms are used to determine the calculation sequence of the functional elements used. The first step in sorting deals with evaluating the simulation configuration by determining the matrix positions of the integrators with their inputs, the block numbers of the previous functional blocks in a counterclockwise signal flow. This results in a list of sorted algebraic functions that allows the calculation for each block as a function of its argument in the determined sequence.

If all of the integration steps in the sorted sequence are finished, the next repetition of the sorted loop will be prepared:

$$n := n + 1 \qquad (5.4)$$

and

$$t_n := t_{n+1}. \qquad (5.5)$$

The simulation runs through the sorted loop until the condition $t_n = t$ and/or another given condition is reached that terminates the procedure. For an algebraic loop, one or more algebraic blocks are connected in a closed loop that contains no

dynamic components such as integrators. Algebraic loops can be eliminated by
opening the loop, which means manipulating the original equation.

Block-oriented simulation systems available beside PSI are:

- *ACSL*: equation-oriented language consisting of a set of arithmetic operators,
  standard functions, a set of special ACSL statements, and a macro capability
  which allows the extension of special ACSL statements; see (URL 1).
- *BORIS*: see (URL 2).
- *DYNAMO*: simulation language and accompanying graphical notation developed within the system dynamics analytical framework; see (URL 3).
- *Modelica*: object-oriented, declarative, multi-domain modeling language for
  component-oriented modeling of complex systems. The free Modelica language
  is developed by the nonprofit Modelica Association for industrial and academic
  users; see (URL 4).
- *ModelMaker*: see (URL 5).
- *SIDAS*: see (URL 6).
- *Simulink*: see (URL 7).
- *VisSim*: microscopic multimodal traffic flow simulation software. The scope of
  applications of VisSIM include evacuation, public transport, urban planning
  over fire protection, traffic planning, traffic engineering, signal timing, simulation visualization, computer animation, etc.; see (URL 8).

5.2.2 Equation-Oriented Simulation Systems

In the interactive block-oriented simulation systems (described in.Sect. 5.2.1), each
block in the flow sheet in sequence (see Fig. 5.2), where its input is given, computes
its output. This approach can be very time-consuming for a certain types of
problems. Thus, equation-oriented modeling is an alternate strategy for solving
flow sheet simulations. Instead of solving each block in sequence, the equation-
oriented simulation approach gathers all model equations together and solves them
at the same time. Therefore, equation-oriented modeling is sometimes called
equation-based or simultaneous equation modeling.

Today, a wide variety of different equation-oriented simulation systems is
available. Most of them use a three-step or five-step model-building process with
the following statements containing the respective structural, data, and control
statements:

- *INITIAL*: appears at the beginning of the program before simulation time moves
  forward and is evaluated only once per run. Calculations in this section involve
  initial conditions of state variables or the initialization of counters.
- *DYNAMIC*: moves forward in simulation time. Within the *DYNAMIC* section,
  the *DERIVATIVE* section moves forward in a continuous manner, controlled by
  the integration algorithm.

- *DERIVATIVE*: contains differential equations and integrations. The simulation system used sorts the equations deciding what has to be calculated in what order.
- *DISCRETE*: are at the same level as *DERIVATIVE* but are activated by *INTERVAL* or *SCHEDULE* statements to describe discrete events.
- *END*: is executed once after simulation time has stopped. When program control transfers out of the *DERIVATIVE* or *DYNAMIC* section in response to the *END* statement, it moves to the beginning of the *END* section. *END* is used for statistical calculations. Also parameters of the model can be changed in the *END* section.

The structural statements transform the mathematical expressions of the real-world system through algebraic statements into functional blocks. The data statements connect the symbolic defined parameters, constants, and initial conditions, with the respective numerical values. The control statements schedule the time dependence of the output variables.

INITIAL contains the parameter values and initial values. *DYNAMIC* contains the model description. *END* terminates the simulation run and contains the parameters with its new values, which may be calculated in a second run.

Equation-oriented simulation tools are:

- *ACSL*: see (URL 1).
- *Aspen*: see (URL 9).
- *ModelMaker*: see (URL 5).
- *Simulink*: see (URL 7).

5.2.3 Summary of Simulation Systems

Based on a foregoing discussion, two types of simulation systems have been introduced and used:

- General-Purpose Simulation Systems/Packages (GPSP): can be used for any application, but might have features for certain one, e.g., manufacturing or process reengineering. There are several GPSP, including:
  - *GPSS*
    - GPSS was one of the first simulation language and *GPSS/H* as successor of GPSS which provides improvements over earlier versions of GPSS.
    - SLX is the successor of GPSS/H which replaces many features of GPSS/H entirely and represents many of GPSS/H features with simpler, more general constructs. SLX is a layered modeling system in which GPSS/H comprises only one of the layers.
  - *SIMSCRIPT III*
    - Process- or event-oriented modeling
    - Animation and graphics in conjunction with SIMGRAPHICS
    - Object-oriented simulation package

- *MODSIM II*
  - Object-oriented language
- *SIMAN V*
  - General-purpose program for modeling discrete and continuous systems.
  - Model frame defines system components such as:
    - Machines
    - Queues
    - Transporters (trucks, AGV, conveyer) and their interrelationships
  - Incorporates ARENA environment.
- *SLAMSYSTEM*
  - Integrated simulation system for Windows based-PCs or OS/2
  - Features are:
    - Multiple networks in a single scenario
    - Output graphics
    - Interface to SimStat
    - OS/2 metafiles for graphic
- Special-Purpose (SPSP) or Application-Oriented Simulation Systems/Packages (AOSP): are designed to be used for a certain class of applications such as manufacturing, health care, or contact centers.
  - Examples for manufacturing are:
    - *AutoMod*
    - *Enterprise Dynamic*
    - *Flexsim*
    - *ProModel*
    - *QUEST*
    - *SIMFACTORY II.5*
    - *WITNESS*
  - Examples for communication networks are:
    - *OPNET Modeler*
    - *QualNet*
  - Examples for process reengineering and services are:
    - *ProcessModel*
    - *ServiceModel*
    - *SIMPROCESS*
  - Examples for health care are:
    - *MedModel*
  - Examples for contact centers are:
    - *Arena Contact Center Edition*
  - Examples for supply chains are:
    - *Supply Chain Builder*
  - Examples for animation (stand alone) are:
    - *Proof Animation*
    - *Systemflow 3D Animator*

In summary, the history of simulation systems is given in Table 5.1.

Table 5.1 History of simulation systems/software

	Generation of simulation systems/software
1955–1960	Simulation was a very expensive and specialized system that was generally used only by large corporations that required substantial capital investments. It requires user programming because there was no user support. Model building was based on the available higher programming languages such as FORTRAN, ALGOL, etc. These models run on the so-called mainframes
1960–1965	The *first generation of simulation systems/software* was launched into the market. They offer a very simple user support through automatically generated computational relations and graphical user interfaces
1965–1970	The *second generation of simulation systems/software* was launched into the market. They are better systems allowing some interactivity
1970–1980	The *third generation of simulation systems/software* was launched into the market which follows the trend that computers were becoming faster and cheaper which extended new possibilities of simulation systems like combined simulation, etc. Moreover simulation was discovered by other industries, although most of the companies were quite still large
1980–1990	The *forth generation of simulation systems/software* was launched into the market with domain-specific and specialized simulators, animation possibilities, and easier model implementation. It became the systems for many companies, notably in the automotive and heavy industries
1990–2000	The *fifth generation of simulation systems/software* was launched into the market, embedding artificial intelligence, model specification, and experimental environments, expanding the possibilities of the systems of the fourth generation especially with much more sophisticated graphic tools that allowed 3D (spatial) and 4D (time) to become a standard. This is the time window where simulation began to mature. Many small- and medium-sized companies embraced the systems, and they began to see its uses at the very early stages of projects
2000–2010	The *sixth generation of simulation software was launched into the market,* embedding object-oriented modeling frameworks; soft-computing methodology like fuzzy sets, neuronal nets, genetic algorithms, evolution theory, probabilistic methods, and virtual and augmented reality environments in simulation, which allow tactile force-feedback interaction; and simulation at the Internet. Simulation become a state-of-the-art tool resident on every systems-analysis computer
2010–today	The *seventh generation of simulation software* was launched into the market, including XML, pipelining concepts, web-based simulation, remote access simulation, and cloud and fog computing simulation. With the rapid advances being made in computers and software, it is very difficult to predict much about the simulation for the distant future

With regard to the manifold of available simulation software, we can consider numerous features when it comes to select simulation software. Therefore, we categorize these features as being in one of the following groups:

- Animation: including built-in animation because in an animation key, elements of the system are represented on the screen by icons that dynamically change position, color, and shape as the simulation model evolves through time.

- Customer support and documentation: including public and customized training opportunities as support service. Good documentation is a crucial requirement for using the simulation software product. Therefore, there should be a detailed description of how each modeling construct works, particularly if its operating procedures are complex.
- General capabilities: including modeling flexibility, meaning the ability to model a system whose operating procedures can have an amount of complexity and ease of use.
- Hardware and software consideration: means for what computer platform the simulation software is available.
- Output report and plots: including standard reports as well as customize reports as well as a variety of graphics such as a histogram, time plots, bar charts or pie charts, and/or correlation plots.
- Statistical features: including a good random-number generator for generating independent observations from a uniform distribution on the interval [0, 1]. The generator should have at least 100 different streams (preferably far more) that can be assigned to different sources of randomness such as interarrival times or service times in a simulation model—the seeds should not depend on the internal clock of the computer. Moreover, the user should be able to set the seed for each stream, if desired (Law 2007).

5.3 Discrete-Event Simulation Systems

With regard to the modeling approaches in transportation systems analysis, modeling, and simulation, desirable simulation software features are important. Discrete-event simulation concerns the modeling of a system as it evolves over time by a representation in which the state variables change instantaneously at separate points in time. These points in time are the ones at which an event occurs, where an event is defined as an instantaneous occurrence that may change the state of the system.

The structure of discrete-event simulation systems is similar to continuous simulation ones, but they contain an event-based control of time allowing the classification of discrete system simulation systems as follows:

- Transaction-oriented simulation software: based on a time-step control, determined through preprogrammed logical conditions related to respective blocks. Language elements are:
  - Transactions
  - Blocks
  - Facilities
  - Queues
  - Pools and storages
  - Logical switches
  - Numerical and logical variables

- – Functions
- – Tables
- • Event-oriented simulation software: based on time-dependent and restricted event-handling language elements
- • Activity-oriented simulation software: based on activity schedules that are started if specific constraints are fulfilled
- • Process-oriented simulation software: based on activation trigger of the following events as specified language elements

Thus, discrete-event system simulation refers to an instantaneous occurrence in time that alters the state of the system: event scheduling and process interaction. In the event scheduling approach, one concentrates on events and their effect on the system state. The system state is a collection of variables, the values of which define the state of the system at a given point in time. Thus, in the event scheduling approach to discrete system simulation, the occurrence of any of the events that specify the model brings about a resultant change in system state. The model must provide the state changes for recording, which is usually managed by collecting one or more types of statistics at the occurrence of each event. A different form of discrete-event simulation is the process interaction approach. A process is a time-ordered collection of events, activities, and delays that are somehow related to an entity. In process-oriented simulations, many processes are usually ongoing simultaneously and involve complex interactions among these many processes (Biles 1996).

The common characteristics of discrete-event simulation systems include a graphical user interface, animation, and automatically collected outputs to measure the simulation system performance. Simulation results are displayed in tabular or graphical form in standard reports and interactively while running a simulation. Outputs from different scenario analyses can be compared graphically or in tabular form. Most discrete-event simulation tools provide statistical analysis that includes confidence intervals for performance measures and comparisons, as well as a variety of other analysis methods. The available discrete-event simulation systems have been developed to meet specific demands.

Discrete-event simulation tools are (Banks et al. 2001):

- • *Arena*
- • *AutoMod*
- • *CSIM*
- • *DEVS*
- • *Extend*
- • *GPSS*
- • *MODSIM*
- • *ProModel*
- • *SIMAN*
- • *SIMFACTORY*
- • *SIMSCRIPT*
- • *SLAMSYSTEM*

- *QUEST*
- *WITNESS*
- Etc.

Arena: Simulation system/software has been developed for simulating discrete and continuous systems as well as combined discrete-continuous simulation and can be used to model business processes and other systems in support of high-level analysis needs. It represents dynamic behavior in a hierarchical flowchart and stores system information in data spreadsheets. Its extension, OptQuest, allows system optimization. Arena's input analyzer automates the process of selecting proper distributions and their parameters to illustrate existing data, such as process and interarrival time. Arena's output analyzer automates the comparison of different design alternatives. The Arena Professional Edition includes the ability to create customized module, which are functionally arranged into a number of templates, and to store them in a new template. The "Basis Process" template contains modules that are used in virtually every model for modeling arrivals, departures, services, and decision logic of entities. The "Advanced Process" template contains modules that are used to perform more advanced process logic and to access external data files in Excel, Access, and SQL databases. Arena also has an option that permits a model to run in real time and to dynamically interact with other processes; this supports applications such as the high-level architecture (HLA) (see Sect. 4.2) and testing of hardware/software control systems (Law 2007).

AutoMod: Simulation system/software has been developed for material handling systems, including vehicle systems, conveyers, automated storage, bridge crane, free conveyors, and kinematics for robots. It contains a full simulation programming language, and its animation can be viewed from any angle and perspective in real time.

Extend: Simulation system/software was developed combining the block diagram approach to model handling with an authoring environment for creating new blocks or complete vertical market simulators. It is process oriented but also capable of continuous and combined modeling.

In some situations, real-world systems include diverse components that require different formalisms for modeling and simulation. This occur when system components are continuous with concentrated parameters that show slow and fast parts or when a system contains a queuing part and a continuous part, which is introduced as a combined systems approach. Using an object-oriented approach, one can simulate combined systems creating objects that simulate system submodels—queuing and continuous—running them concurrently. Hence, submodels of very different kinds can run and interact in the same simulation environment.

AnyLogic: Simulation system/software can be used for combined simulation. It was developed based on complex system modeling theory and working standards in system design and allows system exploration at any desired depth and level of abstraction. The fundamental basis for modeling systems with AnyLogic includes:

- Arbitrary complex behavior logic; timing; topologies, such as ring, chain mesh, etc.; and routing
- Block-based flowchart modeling
- Differential and algebraic equations
- Direct links to data bases and GIS
- HLA support
- MATLAB-SIMULINK-type library
- Modeling in Java, to run models on any Java-enabled platform, or even as applets on web browsers
- Message passing, ports, custom routing
- State chart modeling, to combine discrete and continuous behaviors
- UML for Real Time (UML-RT)

UML-RT in AnyLogic is specifically adapted for the development of complex, event-driven, real-time, real-world systems. Its modeling constructs have rigorous formal semantics that provide for model execution. UML-RT modeling means:

- Explicit structural decomposition
- Clear separation of structure and behavior
- High degree of reusability

UML-RL is a complete working modeling standard. When developing a model in AnyLogic, one develops classes of active objects representing real-world objects. Active objects can encapsulate other active objects to any desired level. There is one designated root class, which describes the model structure. Active objects interact with their surroundings solely through interface objects, which are ports and variables. Ports are used for discrete communication (message passing), optionally using a queue, while variables are used for continuous communication.

To use UML-RT, AnyLogic provides a graphical user interface similar to classical visual development tools for programming languages. The libraries included give that development environment the extension to do discrete, continuous, and hybrid simulation modeling. If licensed by the user separately, Numerical Algorithms Group (NAG) libraries or other numerical methods, not yet implemented in Java, can be added.

The notation on the interface makes active object classes highly reusable. Moreover, the modeler can define the object behavior as a Java method and run it within the active object as a separate thread, the execution of which can be synchronized with other activities through:

- Delay (timeout) method
- WaitEvent (static event) method
- WaitFORMessage(..) method of the PortQueuing class

The innovative core technology together with a remarkable set of features makes AnyLogic an advanced technology solution for a broad range of real-world applications.

This core technology is continuously extended by different libraries, such as MATLAB-SIMULINK, to provide the look and feel of the original SIMULINK simulation environment, or an Enterprise Library, which gives similar possibilities for building traditional discrete-event simulation models.

Any AnyLogic Java class can be integrated into a Java program. On the other hand, external routines can be integrated either by including external classes or using the JNI interface to non-Java libraries.

As already mentioned, AnyLogic includes the optional OptQuest, the most common and widely used optimization software.

As long as there are no platform-dependent libraries needed for a specific simulation model, the precompiled Java-Applet can be put on a web server and used over the Internet without any restrictions, a distinct advantage in distributed development or project management.

ProModel: Simulation and animation system primarily designed to model manufacturing systems. ProModel offers, like Arena, a flowchart-based simulation system for business processes. The modeling elements in ProModel are parts or entities, locations, resources, path networks, routing and processing logistics, and arrivals. With regard to these modeling elements, ProModel can be introduced as a LEAP technology simulation tool in which the model components are embedded in the model as *locations*, *entities*, and *arrival processes* (LEAP).

The first step in LEAP is the definition of *locations*. Locations represent fixed spots in the model like a container yard. At these model places, operations on objects can be made. In the graphical user interface (GUI), they appear as an icon of the layout and can be fitted with a logo and an advertisement. Such a display specifies how many dynamic objects (*entities*) are included in the *location* at the current time, which makes it easier to track the system state during a simulation run. The capacity of a *location* is explicitly set in the model, which means that a *location* at any time, up to a certain number, must contain only *entities*.

In the second step, the definition of *entities* is carried out. *Entities* are those objects that move during the simulation runs through the model or a part of the model and may be changed. *Entities* are assigned a speed and size.

The third step is the definition of *arrivals*. *Arrivals* define when *entities* arrive in the model. Here, it is determined in which *location* this will happen, if and in which distance repetitions occur, how many repetitions of *arrivals* there are, and how many *entities* enter the simulation model per repetition set. Moreover, at *arrival*, a manipulation of the global variable can be defined as well as the initial allocation of *attributes*.

In the final step, the processing, one or more target *locations* are issued from a *location* for an *entity*, which reaches the next *entity*. Leaving a *location* can also change the type of *entity*. Moreover, waiting periods can be specified, conditions examined, and variables and attributes manipulated. If several *locations* are specified as a target, one can be selected based on conditions, meaning that all possible paths an *entity* can use to pass through the model are defined. Should an *entity* leave the model, "EXIT" is specified as the target *location*.

The following use case shows the source code for an attribute query of submodel input as part of processing the location in-arrival container.

```
Distribution of incoming containers based on attribute origin
#Ein container arrives... #zur location in_load_ship, #wenn he should be
 unloaded from a ship
# Distribution of incoming containers based on attribute origin
      #A Container arrives,
      #at Location in_load_ship,
      #if he should be unloaded from a ship
IF origin = 12 THEN
{
 ROUTE 1
}
   #to Location in_to_hh,
   # if he should arrive without shipping in the port of Hamburg
 ELSE IF origin < 6 THEN
{
 ROUTE 2
}
   #at Location in_to_drport,
   #if his transportation part start in Maschen or Rade ELSE IF (origin =
6 OR origin = 7) THEN
{
 ROUTE 3
}
   #at Location in_to_cth,
   #if his transportation path start in Magdeburg
 ELSE IF origin = 8 THEN
{
 ROUTE 4
}  #at Location in_to_dd_container,
   #if his transportation part start in Riesa
 ELSE IF origin = 9 THEN
{
 ROUTE 5
}
   #atLocation in_to_ctw,
   #if his transportation path start inWilhelmshaven
 ELSE IF origin = 10 THEN
{
 ROUTE 6
}
```

Beside *locations, entities, arrivals,* and *processing,* ProModel offers element *attributes, networks, resources, macros,* and *global variables.*

Attributes: are either defined for *locations* or *entities* and can be created as integer or real. If an *attribute* is defined for *entities*, a value can be assigned to each individual *entity*. The value of the *attributes* can be changed during a simulation run.

Networks: build graphs and essentially consist of *nodes* and *paths. Nodes* can have either a certain limited or an unlimited capacity. For a limited capacity, a value must be explicitly specified. Then, the concerned *node* can contain, at any time, up to its maximum only *entities* and *resources. Nodes* can be linked to *locations.* Such connections are required if *entities* in a network pass from one *location* to another. Paths connect two *nodes*. They have a certain length, which is explicitly specified. Both *entities* and *resources* can pass along networks.

Resources: are used in the model to transport *entities* from one *location* to another. *Resources* are linked to a *network*. If an *entity* is assigned to a *resource* when changing from one *location* to another is assigned in the *processing*.

Macros: are placeholders for text parts. A *macro* has a name for identification, which forms the placeholder, and a text, which is inserted and executed during execution of the simulation at the location of the placeholder. *Macros* can be defined as a scenario parameter. For a scenario parameter, the text that occurs during the simulation in the place of the *macros* can be changed before beginning the simulation run. Macros are essential in ProModel to define scenarios. To create scenarios, there must be at least a *macro* that is defined as a scenario parameter. A *macro* that is defined as a scenario parameter is assigned a value for each scenario to be examined. This happens outside the scope of the model so that the model itself is not modified.

Global variables: are variables that can be accessed from any part of the model. This means that their current value can be queried and changed, allowing different model parts to communicate with each other.

Is it also possible to first develop a model in the form of more than one part or in submodels and then merge this into an overall model. The process of putting the model together is called merging. There are two kinds of merging. In a merge as a model, the pasted model is not changed. In a merge as a submodel, however, the names of all model elements, such as locations and global variables, receive a prefix or suffix that is specified during merging. Entities and attributes are excluded from these changes. Submodels can communicate only via global variables so that no further adjustments need to be made after the merge. If it should be possible to move entities between the submodels, then the submodels can be adapted for this purpose; however, the target location must be changed after the merging of the models in the processing of individual rules.

The simulation runs can be traced on the screen, because movements are graphically represented by entities and resources. Also, the current values of global variables and the number of entities in the individual locations can be traced if their representation in the model is created. Through these opportunities, model building is easy to understand; and the black-box nature of the simulation is thus at least partly resolved. This can improve the acceptance of the results for decision makers who have no direct relation to the modeling and simulation methodology.

There are more discrete-event simulation tools available, but we have to restrict ourselves to a selection of tools described which have a wide dissemination. In Sect. 5.8, several textbook and web link references are listed which give a good introduction into the respective simulation tools.

5.4 Object-Oriented Simulation

In the last 10 years, there has been a huge interest in object-oriented simulation. This is probably a result of the strong interest in object-oriented programming. Therefore, object-oriented modeling and simulation is a fast-growing area that provides a structured, computer-supported concept of mathematical and equation-based modeling which emphasizes the static declarative structure of a mathematical model. Thus, in object-oriented simulation, a simulated system is considered to consist of objects, e.g., an entity or a server that interacts with each other as the simulation evolves over time. There may be several instances of certain object types present concurrently during the execution of a simulation. Hence, the object-oriented approach itself can be found in several programming languages for which scalability was a major design goal as well as in object-oriented simulation environments. Ranging to scalability, the most obvious fact of scalability involves efficiency and execution speed, which becomes a problem when modeling requires scaling from only a few objects to hundreds or thousands. The solution to overcome this problem lies in parallel computing which involves partitioning of an object-oriented solution into subsets that can be effectively distributed among the processors (Jefferson and Sowizral 1982).

Objects contain data and have methods. Data describe the state of an object at a particular point in time, while methods describe the actions that the object is capable of performing. Other object instances can only view the data which is called encapsulation.

Meanwhile, a number of simulation software systems, packages, and languages have been developed for object-oriented modeling and simulation such as Modelica, the Modelica extension language MOSILA (Modeling and Simulation Language); TOMAS, a tool for object-oriented modeling and simulation; SIMPLE$_{++}$, a special language, similar to C$_{++}$, that provides facilities for object-oriented programming for object-oriented simulation; and others like OOCSMP, an extension of the old CSMP simulation language, in the sense that CSMP programs can still be compiled and executed by the respective compilers. The extensions make it possible to build compact object-oriented models when the system simulated consists of many similar interactuating components or parts. Hence, a class of objects can be defined by the following syntax:

```
CLASS class-name [: parent-class] (
          data declaration section
          [INITIAL section]
```

```
DYNAMIC [argument-list]
    dynamic section
[method-name [argument-list]
method body
[directive section]
]
```

Instances of a class can be declared with the following syntax:

```
class-name    object-name ( [list-of-attribute-values] )
```

A collection of objects are declared as follows:

```
class-name    collection-name  := object-list
```

Previously defined classes in a new model can be included as follows:

```
INCLUDE    file-name
```

Attributes and invoke methods on objects and/or collections can be referred by the following syntax:

```
object-name.attribute
object-name.method([argument-list])
```

Modelica: A modeling and simulation language which effectively unifies and generalizes previous object-oriented modeling languages. Today's OpenModelica version is an open-source Modelica-based modeling and simulation environment intended for industrial and academic usage. Its long-term development is supported by a nonprofit organization—the Open Source Modelica Consortium (OSMC). The goal with the OpenModelica effort is to create a comprehensive Open Source Modelica modeling, compilation, and simulation environment based on free software distributed in binary and source code form for research, teaching, and industrial usage (https://openmodelica.org/).

The modeling description language, *MOSILA*, is based on Modelica and specified/used in the GENSIM project. From the modeler's perspective, MOSILA is mainly an extension of *Modelica*. Thus, existing models and the Modelica standard library can be directly reused within the GENSIM simulation tool or with a small effort of adaptation. However, the means of expressions in Modelica, particularly for the description of variable model structures, are not powerful enough for using special simulation technologies. Therefore, extensions are added in MOSILA making use of the UML (Unified Modeling Language) standard (Nytsch-Geusen et al. 2005). UML provides, at its basic level, a set of graphic notation techniques to create visual models of object-oriented software systems. At its higher levels, it covers a process-oriented view of a system.

TOMAS: A software package developed for discrete-event simulation of complex control problems in logistic and production environments, which offers maximum flexibility in describing unique control processes. Supporting the design process of such complex systems, the model description is closely connected to modeling techniques used in business and logistics management. Thus, a TOMAS model is described by means of a process-oriented approach, where processes are "normal" object methods instead of threads. TOMAS has been applied in the simulation of Automated Guided Vehicle—systems for container transport and for operational scheduling in enterprise resource planning (ERP) environments (Veeke and Ottjes, 2000). Flow elements (processed elements) perform part of a process in TOMAS by defining a general class, the *TomasElement*. Specific elements can be defined as descendants of a *TomasElement*, and each *TomasElement* may or may not have a process method.

The major features of object-oriented simulation packages are:

- *Inheritance*: means that if one defines a new object type (in the object-oriented notation called a child) in terms of an existing object type (in the object-oriented notation called parents), then the child type inherits all the characteristics of the parent type
- *Polymorphism*: means that different object types with the same ancestry can have methods of the same name, but when invoked may cause different behavior in the various objects (Law 2007)
- *Encapsulation*: used to refer to one of two related but distinct notions such as a language mechanism for restricting access to some of the object's components or a language construct that facilitates the bundling of data with the methods operating on that data (URL 10)
- *Hierarchy*: classical inheritance relationship between classes

5.5   Online Simulation

Online simulation allows users to run simulations from a web browser environment. Whether users are using a Microsoft Windows, Linux, or Apple computer, a smartphone, or a tablet, they will always be able to run a transportation-specific simulation. For this purpose, online simulation offers realistic graphics, authentic sound effects, and comprehensive editing as well as domain-specific simulation software available on the Web, such as:

- Train simulation using different types of railroad engines to deal with different activities running through different types of landscape, etc.
- Aviation simulation using different types of planes approaching different destinations, etc.
- Maritime simulation using different types of ships approaching different types of harbors, etc.

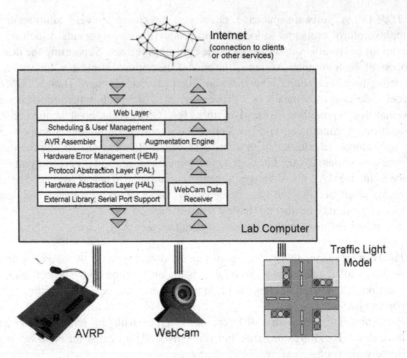

Fig. 5.3 Online simulation environment for a traffic light system (Sitzmann et al. 2014)

Online simulation is mostly allocated to the game industry. Besides gaming, online simulation is also embedded in online transportation systems planning to illustrate transportation-specific scenarios, such as traffic flow behavior as a function of traffic light sequencing where traffic lights change in the following sequence (see Sect. 2.3) whenever a person pushes a button.

$green \rightarrow amber \rightarrow red \rightarrow amber \rightarrow green$

Such an online simulation tool provides interactivity which requires the following two major improvements:

1. A webcam broadcast that shows the events in the so-called residential lab which is extended by simulated animations related to the programming of the traffic light sequencer. So far, the webcam shows a crossroad with traffic lights.
2. In addition to that, the web application shows animated cars and pedestrians, as can be seen in Fig. 5.3.

The animations are simulated on the basis of the feedback signals. First, they can be realized by editing the webcam stream before passing it along to the user (client), by overlaying Document Object Model (DOM) elements synchronized to the webcam image or by using the HTML5 canvas element and Javascript functions. Second, various simulation data or experimental values can be shown in the user interface, e.g., control signals, other feedback data of the hardware module, or experimental setup or source code highlighting the logical equations that are currently used for the control of the traffic light in real time. This feature especially helps to connect theoretical issues with effects in the real world; and thus, it fosters a better understanding of the complexity of the transportation system scenario under test. Another important improvement is providing an e-tutor—which can be realized as a chatbot or through a human tutor connected via the system with the users through a video chat—which helps to avoid problems in understanding the scenario or using the lab. Generally, these issues can be mitigated by a broader use of social software techniques. In terms of the advent in technology, online simulation can also support mobile laboratory experiments embedding e-learning via single source publishing or optimized layout templates for heterogeneous devices and achieving platform independence through the usage of standard web browsers and multilingualism by default.

5.6   ProModel Case Study in Transportation Analysis

The ProModel case study in transportation analysis, modeling, and simulation describes a feasibility study for potential hinterland connections (dry ports) of the Seagate port of metropolitan Hamburg, Germany, based on discrete-event simulations, which deal with modeling of systems in which the state variable changes only at a discrete set of points in time. This type of simulation model is analyzed by numerical rather than by analytical methods.

The freight yard at Maschen and the logistic center at Rade have been chosen as potential dry ports, because the case study is based on the assumption that a tunnel-based connection between the Seagate harbor of metropolitan Hamburg and the dry ports through which the containers are conveyed on carriers is planned. In addition to the carriers, trucks and trains are included as additional means of transportation. The goal of the feasibility study is to investigate the dry port scenario concept to identify whether or not an efficient and logistically successful operation and ecologically sufficient transport of containers between the Seagate harbor port and the hinterland can be achieved to strengthen the competitiveness of the port of Hamburg. Once developed and validated, the ProModel case study in transportation analysis of the dry port model can be used to investigate a wide variety of what-if questions about the real-world hinterland connection workflow. This also allows studying optimization scenarios and/or options at an early stage, before anything physically has been done. Thus, modeling and simulation in the maritime domain can be used to predict the effect of changes with regard to the actual existing implementation of workflows and/or layouts of service stations, etc., at the dry port. It can also be used as a scenario analysis tool to predict the performance of a hypothesis under varying sets of circumstances.

The simulation tool ProModel 2011 (see Sect. 5.3) was used to simulate the container transport. ProModel is a simulation and animation tool with a flowchart-based simulation procedure which allows its use for business processes, such as the container transportation case study for dry port planning and development. The time course of the intrinsic process dynamic can be implemented using a GUI base and simulated considering realistic data, such as container volumes, deletion and transport times, etc. In addition, ProModel's programming language provides for modeling specific situations not covered as standard, built-in choices.

In this case study, three different models were developed whereby different tunnel connections and means of transport modes were adapted to the container transport from the Seagate port to the dry ports.

The important processes investigated for the Hamburg Seagate port are discharging containers from container ships and feeders and the subsequent loading of vessels and feeder ships with containers. These processes run at a certain speed. For the individual terminal, the number of containers and feeder ships, which are loaded and/or unloaded at the same time, is limited by the number of moorings.

Containers that arrived at the container terminal at the Seagate port on container ships are assumed to allow grouping based on their transportation target. Thus, only transport targets from the modeled immovable elements are taken into account. These places are usually only stops within the real transportation chain of containers. After debarkation, from the container ships at the container terminals at the Seagate port, only containers are pursued, which are part of the model of one of the two planned dry port concepts.

To transport the containers from the container terminal to the assumed dry port requires a tunnel entrance at the Seagate port to load the containers onto the carriers and unload them at the dry port. The time required for this loading process and the transportation time through the tunnel depend on the transportation carrier system used. Regardless of which type of carrier is used for the connection between the container terminals at the Seagate and the dry ports, a container only leaves via a container terminal to the assumed tunnel entrance for further transportation. The container can either be loaded immediately onto a carrier or plenty of storage capacity is available, storing the container only for a very short time. During transportation through the assumed tunnel tube, the carrier maintains a minimum safe distance in both directions. For the tunnel concept, a Y tunnel approach is assumed, which allows exits at the dry ports in Maschen and Rade, with the entrance at the Seagate port of metropolitan Hamburg, and vice versa. For safety reasons, the carrier should maintain the minimum safe distance in the respective tunnel tube from these dry ports.

To start with model building, the elements in this case study have to be determined. Moving elements are containers, container ships, feeder, carrier, trains, and trucks. Other relevant elements are the Container Terminal Altenwerder (CTA), Container Terminal Burchardkai (CTB), Container Terminal Eurogate (CTE), Container Terminal Steinwerder (CTS), Container Terminal Tollerort (CTT) (see Sect. 1.6), Container Terminal Hansa Terminal (CTH) in Magdeburg, Container Terminal Wilhelmshaven (CTW), dry ports in Maschen, and Rade, the tunnel system between the Seagate and the dry ports, the tunnel between Container Terminal Wilhelmshaven and the dry port in Rade, and the road network between the container terminals in Hamburg and the container terminal tunnel entrance (Dankers 2012).

The main process in the Seagate port is the unloading of containers from container ships and feeders, as well as the subsequent loading of container vessels and the feeder with containers. These processes are at a certain speed. For the individual terminals, the number of container ships and feeders, which at the same time can be loaded or unloaded, is limited by the number of moorings. All requirements mentioned apply to the container terminal in Magdeburg and Wilhelmshaven, where container ships are considered and in Magdeburg only feeder for dis-embankment or embankment.

Containers that arrived at the terminals in the port of Hamburg on container ships are assumed to be grouped for container flows on the basis of their transport target. In this research study, only transport targets from the amount of modeled immobile elements are taken into account. These places are usually only stopovers in the real container transport. After unloading the container ships in the container terminals in Hamburg, only the container being pursued is part of the model of one of the dry ports in Maschen or Rade or the container terminal in Magdeburg. Transportation of containers between the terminals in Hamburg and the tunnel entrance of Hamburg requires that the containers be loaded onto transport vehicles and unloaded at another place. The time required for these loading processes depends on the transport vehicles used. Regardless of what form the connection between the terminals in the Seagate port and the dry ports is, a container should only be transported from a terminal to the tunnel entrance, if the container can be loaded immediately onto a carrier or plenty of storage of short-term storage is available. During transportation through the tunnel, the carriers in both directions have to maintain a minimum distance. If a Y tunnel concept is accepted, it has to be taken into account that the carrier from Maschen and Rade must adopt a minimum safety distance from these two directions passing along the tunnel (Table 5.2).

Table 5.2 Collection of all attributes used with name and explanation of what is specified by them

Name	Explanation
Origin	Indicates the place of origin of an entity—modeled start of the entity and not the location where the entity enters the model. Manipulation of this attribute can be saved which is considered from this point of origin
Destination	Specifies the destination of an entity. It models the entity target and not the location where the entity leaves the model
Cargo	Specifies the number of entities to be loaded or the current load of an entity. This can be, e.g., the number of containers, which should be loaded or loaded on a container ship
Cargo_max	Specifies the capacity of an entity. In the model, it determines how many containers can be loaded by a feeder
Cargo_max_dd	Specifies the capacity of an entity for the area between Magdeburg and Dresden
Cargo_md	Specifies the number of loaded entities for Magdeburg
Cargo_dd	Specifies number of loaded entities for Dresden and Riesa

Table 5.3 Meaning of numerical values of attributes origin and destination

Number	Explanation
1	CTA
2	CTB
3	CTE
4	CTS
5	CTT
6	Maschen
7	Rade
8	Magdeburg (Hansa Terminal)
9	Dresden/Riesa
10	JadeWeserPort (CTW)
11	Disposition in the terminal; specifies that a container should not be transported either to Magdeburg and Dresden or to a dry port; the model should be left immediately after unloading of a container ship
12	Transportation by a container ship; indicates that a container on a container ship enters a terminal
13	Tunnel entrance; tunnel entrance of the Seagate port of Hamburg

In the present model, attributes are mapped to entities. Depending on the specification of attributes, the entities take different paths through the model. The loading of ships and trains is set by attributes, while the loading is only restricted by feeders by specifying attributes for them. The initial assignment of the attributes is specified by input tables for each entity. The explicit value assignment is made only if the attribute for the corresponding entity during the simulation is of importance.

The attributes, *origin* and *destination,* are specified coded start and destinations as integers. Table 5.3 shows the corresponding illustration.

Macros in ProModel are placeholders for text blocks and are used to define scenarios. Table 5.4 shows a selection of macros with their use and the number replaced by them, which are defined as a scenario parameter.

Containers which are unloaded by container ships in Wilhelmshaven should be taken only into account if their target is the dry port in Rade. This, we considered only the direct transport of the container through the tunnel after Rade. The carrier should maintain a minimum distance in the downhill at Rade and Wilhelmshaven.

Containers which are unloaded by container ships in Wilhelmshaven should be taken into account only if their target is the dry port in Rade. This, we considered only the direct transport of the container through the tunnel after Rade. The carrier should maintain a minimum distance in the downhill at Rade and Wilhelmshaven.

The structure of the submodels of the Seagate port to dry ports overall model is shown in Fig. 5.4 (Dankers 2012).

Figure 5.4 shows the entire model with submodels. Boxes with numbers symbolize the several submodels: 1, Input; 2, Seagate port of Hamburg; 3, Dry port; 4, River Elbe; 5, Hansa Terminal; 6, CTW; 7, JWP. Arrows indicate entity flows between the submodels.

Table 5.4 Overview for selected scenario parameter defined macros with their purpose and the numbers replaced by them

Name	Usage	Number
cta_rate	Specifies the maximum number of containers that can be loaded at the CTA per hour on vessels or unloaded from ships, container ships, as well as feeder ships Achieving this value depends on whether all berths are occupied	600
ctb_rate	Specifies the maximum number of containers that can be loaded at the CTB per hour on vessels or unloaded from ships, container ships, as well as feeder ships. Achieving this value depends on whether all berths are occupied	1,040
ctw_rate	Specifies the maximum number of containers that can be loaded at the CTW per hour on vessels or unloaded from ships, container ships, as well as feeder ships. Achieving this value depends on whether all berths are occupied	960
dry_delta_t	Determines the minimum time interval of two carriers loaded with containers passing along the tunnel system between the Seagate port Hamburg and dry ports in Maschen and Rade. This interval is specified in seconds	5
jwp_delta_t	Determines the minimum time interval of two carriers loaded with containers at least must comply passing along the tunnel system between the Seagate port Wilhelmshaven and dry port in Rade. This interval is specified in seconds	10

The model depicts container flows emanating from the Seagate port of metropolitan Hamburg, as well as those in the opposite direction. Also transports involve Rade and the container terminal of containers between the dry port in Wilhelmshaven, Germany. The model is divided into 7 individually created submodels. After developing, they are adapted with few changes of existing processing rules for the required total model. This leads to a reduction of complexity of the overall model and improved intelligibility and customization.

Figure 5.5 shows the schematic diagram of the submodel input. Circles marked with digits represent the eleven locations, arrows facing transportation lines of entities between locations.

This submodel is an abstract construct to assign initial locations to the containers, container ships, feeders, and trains. It forms the central point of all entities in the model. Each entity type has its own entry point in form of a location. There are four locations where entities enter the model. Located in the upper part of the graphical representation are the submodel locations 1–4. The locations 6–11 are the entities of the submodels which serve to distribute container to dry port and/or terminal submodels. For this redirection of entities in the appropriate submodels, there is a location that serves as a connection point for each additional submodel. Also, the container ships, loaded for the first time with containers in this submodel location 5. Below are listed the identifiers and functions of locations:

1. *In_arrival_train*
2. *In_arrival_ship*

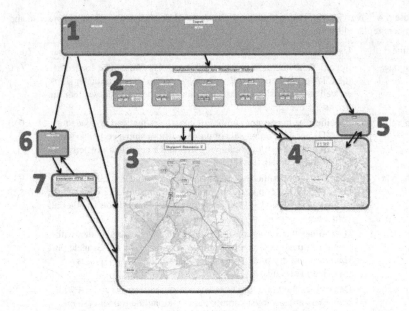

Fig. 5.4 Submodel architecture of the Seagate port to dry ports

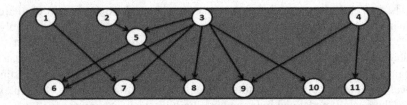

Fig. 5.5 Schematic diagram of submodel input. The *circles* marked with digits represent the 11 locations, *arrows* facing transportation lines of entities between the locations

  3. *In_arrival_container*
  4. *In_arrival_feeder*: arrival of feeder ships
  5. *Load_ship*: loading of ships
  6. *In_to_ctw*: connection with submodel Container Terminal Wilhelmshaven (CTW)
  7. *In_to_dry*: connection with submodel dry port
  8. *In_to_hh*: connection with submodel (Seagate port Hamburg)
  9. *In_to_md*: connection with submodel (Hansa Terminal)
 10. *In_to_dd_c*: connection for container for primarily planned submodel Dresden
 11. *In_to_dd_f*: connection with feeder for primarily planned submodel Dresden

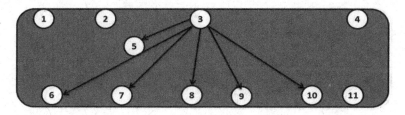

Fig. 5.6 Schematic diagram of submodel input. The *circles* marked with digits represent the 11 locations; the *arrows* face transportation lines of containers between the locations

In this submodel, the attributes *origin* and *cargo* are used. Attribute *origin* maps *entities* in *processing* with the right paths. Loading of container ships with the right number of containers is specified by the attribute *cargo*.

Processing of the used *entities container, container ships, feeder,* and *train* is described in detail in Fig. 5.6. Circles marked with digits represent the eleven locations, arrows facing transportation lines of containers between locations.

Starting from location No. 3 *in_arrival_container*, each container with the attribute *origin* will be assigned to one of the following routes: If *origin* = 12, the container in_arrival_container will immediately be sent to location No.5 *in_load_ship*. If *origin* < 6 applies, the container changes from *in_arrival_container* to in_to_hh No. 8 in the above figure, and for *origin* = 6 or *origin* = 7, they promptly change to location No. 7 *in_to_dry*, for origin = 8 to location No. 9 in_to_md, in *origin* = 9 to location No. 10 *in_to_md*, and for origin = 10 to location No. 6 *in_to_ctw*.

Circles marked with digits in the following figure represent eleven locations; arrows face transportation lines of trains, container vessels, and feeder between the locations.

Each container ship in Fig. 5.7 will change from location No. 2 *in_arrival_ship* immediately to location No. 5 *in_load_ship*. Starting from location No. 5 *in_load_ship*, each container ship is assigned with a numerical value of attribute *origin* to one of the following routes. If *origin* < 6, the container ship promptly changes to location No. 8 *in_to_hh* and for *origin* = 10 to location No. 10 *in_to_ctw*.

Each feeder in Fig. 5.7 will change from location No. 4 *in_arrival_feeder* and is assigned based on a numerical value for attribute *origin* to one of the following routes: if origin = 8, the feeder promptly moves to location No. 9 *in_to_md*, and for origin = 9 to location No. 11 *in_to_dd*. In location No. 1 *in_arrival_train*, each train mode immediately changes to location No. 7 *in_to_dry*.

Now a schematic diagram of the submodel Seagate port of Hamburg will be discussed in which boxes represent the models of container terminals which are Tollerort, Altenwerder, Burchardkai, Eurogate, and Steinwerder. Circles with numbers are again used to represent the locations to which the terminal models are added and arrows represent the transport links between the locations.

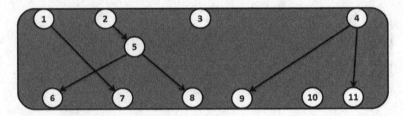

Fig. 5.7 Schematic diagram of submodel input. The *circles* marked with digits represent the 11 locations; the *arrows* face transportation lines of trains, container ships, and feeder between the locations

The individual terminal models are part of the submodel Seagate port of Hamburg for which more tasks are required to describe the functionality of the terminals. These are, along with others, bundling and distribution of entity flows between the submodel Seagate port of Hamburg and the other submodels. For each submodel, there is a relationship in the form of entity flows for each location that serves as a connection point, e.g., container and container ships coming from the submodel input in location No. 1, from where they are distributed to the different terminals. This is done based on the attribute *origin*. In location No. 2, container and feeder streams are bundled and forwarded from the terminals to the submodel's dry port of River Elbe. On the basis of their entity type and the attribute *origin*, they will be forwarded to terminals or locations which act as connection points to the submodel's dry port (No. 3) and River Elbe (No. 4). From the submodel Seagate port of Hamburg, solely containers are directed only to the submodel dry port and only the feeder to the submodel River Elbe.

Of the models developed, there is only one submodel dry port. The different scenarios presented come up with specific model requirements for this submodel. Common to these different submodel versions is that containers from submodel Seagate port of Hamburg, represented by the attribute *origin*, are distributed to the container terminals of the port of Hamburg, so containers from CTA in submodel Seagate port represent arrivals from submodel CTA. From the terminals, containers are allocated to the representations of dry ports in Maschen and Rade and vice versa. Containers that arrive from the direction of the dry ports in the terminal leave the submodel dry port in the direction of the submodel Seagate port of Hamburg. It is assumed that there is a tunnel which begins south of the harbor container terminals and ends directly in the dry port in Rade for all developed submodels. In each tunnel connected with the dry port submodel, only carriers are moving which transport the containers, keeping at least 5 seconds between containers. Moreover, the inflow of containers from the container terminals is limited. Hence, the number of containers from the container terminal cannot exceed the number of already existing containers.

Submodel dry port is based on the assumption that a y-shaped tunnel exists which begins south of the container terminals and separates into two tunnels in the south direction, one leading to Maschen and the other to Rade. This tunnel system is

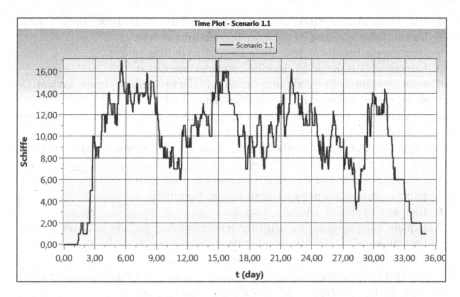

Fig. 5.8 Total of container ships and feeder in terminals Altenwerder, Eurogate, Burchardkai, and Tollerort for the simulation period of 35 days. Hourly averages of the number of container ships and feeder per terminal are recorded for summation (Dankers 2012)

a network model *dry_tunnel*, marked in blue, through which the carrier dry_Carrier moves. Between the container terminals and the south tunnel entrance, the containers are transported on trucks *dry_Truck* on the network transport *dry_inter-terminal,* marked in blue. This network covers the existing road network between the container terminals and the Seagate port of Hamburg tunnel entrance.

To enhance the comparability of the scenarios being investigated, the same number of ships is assumed for all scenarios. The number of ships is based on real data from May 2012. The number of container ships, summarized for terminals Altenwerder, Eurogate, Burchardkai, and Tollerort, is shown in Fig. 5.8. During this period, the five areas show an increased number of incoming ships. These areas lie between 4 and 9 days, 25 to 27 days, 13 and 17 days, 20 and 24 days, and 29 and 32 days. The maximum number of ships occurs on Day 6, Day 16, Day 22, Day 26, and Day 31. The absolute values of the aggregated number of ships range between 0 and 17.

5.7 Exercises

1. Explain what is meant by the term behavior level.
2. Give an example of a behavior level modeling approach.
3. Explain what is meant by the term composite-structure level.
4. Give an example of a composite-structure level modeling approach.
5. Explain what is meant by the term continuous-time system.

6. Give the mathematical description for a second-order continuous-time system.
7. Explain what is meant by the term discrete-time system.
8. Give the mathematical description for a discrete-time system.
9. Explain what is meant by the term computer simulation.
10. Give an example of a computer simulation of a first-order dynamic system
11. Explain what is meant by the term simulation tool.
12. Give example of simulation tools for continuous-time systems and discrete-time systems.
13. Explain what is meant by the term decomposition of a dynamic system of nth order.
14. Give an example of a dynamic system of the third order.
15. Explain what is meant by the term sorting algorithm.
16. Give an example of a sorting procedure.
17. Explain what is meant by multistep model-building process in discrete-event simulation.
18. List and define the five main characteristic statements.
19. To what specific approaches does discrete-event system simulation refer to?
20. Describe the two approaches for discrete-event simulation.
21. Explain what is meant by the term LEAP technology in ProModel.
22. Describe the LEAP approach in ProModel in detail.
23. Explain what is meant by the term object-oriented simulation.
24. Give an example of an object-oriented simulation.
25. Explain what is meant by the term online simulation.
26. Describe how the interactivity in online simulation can be achieved.
27. Explain what is meant by the term dry port.
28. Describe the services a dry port has to offer.
29. Describe the case study example in this chapter in your own words.
30. Describe the results shown in Fig. 5.8 in your own words.

References and Further Readings

Alves G, Marwues A, Paiva C, Nogueira P, Guimaräes P, Couto R, Cherem L, Borba V, Ferreira G, Koch S, Pester A (2013) A remote lab for projectile launch experiments: professional & academic perspectives. iJEP – Volume 3 (Special Issue 2: IGIP2012 Conference), pp 18–22

Andujar JM, Mejias A, Marqezm MA (2011) Augmented reality for the improvement of remote laboratories: an augmented remote laboratory. IEEE Trans Educ 54(3):492–500

Balamuralithara B, Woods PC (2008) Virtual laboratories in engineering education: the simulation remote lab. Comput Appl Eng Educ 17:108–118

Banks J, Carson JS II, Nelson BL, Nicol DM (2001) Discrete-event system simulation. Prentice Hall, Upper Saddle River

Bates AW (2005) Technology, e-learning and distance education. Routledge Publ., London

Biles WE (1996) Discrete-event systems. In: Kheir NA (ed) Systems modeling and computer simulation. Marcel Dekker, Inc., New York, pp 219–277

Bloechle WK, Laughery KR Jr (1999) Simulation interoperability using micro saint simulation software. In: Farrington PA, Nembhard HB, Sturrock DT, Evans GW (eds) Proceedings 1999 Winter Simulation Conference, pp 286–288

Brown S, Rose J (1996) FPGA and CPLD architectures: a tutorial. Des Test Comput IEEE 13(2):42–57

Bunus P, Fritzson P (2000) DEVS-based multi-formalism modeling and simulation in Modelica. In: Waite WF (ed) Proceedings Summer Simulation Conference, pp 91–96

Chapra SC, Canale RP (2006) Numerical methods for engineers. McGraw-Hill Publ., Boston

D'Apice C, D'Auria B, Manzo R, Salerno S (2000) Using mathematica to simulate general queuing systems. In: van Landeghem R (ed) Proceedings 14th European simulation multi-conference, pp 23–25

Dankers I (2012) Decision support for port development planning in metropolitan regions by mathematical modeling and simulation: case study Hamburg metropolitan region (in German). Bachelor thesis, University of Hamburg, Germany

Donald DL (1998) Tutorial on ergonomic and process modeling using quest and IGRIP. In: Medeiros DJ, Watson E, Carson J, Manivannan S (eds) Proceedings 1998 Winter Simulation Conference, pp 297–3022

Fritzson P (2004) Principles of object-oriented modeling and simulation with Modelica 2.1. Wiley, Piscataway

Fritzson P (2014) Principles of object-oriented modeling and simulation with Modellica 3.3: a cyber-physical approach. Wiley, Hoboken

Hullinger DR (1999) Taylor enterprise dynamics. In: Farrington PA, Nembhard HB, Sturrock DT, Evans GW (eds) Proceedings 1999 Winter Simulation Conference, pp 227–229

Itmi M, Huntsinger R, Jarjoui P, Pécuchet JP (2000) A simulation methodology description of container placement using a unique system to control the logistics. In: Waite WF (ed) Proceeding Summer Simulation Conference, pp 63–67

Jefferson D, Sowizral H (1982) Fast concurrent simulation using the time warp mechanism, part I: local control, N-1906-AF. The Rand Corporation, Santa Monica

Kelton WD, Sadowski RP, Sturrock DT (2004) Simulation with Arena. McGraw Hill, 2005

Klee H (2000) Real-time traffic generation for a driving simulator using Matlab. In: Waite WF (ed) Proceeding Summer Simulation Conference, pp 68–73

Ko CC, Chen BM, Chen J (2004) Creating web-based laboratories. Springer, New York

Krahl D (1999) Modeling with extend. In: Farrington PA, Nembhard HB, Sturrock DT, Evans GW (eds) Proceedings 1999 Winter Simulation Conference, pp 188–195

Law AM (2007) Simulation modeling and analysis. McGraw-Hill International Publ.

Leemis LM, Park SK (2006) Discrete-event simulation: a first course. Pearson Prentice Hall, Upper Saddle River

Lowe D, Murray S, Lindsay E, Liu D (2009) Evolving remote laboratory architectures to leverage emerging internet technologies. IEEE Trans Learn Technol 2(4):289–294, ISSN: 1939–1382

Ma J, Nickerson JV (2006) Hands-on, simulated, and remote laboratories: a comparative literature review. ACM Comput Surv 38(3), Article No. 7

Mehta A, Rawles J (1999) Business solutions using WITNESS. In: Farrington PA, Nembhard HB, Sturrock DT, Evans GW (eds) Proceedings 1999 Winter Simulation Conference, pp 230–233

Mohammed F (2009) Blended learning and the virtual learning environment of Nottingham Trent University, Development in eSystems Engineering. In: 2nd international conference, pp 295–299

Nedic Z, Machotka J, Nafalski A (2003) Remote laboratories versus virtual and real laboratories. In: Frontiers in education, 33rd annual meeting, vol 1, pp T3E-1–T3E-6

Nytsch-Geusen C, Ernst T, Nordwig A, Schneider P, Schwarz P, Vetter M, Wittwer C, Holm A, Nouidui T, Leopold J (2005) Advanced modeling and simulation techniques in MOSILAB: a system development case study. In: Proceedings 5th international Modelica conference, pp 63–72

Price RN, Harrel CR (1999) Simulation modeling and optimization using ProModel. In: Farrington PA, Nembhard HB, Sturrock DT, Evans GW (eds) Proceedings 1999 Winter Simulation Conference, pp 208–214

Rohrer M (1999) AutoMod Product Suite Tutorial. In: Farrington PA, Nembhard HB, Sturrock DT, Evans GW (eds) Proceedings 1999 Winter Simulation Conference, pp 220–226

Sadowski D, Bapat V (1999) The Arena product family: enterprise modeling solutions. In: Farrington PA, Nembhard HB, Sturrock DT, Evans GW (eds) Proceedings 1999 Winter Simulation Conference, pp 159–166

Schmidt G, Doli U, Mattes A (2005) MOSILAB: development of a Modelica based generic simulation tool supporting model structural dynamics. In: Proceedings of the 4th international Modelica conference. http://www.Modelica.org/events/Conference2005

Sitzmann D, Möller DPF, Becker K, Richter H (2013) TIO—A software toolset for mobile learning in MINT disciplines. In: Proceedings of ICTERI, Kherson, Ukraine, pp 424–435

Sitzmann D, Moeller DPF, Mücke F (2014) Augmented remote labs for the on-line access to programmable hardware in computer engineering education. In: Proceedings INTEND conference

Sokolowski JA, Banks CM (2009) Principles of modeling and simulation: a multidisciplinary approach. Wiley, Hoboken

Veeke HPM, Ottjes JA (2000) TOMAS: tool for object-oriented modeling and simulation. In: Proceedings of the business and industry simulation symposium, ISBN 1-56555-199-0

Ziegler BP, Praehofer H, Kim TG (2000) Theory of modeling and simulation: integrating discrete event and continuous complex dynamic systems. Academic

Links

(URL 1) http://en.wikipedia.org/wiki/Advanced_Continuous_Simulation_Language
(URL 2) http://www.kahlert.com/web/english/e_boris.php
(URL 3) http://en.wikipedia.org/wiki/DYNAMO_%28programming_language%29
(URL 4) https://modelica.org/
(URL 5) www.modelmakertools.com/
(URL 6) http://link.springer.com/chapter/10.1007%2F978-3-642-706400_24#page-1
(URL 7) www.mathworks.de/products/simulink/
(URL 8) http://en.wikipedia.org/wiki/VisSim
(URL 9) http://www.ualberta.ca/CMENG/che312/F06ChE416/HysysDocs/AspenPlus20041Getting StartedEOModeling.pdf
(URL 10) http://en.wikipedia.org/wiki/Encapsulation_%28object-oriented_programming%29

More Links

http://www.jedec.org/. Downloaded 10 Jan 2014
http://ti-online.org/en. Downloaded 10 Jan 2014
http://www.youtube.com/user/anylogic
http://www.arenasimulation.com/Arena_Home.aspx
http://www.simplan.de/de/software/tools-simulation/automod.html
https://www.extendsim.com/
http://www.mathworks.de/products/simevents/
http://www.modprod.liu.se/workshop_2007/1.46545/tutorial-fritzson.pdf
https://www.promodel.com/
https://www.promodel.com/
http://en.wikipedia.org/wiki/List_of_discrete_event_simulation_software

Transportation Use Cases

This chapter introduces several case studies which have been conducted during a long period of collaboration with Prof. Bernard Schroer, Ph.D., University of Alabama in Huntsville (UAH), who is coauthoring this chapter of this book. Section 6.1 introduces, from a general perspective, critical issues in the design, development, and usage of simulation models of transportation systems. The Coal Terminal case study in Sect. 6.2 refers to the real-world situation at the McDuffie Coal Terminal at the Alabama State Docks in Mobile. The Port of Mobile was interested in increasing the volume of coal to 30 million tons annually and had looked at a number of process improvements to achieve this goal. The objective of the simulation was to evaluate the impact of these process improvements on the volume of coal through the Port. The Container Terminal case study in Sect. 6.3 gives an overview of the container terminal at the Port of Mobile and its intermodal container handling facilities. The Port of Mobile had just completed a major expansion to its container terminal and was very interested in validating the design of the expanded facilities. The objective of the simulation was to identify maximum throughput capacities, especially utilization of the very expensive large cranes used for unloading and loading containers. The Intermodal Container Center in Sect. 6.4 offers an approach to analyzing the operation of an intermodal center and evaluating various operational alternatives before finalizing the design of any planned expansion. Hence, the objective used was to determine if throughput can satisfy anticipated demand and if sufficient resources are available to meet anticipated growth in demand. Port Security Inspection is the topic of the case study in Sect. 6.5. Increased security is having a significant impact on the operation of seaports. The objective of the simulation was to evaluate the impact of various inspection protocols on the operation of the container terminal at the Port of Mobile. The objectives of the Interstate Traffic Congestion case study in Sect. 6.6 were to determine the congestion point as traffic increased and to evaluate adding additional lanes at congestion points. This is an important issue because the growth

© Springer-Verlag London 2014
D.P.F. Möller, *Introduction to Transportation Analysis, Modeling and Simulation*,
Simulation Foundations, Methods and Applications,
DOI 10.1007/978-1-4471-5637-6_6

in automotive vehicle manufacturing in the south has greatly increased traffic on I-65 from Chicago to Mobile, AL. The objective of the simulation was to evaluate the increase in congestion on I-65 south from Birmingham to Montgomery, AL. The objective of the simulation case study in Sect. 6.7 was to evaluate the impact of increased 18-wheel truck traffic on the I-10 tunnel crossing the Mobile River in Downtown Mobile, AL. The case study in Sect. 6.8 focuses on passenger and freight operations at Hamburg Airport to estimate the maximum numbers of passengers and freight that can be dispatched and to identify potential opportunities in process optimization with regard to the expected growth in passenger and freight numbers. In Sect. 6.9, a case study of a project modeling and simulating an Italian airport transportation operation conducted by an international student team is introduced to demonstrate how an international group of students can be motivated to conduct an innovative, advanced project in a very complex area of concentration in modeling and simulation.

6.1   Introduction

Modeling and simulation has a great potential for solving problems in the multimodal transportation systems sector. The power of modeling and simulation in multimodal transportation lies in the three Rs, namely, reductionism, repeatability, and refutation. Reductionism recognizes that a transportation system can be decomposed into a set of components or compartments that mainly follow the laws of engineering. Repeatability or test-retest reliability is a test on the same item, the multimodal transportation system model, and under the same conditions. This means that the multimodal transportation system model under test can be said to be repeatable when its variation is smaller than an agreed-upon boundary or target function. In that case, the multimodal transportation system model under test can be said to be validated by its repeatability; and one can make predictions with this model by refutation of the hypothesis. Against this background, while validation discovers that the chosen assumptions are true, a refutation does the opposite; and it proves something is false in the assumed prediction or hypothesis.

Due to the advancements in transportation engineering and planning for handling complex systems and/or processes, modeling and simulation have become very important tools in the transportation systems sector. Not only do they help in achieving a better understanding of real-world objectives in multimodal transportation, they are also important in planning new transportation systems and determining their impact before a new system has been built. This kind of abstraction from reality is fundamental to all formal methods for design, planning, or performance analysis and prediction. Moreover, modeling and simulation allows engineers, managers, and politicians to accurately analyze multimodal transportation system behavior under various operating conditions. This is why a White House report has identified modeling and simulation as one of the key enabling technologies of the twenty-first century. Its applications are virtually universal.

Modeling and simulation itself can be viewed as an iterative process consisting of successive mathematical model building and computer simulation steps. The process of mathematical model building is based on three separate tasks:

1. Identification and idealization of individual elements of the transportation system, so-called transportation subsystems
2. Identification and idealization of the interaction of individual elements of the transportation subsystems
3. Application of the respective scientific laws involved to abstract the model

The variety of levels of conceptual and mathematical representations is evident by the wide spectrum of models available. Selecting a model depends on the goals and purposes for which it was intended, the extent of a priori knowledge available, data gathered through experimentation and measurements on the real-world system, and estimates of system parameters and system states. Therefore, the multimodal transportation system can be seen as a system that is decomposed to a certain level of detail, as introduced in Sect. 1.4.

Difficulties in developing mathematical models of transportation systems arise because systems are, in general, extremely complex. In addition, sufficient amounts of operating data are often not available. Hence, mathematical model building for a given real-world system first requires the selection of a model structure and then some form of parameter estimation to determine the acceptable model parameter values, if they are not available. For this reason, it may sometimes be better to develop a more simplified model to eliminate intrinsic characteristics of the system, as an overly complicated mathematical model will cause mathematical difficulties.

From a more general point of view, two major facts are important when developing mathematical models of transportation systems:

1. A model is always a simplification of reality but should never be so simple that its answers are not true.
2. A model has to be simple enough that it is easy to use, as described in Sect. 1.4. The success of model building depends on the validity of the assumptions made by the model builder in arriving at the model.

The difficulty, but also the fascination of transportation analysis, modeling, and simulation, derives from the intrinsic complexity of the multimodal transportation systems sector. Therefore, simulation can be used as inexpensive insurance against costly mistakes. Some companies will not launch a major expansion, change a process, or incur a capital expenditure until a detailed analysis using simulation has been completed. Simulation models are plagued with a number of difficulties. Probably the primary difficulty is the time required for model development, with the level of detail the most critical determinant of time required. In many instances, a macro-level model will satisfy the requirements better than a very detailed, complex microlevel model. But the question is, "What level of detail is sufficient?" There are many issues that need to be addressed to ensure a good model design, an accurate simulation model, and confidence in the simulation results. Some of these issues are:

- Design of the modeling or simulation environment consisting of defining the level of model design, submodels, interconnection of submodels, submodel entities, global simulation variables, and submodel attributes.
- Data collection; the level of data collection and use of approximations
- Model verification and validation
- Model starting and stopping conditions
- Point in model execution when model reaches steady state or equilibrium
- Length of simulation to achieve good results or sample size
- Statistical confidence intervals on simulation results
- Analyzing simulation results

The conceptual framework for rapidly modeling transportation systems is shown in Fig. 6.1. In this description, the model consists of a number of submodels that run independently as well as concurrently. Each model has its own data input and entities with specific attributes. Data are shared between the submodels by global variables. The content of these variables can be altered within any submodel, with the new values immediately shared and used by any other submodel. These global variables not only pass data between the submodels but can also be used in logic statements to control the movement and routing of entities, branching logic, updating entity attributes, etc.

A second difficulty is data collection. The more detailed the model, the more detailed the data. Quite often, data do not exist, are only vague, are not measurable, or are very time-consuming to obtain. A simplified, rapid approach to data collection is to ask the appropriate questions through interviews with personnel directly

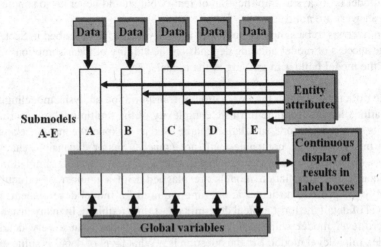

Fig. 6.1 Conceptual framework for rapidly modeling transportation systems

involved with the application. This is not only effective but also a time-saving approach to obtaining data.

When data is constrained, the questions that need to be answered are whether the constraints apply on one side of the distribution or both, and, if so, what the limits on values are. Once these questions have been answered, there are two choices. One is to find a distribution that conforms to these constraints. For instance, the lognormal distribution can be used to model data, such as revenues and stock prices that are constrained to be never less than zero. For data that have both upper and lower limits, one could use the uniform distribution, if the probabilities of the outcomes are even across outcomes or a triangular distribution (if the data is clustered around a central value). In these instances, the triangular distribution is often used as a subjective description of a population when there are only limited sample data and especially where actual data are scarce and the cost of collection is high. For example, if the smallest value, the largest value, and the most likely value are known for a process, then the outcome can be approximated by the triangular distribution. Most personnel engaged in a process can readily give estimates for the minimum, maximum, and most likely values which correspond to the three parameters of the triangular distribution (see Fig. 6.2).

A reasonable assumption is that service times follow a triangular distribution. It is rather easy to ask knowledgeable personnel for the smallest time (parameter a), the most frequent time or mode (parameter c), and the largest time (parameter b) to obtain the needed parameters for the triangular distribution shown in Fig. 6.2. The triangular distribution (probability density function) is a continuous distribution with a mode of c, a mean of $\frac{(a+b+c)}{3}$, and a variance of $\frac{\left(a^2+b^2+c^2-ab-ac-bc\right)}{18}$. The triangular distribution closely resembles the normal distribution if $(c - a) = (b - c)$. However, most data are skewed and more accurately represented by the log normal distribution. The triangular distribution in Fig. 6.2 resembles the log normal since $(c - a) > (b - c)$. It should be noted that log normal distributions can have relatively long tails, which may or may not be desirable in a simulation (Crow and Shimizu 1998).

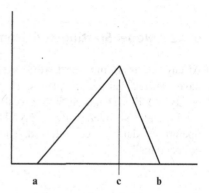

Fig. 6.2 Triangular probability density function

6.1.1   Model Verification and Validation

Model verification is defined as determining if the model is correctly represented in the simulation code. Several of the common techniques for model verification are:

- Using the trace feature in the simulation software to trace an entity through each model segment
- Replacing the distributions with mean values, then running the model, and evaluating the results
- Using the animation feature in the simulation software to observe any abnormalities during model execution
- Running a single entity through the simulation model, computing the entity time in the system, and comparing results to real-world data

Model validation can be defined as determining if the model is an accurate representation of the real-world system. Thus, model validation examines the fit of the model to empirical data (e.g., measurements of the real-world system to be modeled). A good model fit means that a set of important performance measures, predicted by the model, match or agree with their observed counterparts in the real-world system. This kind of validation is only possible if the real-world system or emulation through the model exists and if the requisite measurements can actually be acquired. Any significant discrepancies would suggest that the model developed is inadequate for project purposes and modifications are required. Simulation tools, such as ProModel (ProcessModel 2011), have a label block that displays data generated by the global variables during the simulation. By slowing the simulation down, it is possible to observe these values as the entities move through the simulation. A group of domain experts can then observe the simulation being run to observe the model operation.

Another model validation technique is to replace all model variation with mean values, run an entity through the model, and then use domain experts to compare the results with real-world data.

6.1.2   Model Starting and Stopping Conditions

At any one time, most real-world systems have entities occupying activities and have entities waiting in queues. However, most simulation models are started empty and idle. That is, at time zero, there are no entities in the system, all activities are idle, and all queues are empty. Therefore, some time needs to elapse before operational data can be collected. The question is, "At what point can we start to collect data?"

Fig. 6.3 Mean times
in the system

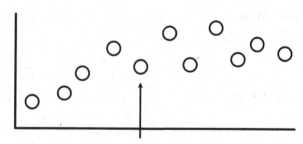

First point that is neither the max nor min
of previous measurements.

6.1.3 Model Reaches Steady State or Equilibrium

A problem which exists in most simulation runs and which is often ignored is
determining when the results of the simulation should be collected from the
simulation run. In most real-world systems, the system has been in operation for
some time before the observer begins collecting measurements; or the observer will
design experiments to ensure that unwanted biases have been removed before
beginning to collect measurements. Likewise, a simulation model must be allowed
to run for a given time in an attempt to remove any bias before measurements are
collected.

Equilibrium is that point in the simulation run when the system has reached
a steady-state condition. One of the most common techniques to determine steady
state is to run the simulation for a given amount of time, collect statistics, reset the
tables, and then repeat the steps for a number of replications.

For example, the simulation model can run for a period of time and data can
be collected. Then the model can be run for another period of time and more data
collected. The sequence can be repeated or replicated for a number of times. As an
example, the average time that an entity is in the system can then be plotted over
time (see Fig. 6.3). Conway's technique is to select the first value that is neither the
maximum nor the minimum of the previous values. For the example in Fig. 6.3,
the fifth value meets this criterion. Therefore, the first four hours of the simulation
are needed to reach steady state. All output tables are reset to zero, and the
simulation is run to collect statistics.

It should be noted that any entities that are in the system after reaching steady
state remain in the system when statistics are collected.

6.1.4 Length of Simulation to Achieve
Good Results or Sample Size

The central limit theorem is a statistical theory that states that given a sufficiently
large sample size from a population with a finite level of variance, the mean of all
samples from the same population will be approximately equal to the mean of the

Table 6.1 Replication results

Replication	Mean time in system Xi (min)
1	1.5
2	2.5
3	2.0
4	2.5
5	1.5

population. Furthermore, all of the samples will follow an approximately normal distribution pattern, with all variances being approximately equal to the variance of the population divided by each sample's size. Let us take samples of size n from the population. As the sample size n increases, the distribution of sample means will approach a normal distribution with mean u and variance $sigma2/n$. For example, let us assume that the simulation model has been run for four replications with the results given in Table 6.1.

Mean time in the system $Xbar$ for the five replications is the sum of Xi/n or $(1.5+2.5+2.0+2.5+1.5)/5 = 2.0$ min. Variance $S^2(n) = sum$ of $(Xi-Xbar)^2/(n-1)$ or $1.00/(5-1) = 0.25$. Then $S(n)$ is the square root of $S^2(n)$ or 0.5.

How many replications N, or samples, do we need to take to be 95 % confident that the estimate for the mean time an entity is in the system is within an error margin of $e = 0.5$ or the mean time in the system is $2.0 + 0.5$? N is computed as

$$N = \left[t\left(n-1, \ 1-\frac{a}{2}\right) * \frac{S(n)}{e} \right]^2$$

where t is from the t tables

$$t\left(5-1, \ 1-\frac{0.05}{2}\right) = 2.776.$$

Then,

$$N = \left[2.776 * \frac{0.5}{0.5} \right]^2 = 7.7 \approx 8 \text{ replications.}$$

If we want to double the accuracy of the mean time in the system, then we use $e = 0.25$. The number of replications is now

$$N = \left[t\left(n-1, \ 1-\frac{a}{2}\right) * \frac{S(n)}{e} \right]^2 = \left[2.776 * \frac{0.5}{0.25} \right]^2 = 30.82 \approx 31 \text{ replications.}$$

Therefore, doubling the accuracy will require a quadrupling of the number of replications from 8 to 31.

6.1.5 Statistical Confidence Intervals on Simulation Results

What confidence do we have in the simulation results? For example, what is the 95 % confidence interval of the mean time an entity is in the system? Let us use the same simulation results in Table 6.1. Then the confidence interval CI can be computed as

$$CI = X\,bar \pm \left[t\left(n - 1,\ 1 - \frac{a}{2} \right) * S_q R_t \left(\frac{S^2(n)}{n} \right) \right] = 2.0 \pm \left[2.776 * S_q R_t \left(\frac{0.25}{5} \right) \right]$$
$$= 2.0 \pm 0.622$$

Therefore, we are 95 % confident that the mean time an entity is in the system is between 1.378 and 2.622 min.

6.1.6 Analyzing Simulation Results

There are a number of points to look for in the simulation results. Any resource with 100 % utilization is a signal for a possible bottleneck. On the other hand, very low resource utilizations are a signal for an excessive number of resources and, possibly, that several resources are not needed.

If at the end of the simulation the queue in front of a station is also the maximum queue content during the simulation, then the system is probably unstable. That is, the queue length will continually increase. As a result, other results will also increase over time.

One method of analyzing the simulation results is to plot the percent change from the baseline results rather than looking just at the actual values. Figure 6.4 is an example showing the percent change of one parameter as compared to the baseline run.

Thus, the use cases introduced in this chapter bring together the theory and the application methods in an attempt to address both the theoretical background and the practical simulation methods.

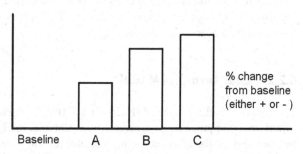

Fig. 6.4 Output analysis as a percent change from the baseline run

6.2   Coal Terminal Simulation

6.2.1   Introduction

The McDuffie Coal Terminal at the Alabama State Docks in Mobile, Alabama, consists of 556 acres and is the largest coal terminal on the gulf coast and the second largest in the USA. Total tonnage through the terminal for 2005 was 15,500,000 tons. Total ground capacity is 2,300,000 tons. Annual throughput capacity is 20,000,000 tons. The Terminal was established in 1976 as an export facility. In 1998, the facility began importing low-sulfur coal for use at power generation plants. The systems and equipment at the coal terminal have evolved over the years resulting in inefficiencies in the operational activities and processes. Thus, equipment and processes, along with customer demand for increased volume, led management to seek opportunities to improve efficiency, productivity, and throughput.

Management of the Alabama State Docks is based on lean management principles, and it was agreed to try these principles at the McDuffie Coal Terminal. The concepts of lean management (NIST 1998; Womack and Jones 1996) include identifying and eliminating inefficiencies, termed waste, in a process. Waste can be categorized as overproduction, inventory, defects, motion, and transportation, waiting, extra processing, and underutilizing people. Some of the types of waste evident in the operations of the coal terminal are:

- Waste of workers—waiting, rework, and inspection
- Waste of machines—setup and breakdown
- Waste of materials—transport and storage
- Waste of information—order processing

These types of waste are non-value-added activities that should be minimized or eliminated. Ideally, coal should arrive at the coal terminal and be immediately dispensed to another transportation mode for delivery to the customer.

Simulation is an excellent tool to explore opportunities for improving processes and minimizing waste. Simulation is valuable in evaluating proposed improvements before significant time and resources are expended. It is critical to understand the impact of change prior to the expending of resources, especially at a large-scale operation such as a coal terminal.

6.2.2   Coal Terminal Model

The coal terminal model used (Harris et al. 2008) is implemented in ProModel and stored on a computer at the University of Alabama in Huntsville connected to the Internet and is, therefore, remotely accessible. The availability of remote access models is of utmost importance for distributing the specific expertise available and integrating theory and practice in a cost-effective way at a high standard of quality in research and education. This is essential to developing and improving researchers' and students' hard and soft skills, thus enabling them to succeed in the highly competitive global job marketplace, a key topic of the USE-eNET: US-Europe e-Learning NETwork project (USE-eNET) in Science and Engineering (Möller et al. 2006).

Hard skills include the knowledge to design and implement simulation models for supervision and control and to integrate them into a transportation modeling and simulation framework. Using simulation means supervision through the front end, the computer simulation tool. ProModel was selected as the simulation package because it is available at many universities and in industry. This allowed the inclusion of comparative studies, scenario planning and analysis, content integration, and modeling and simulation case studies. Because of the vast opportunity for applying modeling and simulation, transportation analysis, modeling, and simulation should emphasize the utilization of the concepts in the respective areas of specialization. This would give researchers and students the opportunity to enhance their problem-solving skills by conducting effective simulation modeling, analysis, and research projects using the simulation tools available.

The conceptual framework of the coal terminal model is based on the submodels shown in Fig. 6.5:

- Ship unloading and loading of coal
- Barge unloading and loading of coal
- Train unloading and loading of coal

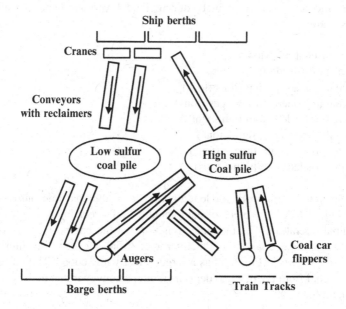

Fig. 6.5 Conceptual framework for coal terminal model

The submodels run independently of one another, each with a different entity. Data are passed between the submodels by a number of global variables. In addition, a number of attributes are assigned to the entities. The variables and attributes control entity movement, branching, and activity operations. The terminal is modeled with two coal piles, or inventory locations. High-sulfur coal arrives on barges and trains and leaves on ships.

Low-sulfur coal arrives on ships and leaves on barges and trains. Modeling resources are:

- Three ship berths
- Three barge berths
- Two cranes for unloading coal onto conveyors
- Conveyors
- Three train slots (maximum number of trains allowed at terminal)
- Two coal car flippers (grabs car and dumps coal onto conveyor)

| Entity | No. of conveyors | |
	Loading	Unloading
Ships	One	Two
Barges	Two	Two
Trains	One	Two

Model entities are ships, barges, trains, and scoops of coal. The model has 29 entity attributes and eight global variables.

The model has a number of ProModel labels that display outputs during the simulation and are used extensively during model verification and validation. These boxes are:

- Low-sulfur coal pile (tons)
- High-sulfur coal pile (tons)
- Total coal unloaded from ships (tons)
- Total coal unloaded from barges (tons)
- Total coal unloaded from trains (tons)
- Total coal loaded onto ships (tons)
- Total coal loaded onto barges (tons)
- Total coal loaded onto trains (tons)

The conceptual framework for loading coal from the coal pile onto a ship is shown in Fig. 6.6. A large piece of equipment called a reclaimer is central to the coal terminal operation. The reclaimer functions as the engine for the conveyor system used to take coal to and deliver coal from the coal inventory piles. The reclaimer shown in Fig. 6.6 has a large wheel with scoops. The wheel spins and collects coal from the pile and deposits it onto a conveyor that delivers the coal to a shipment location.

Fig. 6.6 Conceptual framework for loading coal from coal pile onto a ship

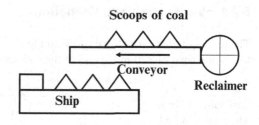

Scoops of coal

Conveyor

Reclaimer

Ship

Table 6.2 Ship unloading/loading submodel

Logic	Activity
GET Ship_Berth	Seize resource
GET 2 Ship_Crane	Seize two resources
Ship_Tons_In = 75,000	Tons to be unloaded
GET 2 ConveyorA	Seize two resources
Tons_OrderedSU = Ship_Tons_In	Tons to be unloaded

6.2.3 Ship Unloading and Loading Submodel

In Table 6.2, the ProModel logic for the ship unloading and loading submodel is shown. The comments next to the boxes and lines are the logic in the action section of activities and the attached routing lines. For example, the following comments next to the activity *Ship_Enters_Terminal* are:

The ProModel logic for the ship coal loading is shown in Table 6.2. At activity *Load_One_Scoop_From_High_S_Pile_Onto_Cnyr*, one scoop of coal is picked up by the reclaimer and placed on *ConveyorB*. Entity *Scoop_Coal* is created and routed to activity *Scoop_Of_Coal_Onto_Ship*. At the same time, the parent entity keeps looping until all of the ordered coal has been placed on *ConveyorB*. During each loop, another entity, *Scoop_Coal*, is created and routed to activity *Scoop_Of_Coal_Onto_Ship*. A conditional test *ScoopsSL*50<Tons_OrderedSL* controls the loop. Once the conditional test has been met (*ScoopsSL*50>=Tons_OrderedSL*), all of the necessary scoops have been loaded onto *ConveyorB* and the parent entity goes to *Wait_Till_Ship_Loaded*.

The activity *Scoop_Of_Coal_Onto_Ship* has a capacity of 5,000 entities, representing the coal capacity of the conveyor. The activity time represents the time for a *Scoop_Coal* to move from the coal pile to the ship. The entities are then batched with the batch quantity equal to the total tons (i.e., scoops) loaded on the ship (*Tons_OrderedSL/50*). The batching is necessary to ensure that all of the scoops for the order have completed the conveyor transfer. The batched entity is then attached with the parent entity, or the ship, to indicate that all of the coal ordered has been loaded onto the ship.

During simulation, an entity in the form of a triangle moves along the *ConveyorB* line to symbolize the movement of a *Scoop_Coal* entity. ProModel continually displays the number of activities in use on the screen to indicate the number of scoops on the conveyor.

6.2.4 Verification and Validation

The model ran for 720 h (30 days) with the following results in the ProModel label boxes at a time of 720 h, the end of the simulation:

Coal	Quantity (tons)
Low-sulfur coal pile	5,315
High-sulfur coal pile	69,960
Low-sulfur coal in from ship	643,815
High-sulfur coal in from barge	539,895
High-sulfur coal in from train	130,065
High-sulfur coal out on ship	600,000
Low-sulfur coal out on barge	538,500
Low-sulfur coal out on train	100,000

Total coal arriving minus total coal going out should equal current coal pile:

Low sulfur coal (tons)		High sulfur coal (tons)	
Arrives via ship	643,815	Arrives via barge	539,895
Departs via barge	−538,500	Arrives via train	+130,065
Departs via train	−100,000	Departs via ship	−600,000
Current coal pile	5,315	Current coal pile	69,960

The total coal unloaded was 1,312,000 tons during 1 month. This equates to 15,600,000 tons annually, which compares favorably to the FY05 tonnage through the terminal of 15,500,000 tons.

6.2.5 Analysis of Results

A number of Kaizen-based approach process improvement events (Imai 1986; Ohno 1988; NIST 1988) were conducted at the coal terminal with the goal of improving operational efficiency and increasing productivity. The results of the Kaizen-based approach identified processes associated with equipment, Total Productive Maintenance (Nakajima 1988), and conveyor operations as two primary areas for improvement. The simulation model was then exercised with various scenarios to address these areas.

The baseline consisted of the following input:

- Time between arrivals
- Arrival capacity

Entity	Arrivals (days)	Capacity (tons)	Type of coal
Ships	3	75,000	Low sulfur
Barges	2	1,500	High sulfur
Trains (100 tons/per car)	1	10,000	High sulfur

- Unloading of one simulation scoop

Method of unloading	Tons per min
Two ship cranes per ship (with two conveyors)	15
Barge auger	15
Train auger	15

- Loading of one simulation scoop

Entity	Method of loading	Tons per min	Departure capacity (tons)	Type of coal
Ship	Reclaimer	50	75,000	High sulfur
Barge	Reclaimer	50	1,500	Low sulfur
Train	Reclaimer	50	10,000	Low sulfur

- Time to move scoop of coal

From	To	Minutes
Ship	Coal pile	10
Coal pile	Ship	10
Barge	Coal pile	6
Coal pile	Barge	5
Train	Coal pile	6
Coal pile	Train	5

- Time for any scoop of coal to be placed on conveyor is one minute
- Space for loadings and unloading

Entity	Space
Ship	Three berths
Barge	Three berths
Trains	Three slots

- Two coal car flippers
- Conveyors

Mode of transportation	No. of conveyors	
	Loading	Unloading
Ships	1	2
Barges	2	2
Trains	1	2

In addition, the simulation started empty and idle; no ships, barges, or trains were initially at the terminal; and all coal piles were empty. The baseline simulation results after running for 720 min are:

- Utilization of resources

Resource	No.	Utilization (%)
Ship berths	3	65
Barge berths	3	54
Train slots	3	96
Ship crane	2	100
Ship unload conveyors	2	100
Ship load conveyor	1	28
Barge unload conveyors	2	52
Barge load conveyors	2	18
Train unload conveyor	2	14
Train load conveyor	2	9
Coal car flipper	2	4

- Time through the terminal (value-added time only)

Entity	Time through terminal (min)	No. through terminal
Ship	6,560	8
Barge	190	359
Train	1,417	10

6.2.5.1 Impact of Process Improvements (Simulation Runs 1–2)

As a result of the Kaizen approach, a number of process improvements were made to the unloading of coal from ships, barges, and trains. Consequently, Table 6.3 outlines the simulation runs conducted to evaluate these improvements:

For example, the barge and train augers have nine buckets, each holding 0.5 tons. At 5.5 rpms the augur can move 1,500 tons per hour. For simulation purposes, this equates to 25 tons per scoop. At 4.4 rpms, the augur can move 1,200 tons per hour. Some of the resource utilizations for each of the runs are given in Table 6.4. As anticipated, the utilizations of the ship and barge conveyors dropped as the scoop size increased. However, utilization of the train conveyor increased.

Table 6.3 Additional simulation runs

Simulation	Crane unload scoop (tons)	Barge unload scoop (tons)	Train unload scoop (tons)
Baseline	15	15	15
Run 1	20	20	20
Run 2	25	25	25

Table 6.4 Resource utilizations (%) for simulation runs

	Simulation		
Resource	Baseline	Run1	Run 2
Ship berths	65	40	35
Barge berths	54	52	63
Train slots	96	90	82
Ship cranes	100	87	70
Ship unload conveyor	100	87	70
Ship load conveyor	28	32	34
Barge unload conveyor	52	42	35
Barge load conveyor	18	18	18
Train unload conveyor	14	22	20
Train load conveyor	9	20	21
Coal car flipper	4	8	8

Table 6.5 Tonnage unloaded and loaded

	Simulation		
Activity	Baseline	Run 1	Run 2
Ship			
Unloaded	643,815	750,000	750,000
Loaded	600,000	750,000	750,000
Barge			
Unloaded	539,895	540,000	534,000
Loaded	538,500	540,000	540,000
Train			
Unloaded	130,065	240,000	250,000
Loaded	100,000	210,000	220,000

The utilization of all loading conveyors increased because of the large scoops at unloading. More coal was unloaded resulting in more coal at the coal piles for loading. Also, there were ships, barges, and trains available for loading which resulted in greater conveyor utilization.

The tonnage unloaded and loaded is given for each simulation run in Table 6.5. As scoop size increased, less time was required to unload a ship, barge, or train. Likewise, tonnage unloaded increased provided there were sufficient coal arrivals.

In Table 6.6, the times in the system for the Baseline Simulation and Simulation Runs 1–2 are shown.

As anticipated, the time in the system decreased as the unload scoop size increased. For example, the time to unload and load a ship was reduced from 6,560 min to 4,560 min, a 30 % reduction. The value-added time for barges was reduced from 190 min to 150 min, a 21 % reduction. The value-added time for trains was reduced from 1,417 min to 1,150 min, a 19 % reduction.

As anticipated, the time in the system decreased as the unload scoop size increased. For example, the time to unload and load a ship was reduced from 6,560 min to 4,560 min, a 30 % reduction. The value-added time for barges was

Table 6.6 Time in the system

	Simulation		
	Baseline	Run 1	Run 2
Scoop size (tons)	15	20	25
Ships			
Number	8	9	9
Time in system[a]	9,106	5,368	4,714
Value-added time in system[a]	6,560	5,310	4,560
Barges			
Number	359	359	353
Time in system[a]	246	248	412
Value-added time in system[a]	190	165	150
Trains			
Number	10	21	22
Time in system[a]	16,233	7,314	5,777
Value-added time in system[a]	1,417	1,250	1,150

[a]Time is given in minutes

reduced from 190 min to 150 min, a 21 % reduction. The value-added time for trains was reduced from 1,417 min to 1,150 min, a 19 % reduction.

In addition to the aforementioned modeling and simulation scenario analysis, many more scenarios can be investigated, such as baseline modification in which the initial baseline model starts empty and idles with all of the coal piles empty. Then the revised model (Simulation Run 3) can be started with 25,000 tons in each coal pile. Another scenario can take into account an increase in ship arrivals. In this scenario, a low-sulfur coal pile is assumed. Therefore, increasing barge throughput requires more low-sulfur coal arriving from ships. The ship unload conveyor utilization is 70 %; and the ship berth utilization is only 55 %, indicating that additional low-sulfur coal could be added to the coal pile (see Simulation Run 3). Thus Simulation Run 4 has ships arriving every 2 days rather than every 3 days.

6.2.6 Conclusion

In summary, the following conclusions can be drawn:

* The simulation model was rapidly constructed using ProModel. On the other hand, model verification and validation was rather lengthy. The use of ProModel labels greatly improved the validation and verification process.
* The semicontinuous flow environment of a coal processing facility was simulated in the discrete event platform by using the "scoop" configuration and a conditional test for aggregate volume.
* ProModel logic was constructed to simulate coal movement on conveyors. However, another package, such as ProModel, could more visually model the coal handling conveyors.

- The use of Kaizens to target the analysis was an excellent approach to obtaining significant impacts from the simulation model. The ship, barge, and train unloading functions were identified as areas for improvement at the coal terminal. Runs 1 and 2 evaluated the greater conveyor loading capacities.
- The coal terminal model appears to be very sensitive to interactions between the unloading and loading of ships, barges, and trains.
- Low utilization of resources does not necessarily indicate areas for improvement. Other constraints may be the cause. For example, if the time between ship arrivals is three days and a ship can be unloaded in two days, then there is a day that the ship berths and ship cranes and supporting conveyors are idle.
- Based on Run 4, the coal terminal can unload 21 M tons and load 19 M tons annually. Because of the nearly 100 % utilization of several of the resources, it appears that the goal of 30 M tons annually may not be possible without an equipment upgrade.

6.3   Container Terminal Simulation

6.3.1 Introduction

Over 90 % of cargo currently transported worldwide is shipped as containerized cargo (Moffatt and Nichol 2002). For example, container port traffic for selected ports in the USA. for the year 2000 is shown in Table 6.7.

As supply chains become more global and the use of containerized cargo increases, the ports throughout the USA. are improving operations and undergoing major expansions.

The Alabama State Port Authority is currently enhancing container and intermodal operations at the Alabama State Docks in Mobile, Alabama (UAH 2005). Figure 6.7 is an overview of the Mobile Container Terminal and intermodal container handling facility. The shipping terminal will include 92 acres with 2,000 feet of berthing space dredged to a depth of 45 feet for two berths. A grade-separated roadway will connect the container terminal with an intermodal terminal and value-added warehousing and distribution area (Moffatt and Nichol 2002). The container operation will consist of 57 acres and will be able to accommodate unit container trains that will pick up or off load containers from the terminal warehousing and value-added areas. Trains up to 8,000 feet in length will be able to serve the facility without blocking rail traffic on the main line.

Table 6.7 Twenty foot equivalents (TEU) volume at selected ports

Port	Volume (millions)
Los Angeles	4.9
Long Beach	4.6
Charleston	1.6
Houston	1.1
Savannah	0.9

Fig. 6.7 Overview of
container and intermodal
facility

The Mobile Container Terminal is operated by Mobile Container Terminal
LLC, a joint venture between APM Terminals North America (a subsidiary of
Maersk, Inc.) and Terminal Link (a division of CMA CGM). Mobile Container
Terminal provides terminal customers with access to global networks covering all
possible trade routes to and from the Port of Mobile. APM operates and manages
the terminal. The state dock officials were very interested in validation of the design
capacities of the container terminal. Of special interest were the utilization of
the berths, cranes, and stackers and the maximum container throughput of the
terminal.

6.3.2 Container Terminal Model

The conceptual framework of the container terminal model is shown in Fig. 6.8
(Harris et al. 2007). The model has five submodels:

1. Ship unloading and loading of containers
2. Train unloading and loading of containers
3. Truck unloading and loading of containers
4. Movement of containers from ship dock to container yard
5. Movement of containers from container yard to ship dock

The terminal model has two container inventory locations: (1) the storage
of containers from ships that are to be loaded onto trains and trucks and

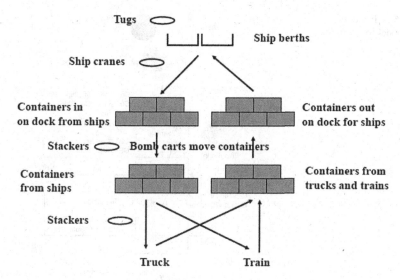

Fig. 6.8 Conceptual framework for container terminal model

(2) the storage of containers from trains and trucks that are to be loaded onto ships. Model resources are:

Resource	Quantity
Tugboats	2
Ship berths	2
Ship cranes	2
Train slots	2
Truck slots	10
Stackers	8
Trailers (bomb carts)	20
Slots for moving containers from ships to container yard	10
Slots for moving containers from container yard to ship area	10

Entities in the model are ships, trains, trucks, and containers. Two other entities for moving containers between the storage area and the dock are *Move_Order* and *Move_Order2*. The model has 13 entity attributes and 10 global variables.

6.3.3   Ship Unloading and Loading of Container Submodel

The ProModel logic for the ship unloading and loading of containers is shown in Fig. 6.9. Many of the activities have considerable logic in the action section.

Fig. 6.9 ProModel
submodel for ships

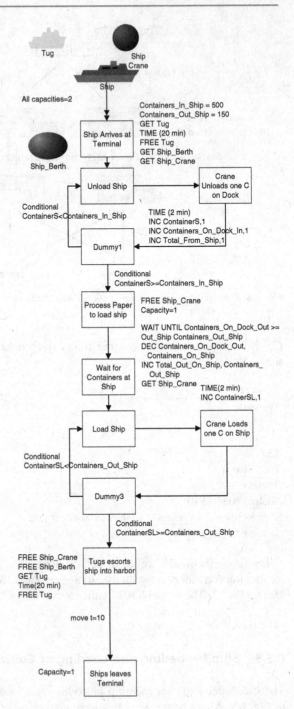

The action logic for the activity *Ship_Arrives_At_Terminal* is assumed to be as follows:

Containers_In_Ship = 500	Set container in count
Containers_Out_Ship = 150	Set container out count
GET Tug	Get resource
TIME (20 min)	Time for tug to position ship
FREE Tug	Free resource
GET Ship_Berth	Occupy resource
GET Ship_Crane	Get resource

The action logic for activity *Crane_Unloads_One_C_On_Dock* is:

TIME (2 min)	Time to unload one container
INC ContainerS,1	Increase counter by 1
INC Containers_On_Dock_In,1	Increase counter by 1
INC Total_From_Ship,1	Increase counter by 1

The branching for the two routings at activity Dummy1 is:

- Conditional: ContainerS>=Containers_In_Ship All containers have been unloaded
- Conditional: ContainerS<Containers_In_Ship Not all containers unloaded; continue to loop

At the routing from *Unload_Ship* to *Crane_Unloads_One_C_On_Dock*, the entity Ship is changed to the entity Container. The graphic for the container is a yellow rectangle. The entity continues to loop until all containers have been unloaded. After activity Dummy1, the entity Container is changed back to the entity Ship.

6.3.4 Verification and Validation

The model was verified by removing all of the variability and using only constants. In addition, ProModel has a "label block" option that displays data from the global variables during the simulation. By reducing the simulation speed, it is possible to observe these values as entities move through the simulation. The values of these labels after running the model for 1,440 h, or 60 days, were:

Containers	Values
Unloaded from ships	10,000
Unloaded from trains	6,000
Unloaded from trucks	1,440
Loaded onto ships	3,000

(continued)

Containers	Values
Loaded onto trains	6,000
Loaded onto trucks	1,440
On dock unloaded from ships	0
On dock waiting to be loaded onto ships	4,400
In container yard from ships	250
In container yard from trains and trucks	0

Containers	Values (tons)	Containers	Values (tons)
Unloaded from ships	10,000	Unloaded from trains	6,000
Loaded onto trains	−6,000	Unloaded from trucks	1,440
Loaded onto trucks	−1,440	Loaded onto ships	−3,000
On dock unloaded from ships	−0	On dock waiting to be loaded onto ships	−4,440
In yard from ships	2,560	In yard from trains and trucks	0

Model validation was not possible since the Mobile Container Terminal is still under construction. However, it was possible to use data from the existing container facility for the service times and to visually observe the operations of the terminal during the simulation.

6.3.5 Analysis and Results

The baseline consisted of the following inputs:

- Arrivals and departures

Mode of transport	Time between arrivals	Arrival capacity (containers)	Departing capacity (containers)
Ship	3 days	500	150
Train	1 day	100	100
Truck	1 hour	1	1

- Loading and unloading

Movement	Minutes
Tug to position or remove ship at berth	20
Crane or stacker to unload or load a container from ship, train, or truck	2
Stacker to load or unload container at ship dock or container yard	2
Bomb carts to move container from ship dock to container yard or from container yard to ship dock	5

- Resources

Resource	Number
Ship berths	2
Tugs	2
Cranes	2
Truck slots for loading and unloading	10
Train slots for loading and unloading	2
Bomb carts for loading and moving containers simultaneously from dock to container yard	10
Bomb carts for loading and moving containers simultaneously from container yard to dock	10
Stackers shared for unloading and loading bomb carts, trains, and trucks	8

In addition, the simulation started empty and idle; no ships, trains, or trucks were initially at the terminal and the container yard was empty. The model ran for 60 days or 1,440 h. The baseline simulation results were:

- Utilization of resources

Resource	No.	Utilization (%)
Tugs	2	1
Ship berths	2	22
Ship cranes	2	22
Bomb carts	20	11
Stackers	8	18

- Average time through terminal

Entity	Time through terminal (min)	Value-added time through terminal (min)	No. through terminal
Ship	2,012	1,347	20
Train	623	409	60
Truck	33	16	1,440

6.3.5.1 Reducing Time Between Arrivals (Simulation Runs 2–3)

Since one of the primary objectives of the project was to determine throughput capacity of the container terminal, several additional simulation model runs were made with a continual reduction in times between the arrivals of ships, trains, and trucks (Table 6.8). All other inputs, as described in the baseline simulation section, were not changed.

The utilization of resources after running the model for 60 days, or 1,440 h, is shown in Table 6.9. As anticipated, the utilizations increased as the time between arrivals of ships, trains, and trucks decreased. However, for Simulation Run 3, the

Table 6.8 Additional
simulation runs

Simulation	Time between arrivals (min)		
	Ships	Trains	Trucks
Baseline	4,320	1,440	60
Run 1	2,880	960	40
Run 2	1,440	480	20
Run 3	720	240	10

Table 6.9 Resource
utilization (%) for runs

Resource	Utilization by simulation (%)			
	Baseline	Run 1	Run 2	Run 3
Ship tugs (2)	1	1	1	2
Ship berths (2)	22	33	67	99
Ship cranes (2)	22	33	67	99
Bomb carts (20)	11	16	33	76
Stackers (8)	18	27	54	99

Table 6.10 Average time at the terminal

	Simulation			
	Baseline	Run 1	Run 2	Run 3
Ships				
Quantity	20	30	59	8
Time in terminal (min)	2,012	2,012	2,012	12,999
Value-added time in terminal (min)	1,347	1,347	1,347	1,347
Trains				
Quantity	60	90	179	352
Time in terminal (min)	623	623	623	1,488
Value-added time in terminal (min)	409	409	409	409
Trucks				
Quantity	1,440	2,160	4,319	8,637
Time in terminal (min)	33	33	33	40
Value-added time in terminal (min)	16	16	16	16

utilization of the resources approached 100 %. This implies that the terminal
reached capacity somewhere in the decreasing of the time between arrivals from
Simulation Run 2 and Simulation Run 3.

Table 6.10 shows the average times each type of entity was at the terminal.
As expected, average value-added times remained the same for all runs because no
service times were changed—only the time between arrivals.

However, this is not true for the average time an entity was in the system or
terminal. Note that the times in the system for Baseline Simulation, Simulation
Run 1, and Simulation Run 2 were identical, implying that the terminal has
additional capacity. However, the Simulation Run 3 time in the system for ships
and trains greatly increased. For Simulation Run 1 and Simulation Run 2, average
time in the system for ships was 2,012 min; and for Simulation Run 3, the average

Table 6.11 Container activity

Container activity	Simulations			
	Baseline	Run 1	Run 2	Run 3
Containers unloaded				
Ship	10,000	15,000	29,973	44,090
Train	6,000	9,000	18,000	35,978
Truck	1,440	2,160	4,320	8,640
Containers loaded				
Ship	3,000	4,500	8,850	13,200
Train	6,000	9,000	18,000	35,300
Truck	1,440	2,160	4,320	8,639
Containers in yard				
From ship	2,560	3,840	7,649	69
From train and truck	0	0	0	271
Containers on dock				
In from ship	0	0	0	72
Out from train and truck	4,440	6,660	13,470	31,137

time was 12,999 min. This large increase for Simulation Run 3 can be attributed to the delay of 9,860 min for a ship to obtain one of the two berths. For Simulation Run 3, the average time in the system for trucks increased from 33 min to 40 min. A plausible explanation may be that trucks only need one container to exit the terminal.

The number of containers unloaded, loaded, and in the container yard after running the model for 60 days is shown in Table 6.11.

6.3.5.2 Empty Train Arrivals (Simulation Runs 4–6)

Simulation Model Runs 1–3 had a large number of containers in the yard and on the dock at the end of the simulation. For example, in Simulation Run 2, 7,649 containers were in the container yard from ships and 13,470 containers were on the dock from trains and trucks. These numbers would continue to increase with longer simulation runs. As a result, the system is suspected to be in an unstable condition since these container capacities would continue to increase.

The initial baseline model started empty and idle and with the container yard empty. Simulation Runs 4 through 6 included additional ProModel logic for the arrival of empty trains with an outgoing capacity of 200 containers. The number of containers incoming and outgoing for the other entities, namely, ships, full trains, and trucks, did not change. The time between arrivals for these simulation runs is given in Table 6.12. The goal is to reduce the large buildup of containers in the container yard and on the dock.

The number of containers unloaded, loaded, and in the yard for Simulation Runs 4–6 after running the model for 60 days is given in Table 6.13. The results indicate that the addition of the logic for the arrival of empty trains greatly reduced the number of containers waiting in the terminal. These numbers were the lowest for Simulation Run 4.

Table 6.12 Time between arrivals, Simulation Runs 4–6 (all times in minutes)

	Simulation		
Entity	Run 4	Run 5	Run 6
Ships	1,440	1,440	1,440
Full trains	1,440	1,440	1,440
Empty trains	720	900	1,080
Trucks	20	20	20

Table 6.13 Container activity for Simulation Runs 4–6

	Simulation		
Container activity	Run 4	Run 5	Run 6
Containers unloaded from			
Ship	29,973	29,973	29,973
Train	6,000	6,000	6,000
Truck	4,320	4,320	4,320
Containers loaded onto			
Ship	8,850	8,850	8,850
Train (including empty trains arrivals)	25,600	25,200	22,000
Truck	4,320	4,320	4,320
Containers in yard			
From ship	49	449	3,649
From train and truck	0	0	0
Containers on dock			
In from ship	0	0	0
Out from train and truck	1,470	1,470	1,470

Table 6.14 Total container activity for Simulation Runs 4–6

	Simulation		
Container activity	Run 4	Run 5	Run 6
Containers unloaded	40,293	40,293	40,293
Containers loaded	38,770	38,370	35,170
Containers in terminal	1,519	1,919	5,119

Table 6.14 reveals the total container activity during the 60-day simulation. It appears that of the three simulation runs, Simulation Run 5 may be the best after considering the total times through the system (see Table 6.15). Simulation Run 4 had an average time in the system for trains of 7,372 min as compared to 802 min for Simulation Run 5.

The results in Table 6.15 (Simulation Run 4) suggest other problems. For example, the average time in the terminal for trains increased to 7,372 min with a value-added time of 407 min as compared to 623 min for Simulation Run 2. The container yard had 49 containers from ships. Therefore, on average, a train waited 6,965 min either for sufficient containers to load the train or the activity capacity of two was filled (able to load only two trains at a time). At the end of the simulation, 24 trains were waiting for containers or an available activity to load the train.

Table 6.15 Average times at terminal

Entity	Quantity	Time in terminal (min)	Value-added time (min)
Simulation Run 4			
Ships	59	2,012	1,347
Trains	152	7,372	407
Trucks	4,319	33	16
Simulation Run 5			
Ships	59	2,012	1,347
Trains	156	802	407
Trucks	4,319	33	16
Simulation Run 6			
Ships	59	2,012	1,347
Trains	140	646	407
Trucks	4,319	33	16

For Simulation Run 5, the average time in the terminal was 802 min with the same value-added time of 407 min. The container yard had 449 containers from ships. For Simulation Run 6, the average time in the terminal was 646 min with the same value-added time of 407 min. However, the container yard had 3,649 containers from ships.

6.3.6 Conclusion

The large container buildup in the terminal for Simulation Runs 1–3 indicates a need to further balance the entity arrivals to give a more accurate estimate of container throughput. Simulation Runs 4 through 6 include logic for the arrival of empty trains. With an empty train arriving every 900 min (Simulation Run 5), the container yard was reduced to 449 containers and 1,479 containers on the dock (after running the model for 1,440 h). The terminal capacity was 240,000 containers for Simulation Run 5 with about 2,000 containers still in the terminal.

In conclusion, it was found that the simulation of a container facility can provide insight for the initiation of operational improvements needed to increase freight throughput and velocity

6.4 Intermodal Container Terminal Simulation

6.4.1 Introduction

Over 90 % of cargo currently transported worldwide is shipped as containerized cargo. In 2000, container port traffic at the three busiest ports in the USA was 4.9 million TEUs at Los Angeles, 4.6 million TEUs at Long Beach, and 1.6 million

TEUs at Charleston. The volume of containerized cargo essentially doubled at the ports of Los Angeles and Long Beach and greatly increased at most other ports.

As the use of containerized cargo increases, the ports throughout the USA are improving operations and undergoing major expansions. The increase in containerized cargo is also impacting inland intermodal centers. Simulation offers an inexpensive approach to analyzing the operations of an intermodal center and to evaluating various operational alternatives before finalizing the design of any planned expansion. The two primary questions answered by the simulation in this case study are: (1) Can container throughput satisfy anticipated demand? (2) Are resources sufficient to support anticipated growth in demand?

6.4.2   Intermodal Center

The International Intermodal Center (IIC) is located at Huntsville International Airport between Huntsville and Decatur, Alabama, on Interstate 565 approximately 10 miles from Interstate 65, which is designated as a Freight Significant Corridor by the Federal Highway Administration. The IIC is served by CSX Railroad and operates its own Class 3 Rail Service to move container car pulls to and from the main line. It should be noted that no containers such as those on trains or trucks go onto cargo aircraft. Cargo aircraft use air cargo boxes that are configured to the shape of the aircraft. These air cargo boxes may or may not contain pallets and are unpacked before the contents are cross-docked and put onto trucks. For simulation purposes the air cargo box is considered as a unit of freight (a simulation entity) with the content placed directly onto a truck. Similarly, no containers that arrive by truck or train are put on an aircraft. Air cargo is delivered to the intermodal center by truck as pallets or cases. For simulation purposes these pallets or cases that arrive on a truck are considered a container. Very infrequently, air cargo arrives by train as pallets or even containers. For simulation case studies, these pallets or containers are considered a unit of freight (a simulation entity). In the literature, many articles can be found dealing with the application of simulation in intermodal freight movement and logistics, as summarized in Sect. 6.4.9.

A diagram of the movement of containerized freight through the IIC is given in Fig. 6.10. Freight moves from airplanes to trucks, trucks to airplanes, trucks to rail, rail to trucks and can go from rail to airplane (but this is very rare).

Table 6.16 summarizes the arrival and departure of containers on airplanes, trains, and trucks. For example, containers or cargo boxes that arrive on airplanes only depart on trucks.

6.4.3   Simulation Model

In Fig. 6.1 a representation of the conceptual framework for the model development is given. The conceptual framework consists of a number of independent, but linked, submodels. Each model has its own data input and entities with specific

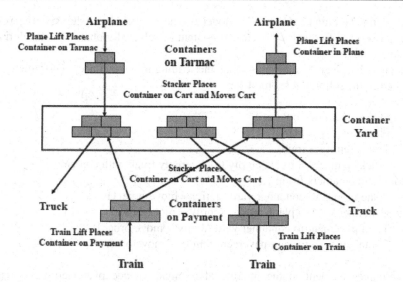

Fig. 6.10 Container flow at Huntsville International Intermodal Center

Table 6.16 Container arrives and departures at International Intermodal Center[a]	Containers depart on		
Containers arrive on	Airplane	Train	Truck
Airplane	No	No	Yes [b]
Train	Yes[c]	Yes[d]	Yes[d]
Truck	Yes[e]	Yes[f]	No

[a]Container, or unit of freight, in the simulation can be a cargo box, pallet, or case, or TEU
[b]Cargo box arriving on airplane and leaving on truck
[c]Pallet or case arriving by train and leaving on airplane
[d]TEU arriving by train and leaving by truck or train
[e]Pallet or case arriving on truck and leaving on airplane
[f]TEU arriving on truck and leaving on train

attributes. Within the conceptual framework, data are shared between the submodels through global variables. The content of the global variables can be altered within any submodel with the new values immediately shared and used by any other submodel. These global variables not only pass data between the sub-models but can also be used in logic statements to control the movement and routing of entities and branching logic and updating entity attributes.

To assist in model verification and validation, the conceptual framework includes a set of output blocks that display the current values of the global variables during the running of the simulation. These values are generally overlaid on top of the simulation model so the user can observe the movement of entities as well as any bottlenecks. ProModel was selected for implementation of the conceptual framework. The building blocks in ProModel were ideal for constructing the submodels. ProModel has four building blocks: activities, entities, resources, and stores. Within each block and for each routing option (connecting line), there is the capability of adding complex logic. Global variables and entity attributes can be easily defined within ProModel. ProModel also has a label block (shown in Fig. 6.1) that can be used to display the current content of selected global variables during the simulation.

Translating the container unloading and loading into the conceptual framework resulted in the following submodels:

- Unloading/loading containers:
  - Planes (entity = plane)
  - Trains (entity = train)
  - Trucks (entities = truck, empty truck, empty truck with container)
- Moving containers from:
  - Plane tarmac to container yard (entity = move order1)
  - Container yard to plane tarmac (entity = move order2)
  - Train pavement to container yard (entity = move order3)
  - Container yard to train pavement (entity = move order4)

Resources resident in the intermodal terminal include plane terminals, train terminals, truck slots, plane lifts, train lifts, stackers, and carts. The plane and train lifts are similar to forklifts or side loaders for unloading and loading containers. The model has 13 entity attributes, 20 global variables, 64 activity blocks, and 9 entity blocks. Figure 6.11 displays the ProModel for the plane unloading/loading submodel. The comments next to the blocks and lines are the imbedded logic within the ProModel blocks and connecting lines.

The simulation model was run for 1,440 h, or 180 eight-hour days, which closely equates to 6 months. The values that were recorded in the ProModel label boxes at the end of the simulation are shown in Table 6.17.

The simulation model was run for 1,440 h, or 180 eight-hour days, which closely equates to 6 months. The values that were recorded in the ProModel label boxes at the end of the simulation are shown in Table 6.17.

The full containers in (8,460) minus the full containers out (8,011) should equal the containers in the intermodal center at the end of the simulation (449). The actual lifts, either a container load or unload, at the intermodal center for 2005 were 34,410. The simulation results for 6 months were $9,180 + 8,267 + 913$, or 18,360 lifts (see Table 6.17). On an annual basis, this equates to 36,720 lifts, which compares favorably to the actual lifts of 34,400 in 2005.

6.4.4 Experimental Design

The experimental design is given in Table 6.18. The current intermodal center operations are defined in Baseline Simulation Run 1. Each of the following simulation runs was based on the output from the previous run. Resources were reduced for each successive simulation run and defined as Simulation Runs 2–10. Each simulation run had fewer resources allocated, reducing the number of plane and train terminals, truck slots, plane and train lifts, stackers, and carts, and was evaluated against the Baseline Simulation Run 1.

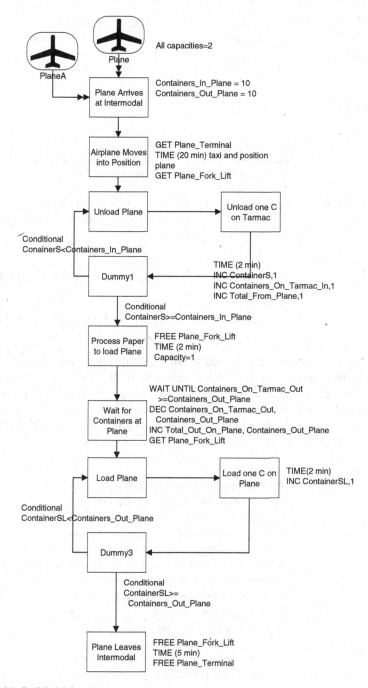

Fig. 6.11 ProModel for airplanes

Table 6.17 Label block values from simulation

Activity	Plane	Train	Truck	Total
Full containers in	1,800	4,500	2,160	8,460
Full containers out	1,800	3,600	2,611	8,011
Containers waiting for	200	198	51	449

Table 6.18 Experimental design

Simulations	Description
Baseline Run 1	Current intermodal center operations
Runs 2–10	Multiple simulation runs reducing the number of resources from Baseline Simulation Run 1 based on the output of the previous simulation run
Run 11	Increased number of entity arrivals in Simulation Run 10
Runs 12–15	Multiple runs reducing the number of resources in Simulation Run 11

Table 6.19 Movement of containers

	Containers out		
Containers in	Airplane	Truck	Train
Airplane		100 %	
Truck	40 %		60 %
Train	25 %	55 %	20 %

The number of plane, train, and truck entity arrivals was increased from Simulation Run 10 to Simulation Run 11. Simulation Runs 12–15 evaluated Simulation Run 11 with fewer resources by continuing to decrease the number of plane and train terminals, truck slots, plane and train lifts, stackers, and carts.

The input data for Baseline Simulation Run 1 are given in Tables 6.19 and 6.20. In addition, the baseline input consisted of:

Quantity/time	Activity
2	Plane terminals for unloading/loading containers
3	Train terminals for unloading/loading containers
20	Maximum trucks in intermodal center at one time
2	Lifts for unloading/loading containers from planes
2	Lifts for unloading/loading containers from trains
8	Stackers for unloading/loading containers from trucks and onto carts
20	Carts for moving containers throughout the center
2 min	Unload/load a container from plane, train, or truck
20 min	Position a plane at a terminal
20 min	Position a train at a terminal
5 min	Position a truck for unloading/loading
2 min	Process paperwork to load a plane, train, or truck
5 min	For plane, train, or truck to exit intermodal center
2 min	Unload/load a cart
5 min	Move a cart between a plane, train, or truck and the container yard

Table 6.20 Entities

Entity	Time between arrivals (min)	Average containers in	Average containers out	Truck leaves with no containers	Truck leaves with container	Truck leaves with full container
Airplane	480	10	10			
Train	960	50	40			
Truck with full container	40	1		10 %	9 %	81 %
Empty truck	240	0				100 %
Truck with empty container	120	1		10 %	9 %	81 %

Table 6.21 Entities through intermodal center for Baseline Simulation Run 1

Entity	Quantity through IIC	Average time in IIC (min)	Average value-added time (min)	Average wait time (min)
Airplanes	180	94	67	27
Trains	90	326	207	119
Trucks	2,160	43	14	29
Empty trucks	360	36	12	24
Truck with empty container	720	39	14	25

Triangular distributions were used for all of the service times. These times remained the same as well as the percentage routings of the entities for all of the simulation runs. The only changes in the data were the number of resources and the time between arrivals of the plane, train, and truck entities. Triangular distributions are three parameter distributions T(a, b, c) where a is the smallest value, b the mean value, and c the largest value. In collecting data, it is rather easy to ask staff working at an operation for the minimum, most likely, and maximum values for any variable. A triangular distribution is a close approximation to the normal distribution with the exception of the infinite tails for the normal distribution.

The results of Baseline Run 1 are given in Tables 6.21, 6.22 and 6.23. Table 6.21 presents the number of entities that pass through the intermodal center after running the simulation for 1,440 h or 6 months. The average entity wait times were relatively low, indicating adequate resources to unload and load all entity arrivals. These waiting times include the time an entity waited on the appropriate container or waited on an available resource.

Table 6.22 Utilization of resources for Baseline Simulation Run 1

Resource	Quantity	Utilization (%)
Airplane terminals	2	9
Train terminals	3	11
Truck slots	20	7
Plane lifts	2	6
Train lifts	2	14
Stackers	8	13
Carts	20	7

Table 6.23 Container activity for Baseline Simulation Run 1

Entity	Containers unloaded	Containers loaded	Containers in process
Airplanes	1,800	1,800	200
Trains	4,500	3,600.	198
Trucks	2,160	2,611	51
Trucks with empty containers	720	256	464
Total	9,180	8,267	913

Table 6.22 shows the percent utilization of the resources. The relatively low utilizations indicate that there is an excess of resources at the intermodal center. It should be noted that the average utilization might be misleading. For example, when a plane arrives, the plane terminal and plane lifts are busy until the plane is unloaded and loaded. Once the plane leaves the intermodal center, these resources become idle.

Table 6.23 provides a summary of the container activity during the simulation run. A total of 9,180 containers were unloaded, 8,267 containers were loaded, and 913 containers were still at the intermodal center. Of these 913 containers, 200 containers were on the tarmac waiting to be loaded onto a plane, 198 containers were on the pavement waiting to be loaded onto a train, 51 containers were in the container yard waiting to be loaded onto a truck, and 464 empty containers were in the container yard.

6.4.5 Removal of Resources from Baseline Simulation Run 1

The relatively low utilization of the resources in Table 6.22 for Baseline Simulation Run 1 indicates that there is an excess of resources at the intermodal center. Therefore, a number of additional runs were made with each run reducing a resource by one. For example, Simulation Run 2 reduced the number of plane terminals from two to one; and the simulation results were compared with Baseline Simulation Run 1. The next two runs reduced the number of train terminals from three to two and then to one. In successive simulation runs, the truck slots were reduced from 20 to 15 and then to 12. The plane lifts were reduced from two to one

Table 6.24 Resources for Simulation Run 10

Resource	Quantity
Airplane terminals	1
Train terminals	1
Truck slots	12
Plane lifts	1
Train lifts	1
Stackers	6
Carts	10

Table 6.25 Entities through intermodal center for Simulation Run 10

Entity	Quantity through intermodal	Average time in intermodal (min)	Average value-added time (min)	Average wait time (min)
Airplanes	180	99	67	32
Trains	90	371	207	164
Trucks	2,160	28	14	14
Empty trucks	360	21	12	9
Trucks with empty container	720	27	14	13

Table 6.26 Utilization of resources for Simulation Run 10

Resource	Quantity	Utilization (%)
Airplane terminals	1	20
Train terminals	1	36
Truck slots	12	5
Plane lifts	1	13
Train lifts	1	28
Stackers	6	17
Carts	10	15

and the train lifts from two to one. The stackers were reduced from eight to six. Carts were reduced from 20 to 15 and then to 10. The final Simulation Run 10 had the resources shown in Table 6.24. A further reduction in the resources greatly increased the entity times at the intermodal center.

The results for Simulation Run 10 are presented in Tables 6.25, 6.26 and 6.27. Table 6.25 shows the entities through the intermodal center after running the simulation model for 1,440 h. The intermodal center resources in Simulation Run 10 were reduced by one-half. Such a reduction in resources is significant because of the corresponding reductions that would be experienced in repairs, maintenance, and overall operating costs. The quantities through the intermodal center were very close to the quantities for Baseline Simulation Run 1 even after a number of resources were removed from the model. The average wait time for the train entity increased from 119 to 164 min. The other entity wait times did not significantly increase.

Table 6.27 Container activity for Simulation Run 10

Entity	Containers unloaded	Containers loaded	Containers in process
Airplanes	1,800	1,800	196
Trains	4,500	3,600	188
Trucks	2,160	2,611	65
Trucks with empty containers	720	270	450
Total	*9,180*	*8,281*	*899*

Table 6.26 gives the utilization of the resources after running the simulation model for 1,440 h. As expected, the utilizations increased since there were fewer resources in Simulation Run 10 as compared to Baseline Simulation Run 1. Train terminal utilization increased from 11 % for Baseline Simulation Run 1 to 36 % for Simulation Run 10. Train lift utilization increased from 14 % for Baseline Simulation Run 1 to 28 % for Simulation Run 10.

Table 6.27 presents the container activity during the simulation run. The total unloaded and loaded containers were almost identical to the results of Baseline Simulation Run 1.

6.4.6 Increase in Entity Arrivals

One of the issues this case study is attempting to answer is the impact of additional entity arrivals on overall intermodal center operations. There is the possibility of a plane arriving every week from Asia. Currently, a plane arrives daily from Europe. The addition of the plane from Asia would require additional truck and train arrivals to move the containers out and to bring additional containers in for shipment by plane. Table 6.28 gives the increase in entity arrivals (Simulation Run 11). The bold values in the table reflect the changes to the entity input data originally presented in Table 6.20.

The increase in entity arrivals should require additional resources. Even so, the number of resources was kept basically the same as Baseline Simulation Run 1 (see Table 6.29), with the exception of train slots that were reduced to two rather than the original three.

The results for Simulation Run 11 after running the model for 1,440 h are given in Tables 6.30, 6.31 and 6.32. Table 6.30 provides the entity times at the intermodal center. The average wait times remain relatively low even after an increase in the arrivals of plane, train, and truck entities and were very similar to Baseline Simulation Run 1.

Table 6.31 gives the utilization of the resources for Simulation Run 11. Note that the resources increased because of the increase in the entity arrivals. However, the utilizations were still relatively low. The utilization of the train terminals increased from 11 % for Baseline Simulation Run 1 to 21 % for Simulation Run 11. This was

Table 6.28 Increase in entity arrivals for Simulation Run 11

Entity	Time between arrivals (min)	Average containers in	Average containers out		
Airplanes-Europe	480	10	10		
Airplane-Asia	2,400	10	10		
Train	720	50	40		
			Truck leaves with no container	*Truck leaves with container*	*Truck leaves with full container*
Truck with full container	30	1	10 %	9 %	81 %
Truck empty	240	0			100 %
Truck with empty container	120	1	10 %	9 %	81 %

Table 6.29 Resources for Simulation Run 11

Resource	Quantity
Airplane terminals	2
Train terminals	2
Truck slots	20
Plane lifts	2
Train lifts	2
Stackers	8
Carts	20

Table 6.30 Entities through intermodal center for Simulation Run 11

Entity	Quantity through intermodal	Average time in intermodal (min)	Average value-added time (min)	Average wait time (min)
Airplanes-Europe	180	93	67	26
Airplane-Asia	36	93	67	26
Trains	120	312	207	105
Trucks	2,879	29	14	15
Empty trucks	360	21	12	9
Truck with empty container	720	27	14	13

Table 6.31 Utilization of resources for Simulation Run 11

Resource	Quantity	Utilization (%)
Airplane terminals	2	11
Train terminals	2	21
Truck slots	20	6
Plane lifts	2	8
Train lifts	2	18
Stackers	8	16
Carts	20	10

Table 6.32 Container activity for Run 11

Entity	Containers unloaded	Containers loaded	Containers in intermodal
Planes	2,160	2,160	544
Trains	6,000	4,800	257
Trucks	2,880	3,235	44
Trucks with empty containers	720	311	409
Total	11,760	10,506	1,254

anticipated since the train arrival times were lowered from 960 min for Baseline Simulation Run 1 to 720 min for Simulation Run 11, and the number of train slots was reduced by 33 %. The increase in the truck arrivals did not increase truck slot utilization because of the large number of slots.

Table 6.32 gives the container activity for Simulation Run 11. As a result of an increase in the entity arrival rates, the container throughput increased to 23,520 (+28 %) from 18,360 for Baseline Simulation Run 1.

6.4.7 Removal of Resources from Revised Model

The relatively low utilization of resources indicates an excess of resources available. Therefore, a number of additional simulation runs were made with each simulation run reducing a resource by one. Additional simulation runs (Simulation Runs 12–15) were conducted, with the final Simulation Run (15) utilizing the resources given in Table 6.33. Further removal of resources greatly increased the entity times at the intermodal center.

The results for Simulation Run 15 are given in Tables 6.34, 6.35 and 6.36 after running the simulation model for 1,440 h. Table 6.34 presents the entities through the intermodal center for Simulation Run 15. Basically, the average entity times remained the same as for Simulation Run 11. The exception was the time in the terminal for Airplanes-Europe that increased to 111 min from 99 min for Simulation Run 11. This increase in time can be attributed to the increase in entity Airplane-Asia.

Table 6.33 Resources for Simulation Run 15

Resource	Quantity
Airplane terminals	1
Train terminals	2
Truck slots	12
Plane lifts	1
Train lifts	2
Stackers	8
Carts	12

Table 6.34 Entities through intermodal center for Simulation Run 15

Entity	Quantity through intermodal	Average time in intermodal (min)	Average value-added time (min)	Average wait time (min)
Airplanes-Europe	180	111	67	44
Airplane-Asia	36	93	67	26
Trains	120	312	207	105
Trucks	2,879	32	14	18
Empty trucks	360	24	12	12
Truck with empty container	720	30	14	16

Table 6.35 Utilization of resources for Simulation Run 15

Resource	Quantity	Utilization (%)
Airplane terminals	1	23
Train terminals	2	21
Truck slots	12	7
Airplane lifts	1	16
Train lifts	2	18
Stackers	8	16
Carts	12	16

Table 6.36 Container activity for Simulation Run 15

Entity	Containers unloaded	Containers loaded	Containers in process
Planes	2,160	2,160	412
Trains	6,000	4,800	414
Trucks	2,880	3,219	35
Trucks with empty containers	720	321	399
Total	11,760	10,500	1,260

The utilization of resources for Simulation Run 15 is provided in Table 6.35. The utilizations increased from Simulation Run 11 due to the reduction in the number of resources available. However, the rate of increase was not as large. The utilization for the plane terminal increased from 11 % for Simulation Run 11 to 23 % for Simulation Run 15 because of the plane arrivals from Asia. Likewise, the plane lifts increased from 8 % for Simulation Run 11 to 16 % for Simulation Run 15. The other resources did not increase significantly because the number of resources was similar to Simulation Run 11.

The resulting container activity for Simulation Run 15 is given in Table 6.36. The containers unloaded and loaded remained identical to Simulation Run 11 (23,520 lifts for both Simulation Runs 11 and 15).

6.4.8 Planning for Additional Growth

Lift capacity reached 47,040 annually for Simulation Run 15 with fewer resources than in the Baseline Simulation Run 1. The estimated 2007 number of container lifts is 45,000. Therefore, it may be possible to increase the number of lifts by restoring those resources in Simulation Runs 11 and 15.

The results of the previous simulation runs suggest that the intermodal center has the capacity for additional container throughput, especially because of the relative low utilization of resources. As a result, the experimental design shown in Table 3 was amended to include an additional simulation run, Simulation Run 16. Table 6.37 provides the revised time between arrivals for Simulation Run 16.

The results of Simulation Run 16 are given in Tables 6.38, 6.39 and 6.40 after running the simulation for 1,440 h. Table 6.38 shows the entity times at the intermodal center. As anticipated, the quantity through the intermodal center

Table 6.37 Increase in entity arrivals (Simulation Run 16)

Entity	Time between arrivals (min)	
	Run 16	Run 15
Airplanes-Europe	360 (6 h)	480
Airplane-Asia	1,200 (20 h)	2,400
Train	480 (8 h)	720
Truck with full container	20	30
Empty truck	240 (4 h)	240
Truck with empty container	120 (2 h)	120
Resources		
Plane terminals	2	1
Train terminals	3	2
Truck slots	20	12
Plane lifts	2	1
Train lifts	2	2
Stackers	8	8
Carts	20	12

Table 6.38 Entities through intermodal center for Simulation Run 16

Entity	Quantity through intermodal	Average time intermodal (min)	Average value-added time (min)	Average wait time (min)
Airplanes-Europe	240	94	67	27
Airplane-Asia	59	93	67	26
Trains	180	307	207	100
Trucks	4,319	28	14	14
Empty trucks	360	21	12	9
Truck with empty container	720	27	14	13

Table 6.39 Utilization of resources for Simulation Run 16

Resource	Quantity	Utilization (%)
Airplane terminals	2	16
Train terminals	3	21
Truck slots	20	8
Airplane lifts	2	11
Train lifts	2	28
Stackers	8	24
Carts	20	15

Table 6.40 Container activity for Simulation Run 16

Entity	Containers unloaded	Containers loaded	Containers in intermodal
Planes	2,990	2,990	966
Trains	9,000	7,200	447
Trucks	4,320	4,370	336
Trucks with empty containers	720	460	260
Total	*11,030*	*15,020*	*2,009*

increased for planes, trains, and trucks. Throughput for planes increased from 216 for Simulation Run 15 to 299 (+38 % increase) for Simulation Run 16, for trains from 120 for Simulation Run 15 to 180 (+50 % increase) for Simulation Run 16, and for trucks from 2,879 for Simulation Run 15 to 4,319 (+50 % increase) for Simulation Run 16.

Even with this increase in throughput, the average time an entity was at the intermodal center remained constant. The average time for Airplanes-Europe was 111 min for Simulation Run 15 and 94 min for Simulation Run 16, for Airplane-Asia 93 min for Simulation Run 15 and 93 min for Simulation Run 16, for trains 312 min for Simulation Run 15 and 307 min for Simulation Run 16, and for trucks 32 min for Simulation Run 15 and 28 min for Simulation Run 16. These results indicate that the additional resources were more than adequate to increase throughput while not causing longer delays for entities to exit the intermodal center.

Table 6.39 presents the utilization of the resources for Simulation Run 16. The train lifts and stackers were utilized more as compared to Simulation Run 15 (see Table 6.35). Simulation Run 15 had 2 train lifts that were utilized 18 %, while for Simulation Run 16, 2 train lifts were utilized 28 %. Simulation Run 15 had 8 stackers that were utilized 16 %, while for Simulation Run 16, the 8 stackers were utilized 24 %. On the other hand, Simulation Run 15 had only 1 plane terminal that was utilized 23 %, while 2 plane terminals for Simulation Run 16 were utilized 16 %. Stacker utilization increased from 16 % for Simulation Run 15 to 24 % for Simulation Run 16.

Table 6.40 provides the container activity for Simulation Run 16. Simulation Run 16 had a significant increase in lifts as compared to Simulation Run 15. Total containers unloaded were 17,030 (+45 %) as compared to 11,760 for Simulation Run 15. Total containers loaded were 15,020 (+44 %) as compared to 10,500 for Simulation Run 15. The number of containers in the intermodal center at the end of the simulation was 2,009 (+59 %), up from 1,260 for Simulation Run 15.

6.4.9   Conclusion

Table 6.41 gives a comparison of the results from Baseline Simulation Run 1, Simulation Run 10, Simulation Run 11, Simulation Run 15, and Simulation Run 16. In summary, the following conclusions may be drawn:

- The current throughput (34,400 lifts in 2005) of the intermodal center can be met with considerably fewer resources than originally estimated for the Baseline Simulation Run 1 and with no reduction in container throughput (Simulation Run 10). The resources for Simulation Run 10 were 1 airplane terminal, 1 train terminal, 12 truck slots, 1 airplane lift, 1 train lift, 6 stackers, and 10 carts. Annual lifts for Simulation Run 10 were 36,720.
- The reduction in truck slots from 20 for the Baseline Simulation Run 1 to 12 for Simulation Run 10 indicates that only 12 trucks need to be inside the intermodal center at a time. This requires considerably less space and, possibly, fewer personnel.
- The container throughput can be increased considerably without any deterioration in entity times at the terminal. For Simulation Run 15, the container throughput reached 47,040 lifts annually, up from 36,720 for Simulation Run 11. Consequently, entity times at the intermodal center remained relatively constant. For example, the average airplane entity time was 93 min for Simulation Run 11 and 111 min for Run 15. The average train entity time was 312 min for Run 11 and 312 min for Simulation Run 15. The average truck entity time was 29 min for Simulation Run 11 and 32 min for Simulation Run 15.
- Resource utilization, after reducing the number of resources, was still relatively low. However, when the resources, such as stackers, were reduced below eight, the average entity times increased significantly because of higher waiting times for either a resource or a container. Future research may be warranted using Overall Equipment Effectiveness (OEE) instead of equipment utilization as a measure.

Table 6.41 Summary results

	Simulations				
	Baseline Run 1	Fewer resources (Run 10)	More entity arrivals (Run 11)	Fewer resources (Run 15)	More entity arrivals (Run 16)
Annual container lifts (estimate)					
Unloaded	18,360	18,360	23,50	23,520	34,060
Loaded	16,534	16,562	21,02	21,000	30,040
Container in yard	1,826	1,798	2,508	2,520	4,018
Total lifts	26,720	36,720	47,00	47,040	68,118
Annual entities through intermodal center (estimate)					
Planes-Europe	360	360	360	360	480
Planes-Asia	NA	NA	72	72	118
Trains	180	180	240	240	360
Trucks	4,320	4,320	5,758	5,758	8,638
Empty trucks	720	720	720	720	720
Truck with empty container	1,440	1,440	1,440	1,440	1,440
Average time in intermodal center (min)					
Planes-Europe	94	99	93	111	94
Planes-Asia	NA	NA	93	93	93
Trains	326	371	312	312	307
Trucks	43	28	29	32	28
Empty trucks	36	21	21	24	21
Trucks with empty containers	39	27	27	30	27
Resources/utilization (%)					
Plane terminals	2/9	1/20	2/11	1/23	2/16
Train terminals	3/11	1/36	2/21	2/21	3/21
Truck slots	20/7	12/5	20/6	12/7	20/8
Plane lifts	2/6	1/13	2/8	1/16	2/11
Train lifts	2/14	1/28	2/18	2/18	2/28
Stackers	8/13	6/17	8/16	8/16	8/24
Carts	20/7	10/15	20/10	12/16	20/15

- Simulation Run 16 indicated that considerably more container traffic was possible with the existing resources from Baseline Simulation Run 1. Simulation Run 16 indicates that these resources can process 68,118 lifts annually. This is a 51 % increase over the projected 2007 container traffic of 45,000 lifts.

- Resource utilization is not a good measure of the utilization of resources during the simulation. For example, when a train arrives at the train terminal, the train lifts are 100 % busy. Then after the train exits the intermodal center, these resources are idle. As a result, the average utilization is low.
- There is considerable interaction between the various submodels. Consequently, decreasing the time between arrivals of one entity might not increase container throughput. In fact, just the opposite might occur because the resources are now busy unloading an entity instead of loading another entity.

6.5 Port Security Inspection

6.5.1 Introduction

Increased security is having a significant impact on the operations of seaports resulting in longer times that ships, trains, and trucks are at container terminals. Ports are wrestling with various inspection procedures and installing equipment to minimize the container inspection times. Therefore, simulation offers an excellent approach to evaluating the impact of port activities, such as container inspections, on terminal operations in advance (Harris et al. 2008). Hence, this case study introduces a simulation model to determine the impact of various inspection scenarios on the operation of the seaport container terminal at the Alabama State Docks in Mobile, Alabama.

An approach integrating modeling and simulation with data fusion for analyzing different security aspects to improve container selection based on risk evaluations was proposed by Bocca et al. (2005) and Lewis et al. (2002). The proposed approach was aimed at understanding the balance between the number of containers undergoing security inspection and the cost of departure delays of outbound vessels and the port cost measured by the number of containers moved. Another paper (Honsi et al. 2005) focused on the training validation of an emergency response plan with traffic flow integration in the event of a disaster. A simulation model to investigate the effects of enhanced security measures on traffic flow in and around selected port gates was developed by Chatterjee (2006). Estimates were made on the length of truck queues, delays, and route alternatives. A paper published by Kerr (2006) developed a framework for managing freight data electronically via the Internet, while Koch (2007) created PortSim, a port security simulation and visualization tool. The tool allows a user to investigate special parameters to determine the impact of those parameters on port operations and costs. An analysis tool for safety and security developed by Berkowitz and Bragdon (2006) depicts air-land-seaport access and potential vulnerabilities in a virtual real-time format. This tool allows for the development of surface and underwater scenes in order to evaluate incident response training and transportation security systems. Sandia National Laboratory developed several simulation models to assist ports in the conduct of cost/benefit tradeoffs of various security measures, such as increased inspections or more scanners, on the movement of containers through a port.

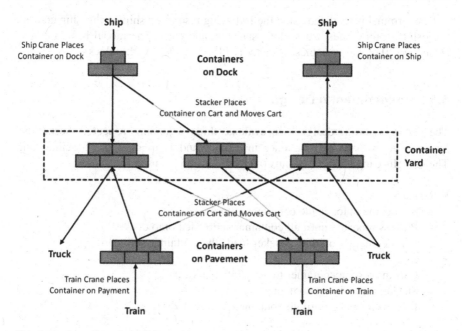

Fig. 6.12 Overview of container traffic

6.5.2   Simulation Model

The ProModel-based simulation model in this case study was initially written to determine the impact of various inspection sampling plans on the container through-put at the Alabama State Docks. The initial model was constructed following the previously presented conceptual framework. This conceptual framework consists of a number of submodels that run independently of one another. Each model has unique data input and entities defined by specific attributes. Data are shared between the submodels utilizing global variables. The content of the global variables can be altered within any submodel with the new values immediately shared with any other submodel. These global variables can also be used in logic statements to control the movement and routing of entities, branching logic, and updating of entity attributes. ProModel has the following submodels.

- Loading/unloading containers:
  - Ship (entity = ship)
  - Train (entity = train)
  - Truck (entity = truck, empty truck and empty truck with container) (Fig. 6.12):
- Movement containers from:
  - Ship dock to container yard (entity = move order1)
  - Container yard to ship dock (entity = move order2)
  - Train pavement to container yard (entity = move order3)
  - Yard to train pavement (entity = move order4)

The terminal is modeled using the following resources: ship berths, ship cranes, train slots, train cranes, truck slots, stackers, and carts. The model has 13 entity attributes, 20 global variables, 70 activity blocks, and 10 entity blocks.

6.5.3 Experimental Design

The experiment was designed to evaluate three unloading/inspecting scenarios. *Scenario A* involves no container inspection and is used as the baseline run. The modeling logic used for ships (and similar logic is used for trains) is:

- Unload
  - Ship crane unloads the container onto dock.
  - Process repeated until all containers unloaded onto dock.
  - After container on dock stacker loads the container onto cart.
- Load
  - Cart moves the container to the container yard.
  - Stacker unloads the container.
  - Process repeated until all containers are moved.

Scenario B consists of unloading a container and then immediately inspecting the container before another container is unloaded. This protocol forbids unloading of another container until the prior container is inspected. The ship logic is:

- Unload
  - Security inspector checks ship's paperwork.
  - Ship crane unloads the container onto dock.
  - Security inspector inspects the container.
  - Ship crane then unloads another container onto dock (unloading is not continued until previous container has been inspected).
  - Process repeated until all containers unloaded.
- Load
  - After container is inspected, stacker loads container onto cart
  - Cart moves the container to the container yard.
  - Stacker unloads the container.
  - Process repeated until all containers are moved.

Scenario C involves the inspection of containers independent of unloading a ship. Unloading containers continues unabated with inspections performed prior to moving the container to the container yard. The ship logic is:

- Unload
  - Security inspector checks ship's paperwork.
  - Ship crane unloads the container onto dock (continue unloading even though containers are not inspected).
  - Process repeated until all containers unloaded.

Table 6.42 Experimental design

Simulation	Scenario	Description
Run 1	A	No container inspection (baseline run)
Run 2	B	100 % inspection of incoming containers
Run 3	B	80 % inspection of incoming containers
Run 4	B	60 % inspection of incoming containers
Run 5	C	Container inspection independent of unloading

- Load
  - After container on dock security inspector inspects container.
  - Stacker places the container on cart.
  - Cart moves the container to the container yard.
  - Stacker unloads the container.
  - Process repeated until all containers are moved.

Table 6.42 presents the experimental design. Scenario A is the Baseline Simulation Run with no container inspection. Three simulation runs are used with Scenario B. An inspection rate of 100 % is used in Simulation Run 2, 80 % in Simulation Run 3, and 60 % in Simulation Run 4. In Scenario C, the inspection is decoupled from container unloading; and all containers are inspected independent of unloading from a ship.

6.5.3.1 Scenario A: No Container Inspection
The input data for Simulation Run 1 are shown in Table 6.43. In addition, the input data consisted of:

- Two ship berths for unloading and loading containers
- Two train terminals for unloading and loading containers
- Twenty truck slots (maximum number of trucks in terminal at one time)
- Two ship cranes for unloading and loading containers from planes
- Two train cranes for unloading and loading containers from trains
- Twelve stackers for unloading and loading containers from trucks and onto and off carts
- Twenty carts for moving containers throughout the terminal
- Two minutes to unload or load a container from plane, train, or truck
- $T(15,20,25)$ minutes to position a ship at a terminal (T = triangular distribution)
- $T(15,20,25)$ minutes to position a train at a terminal
- $T(4,5,6)$ minutes to position a truck for unloading or loading
- Two minutes to process paperwork to load a plane, train, or truck
- $T(4,5,6)$ minutes for plane, train, or truck to exit terminal
- Two minutes to unload and load a cart
- $T(4,5,6)$ minutes to move a cart between a plane, train, or truck and the container yard

Table 6.43 Entity parameters

Entity	Time between arrivals (min)	
Ship	T(1320;1440;1560)	
Train	T(420;480;540)	
Empty train	T(2080;2320;2560)	
Truck with full container	T(54,60,66)	
Empty truck	T(90,120,150)	
Truck with empty container	T(180,240,300)	
Entity	Containers in	Containers out
Ship	T(400,450,500)	T(200,250,300)
Train	T(90,100,110)	T(90,100,150)
Empty train	0	T(90,100,150)
Truck with full container	1	81 % leave with container
		10 % leave with no container
		9 % leave with empty container
Empty truck	0	1
Truck with empty container	1	81 % leave with container
		10 % leave with no container
		9 % leave with empty container

Table 6.44 Scenario results

Entity	Time between arrivals (min)	
Ship	T(1320,1440,1560)	
Train	T(420,480,540)	
Empty train	T(2080,2320,2560)	
Truck with full container	T(54,60,66)	
Empty truck	T(90,120,150)	
Truck with empty container	T(180,240,300)	
Entity	Containers in	Containers out
Ship	T(400,450,500)	T(200,250,300)
Train	T(90,100,110)	T(90,100,150)
Empty train	0	T(90,100,150)
Truck with full container	1	81 % leave with container
		10 % leave with no container
		9 % leave with empty container
Empty truck	0	1
Truck with empty container	1	81 % leave with container
		10 % leave with no container
		9 % leave with empty container

The simulation model runs 1,440 h, or 180 eight-hour days, which is 6 months. The results of Simulation Run 1 are presented in Table 6.44. Several interesting observations can be made with reference to Simulation Run 1:

• The utilization rate is relatively high for ship berths and cranes at 68 % and 67 %, respectively.

- The utilization rate is relatively high for train slots and train cranes at 71 % and 72 %, respectively.
- The utilization rate is very low for tugs, at 1 %, indicating one less tug may be possible.
- The utilization rate is very low for truck slots (the maximum number of allowed trucks in the terminal at one time is 20). It may be possible to reduce this resource, freeing up space for other terminal operations.
- There are currently more stackers than required since the average utilization is 34 %.

6.5.3.2 Scenario B: Container Sampling, Simulation Runs 2–4

In Simulation Runs 2, 3, and 4, inspection times were modified as shown below:

- T(10,15,20) minutes for inspector to check paperwork before unloading of containers from ship or train
- T(2,3,4) minutes for inspector to check paperwork and container from the truck
- Three minutes for inspector to inspect a container from a ship, train, or truck
- Five inspectors available to inspect the containers

Tables 6.45, 6.46 and 6.47 present the results for Simulation Runs 2, 3, and 4 with the associated varying percentages of container inspections. The simulation models run for 1,440 h or 180 eight-hour days.

Ships processed through the terminal dropped from 59 with no container inspection in Simulation Run 1 to 53 with 100 % inspection in Simulation Run 2, a 10 % reduction in throughput. Likewise, the time a ship stayed in the terminal showed a significant increase from 2,013 min with no container inspection to 7,258 min with 100 % inspection, an increase of 260 %. Trains processed through the terminal dropped from 180 with no container inspection to 166 with 100 % inspection, a 7 % reduction. The time a train spent at the terminal increased from 684 min with no container inspection to 3,952 with 100 % inspection, an increase of 477 %. The truck throughput remained relatively constant (1,440 to 1,441); however, the time a truck was in the terminal increased 96 % from 26 min with no container

Table 6.45 Scenario B results: entity time at terminal

Sampling entity	Simulation Run 2, 100 %		Simulation Run 3, 80 %		Simulation Run 4, 60 %	
	Quantity	Time (min)	Quantity	Time (min)	Quantity	Time (min)
Ships	53	7,258	57	3,677	59	2,78
Trains	166	3,952	176	1,966	179	987
Empty trains	37	593	37	532	38	502
Trucks	1,441	51	1,439	34	1,439	32
Empty trucks	724	42	720	23	717	21
Trucks with empty containers	358	45	361	28	360	27

Table 6.46 Scenario B results: resource utilization

Resources	Quantity	Simulation Run 2	Simulation Run 3	Simulation Run 4
Sampling		100 %	80 %	60 %
		Utility	Utility	Utility
Ship berths	2	99 %	99 %	90 %
Ship cranes	2	98 %	98 %	93 %
Tugs	2	1 %	1 %	1 %
Train slots	2	99 %	99 %	94 %
Train cranes	2	96 %	95 %	91 %
Truck slots	2	6 %	4 %	3 %
Stackers	12	32 %	34 %	35 %
Carts	20	48 %	51 %	53 %
Inspectors	5	56 %	54 %	49 %

Table 6.47 Scenario B results: full container throughput

Entity	Simulation Run 2	Simulation Run 3	Simulation Run 4
Sampling	100 %	80 %	60 %
0 Ship			
In	21,707	23,528	24,309
Out	13,454	14,192	14,631
Yard	924	886	852
1 Train			
In	16,718	17,710	18,094
Out	21,842	23,011	23,511
Yard	1,184	1,879	1,956
2 Trucks			
In	1,441	1,440	1,440
Out	2,176	2,169	2,147
Yard	279	526	729
Total			
In	39,866	42,678	43,843
Out	37,472	39,372	40,289
Yard	2,384	3,291	3,537

inspection to 51 min with 100 % inspection. The containers processed through the terminal decreased from 43,729 with no container inspection in Simulation Run 1 to 39,866 with 100 % inspection, an 8 % decrease in throughput.

Ships processed through the terminal increased 11 % from 53 with 100 % inspection in Simulation Run 2 to 59 with 60 % inspection in Simulation Run 4. Note that 59 ships with no inspection were processed through the terminal in the baseline run. The time for a ship at the terminal decreased 61 % from 7,258 min for 100 % inspection to 2,778 min with 60 % inspection. It is interesting to note that the time a ship was in the terminal with no inspection was 2,013 min.

The results for trains were similar to that for ships. Trains through the terminal increased 7 % from 166 with 100 % inspection in Simulation Run 2 to 179 with 60 % inspection in Simulation Run 4. The trains processed through the terminal were 180 with no inspection in Simulation Run 1. The time a train spent at the terminal decreased from 3,952 min with 100 % inspection to 987 min with 60 % inspection, a 75 % reduction. With no inspection in the Baseline Simulation Run, the time a train spent in the terminal was 684 min.

The utilization of resources remained fairly constant during the simulation runs with varied sampling rates. The total quantity of containers unloaded increased from 39,866 in Simulation Run 2 with 100 % inspection to 43,843 in Simulation Run 4 with 60 % inspection, a 9 % increase.

6.5.3.3 Scenario C: Container Inspection After Unloading Simulation Run 5

Table 6.48 shows the results for Simulation Run 5 for scenario C. The simulation models run for 1,440 h or 180 eight-hour days. Surprisingly the results were identical to the Baseline Simulation Run with no container inspection. Also surprising was that the security inspection did not delay the loading of containers onto ships, trains, and trucks. Unfortunately ProModel does not have the necessary detail logic to uniquely identify a container in the terminal and to assign the container for loading on a specific entity. As a result, as long as the containers are in the container yard, the loading continues.

Table 6.48 Scenario C results

Entities through terminal	Quantity	Time (min)	Value-added time (min)
Ships	59	2,007	1,352
Trains	180	695	450
Empty trains	38	430	255
Trucks	1,442	33	20
Empty trucks	719	21	12
Truck with empty container	360	27	14
Resource	Quantity	Utilization	
Ship berths	2	67 %	
Ship cranes	2	67 %	
Tugs	2	1 %	
Train slots	2	72 %	
Train cranes	2	72 %	
Truck slots	20	3 %	
Stackers	12	34 %	
Carts	20	52 %	
Inspectors	5		
Containers	Unloaded	Loaded	At terminal
Ships	23,669	14,786	674

6.5.4   Analysis

Figures 6.13, 6.14, 6.15 present bar chart graphs of the time ships, trains, and trucks, respectively, spent in the terminal. The total time entities spent in the terminal for Scenario B were considerably greater than for Scenario A Simulation Run 1. However, with Scenario C, the time in the terminal and the quantity of containers unloaded were almost identical to the Baseline Simulation Run conditions.

   Figure 6.15 presents the quantity of containers unloaded at the terminal for Simulation Runs 1–5.

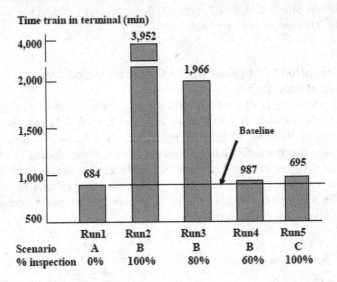

Fig. 6.13 Time ship was in terminal

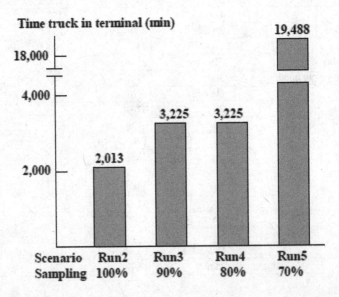

Fig. 6.14 Time truck was in terminal

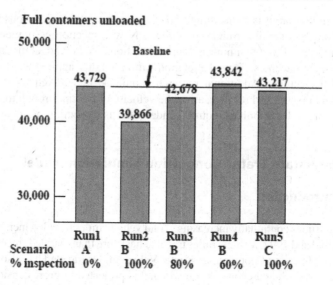

Fig. 6.15 Containers unloaded

6.5.5 Conclusion

In summary, the following conclusions may be drawn:

- Any inspection plan for containers that includes inspection as a part of the unloading operation, such as that described in Scenario B, increased the time for entities at the terminal. For example, 100 % inspection of all incoming containers increased the time a ship was at the terminal by 260 %, a train by 477 %, and a truck by 96 % (Simulation Run 2 versus no inspection for Simulation Run 1). A 60 % sampling plan of incoming containers increased the time a ship was at the terminal by 38 %, a train by 44 %, and a truck by 20 % (Simulation Run 4 versus no inspection for Simulation Run 1).
- Decoupling the container inspection from the unloading of the container minimized the impact of the inspection. The inspection Scenario C for Simulation Run 5 resulted in entity times identical to the Baseline Simulation Run 1 with no inspection. The time a ship was at the terminal was 2,007 min for Simulation Run 5 as compared to 2,013 min for Simulation Run 1. The time a train was at the terminal was 695 min as compared to 684 for Simulation Run 1. The time a truck was at the terminal was 33 min as compared to 26 min for Simulation Run 1. It can be assumed that the decoupled inspection process might require resources similar to the in-process inspections described in Scenario B.
- The ProModel model that was previously developed to simulate a container intermodal center was easily and rapidly modified to include the container inspection logic.

In conclusion, container inspection scenarios are critical in minimizing delays at the container terminal. It is obvious that any sampling protocol must be decoupled as much as possible from the actual unloading of containers.

Using simulation, it is rather simple to evaluate the impact of various sampling scenarios on the overall terminal operations. New inspection equipment is constantly being introduced that improves the inspection process and, at the same time, reduces inspection times. Thus, simulation can be readily applied to evaluate the impact of inspections. Additional research should be undertaken where specific containers can be tracked for measuring the velocity of freight through the terminal and determining the resources required under various inspection protocols.

6.6   Interstate Traffic Congestion Simulation Model

6.6.1   Introduction

Interstate traffic is continually increasing. Total vehicle miles traveled increased 63 % between 1980 and 1997. Vehicle miles traveled has more than doubled since 1970 and has exceeded the rate of population growth. Vehicle miles traveled has also outpaced employment growth and economic growth and is projected to grow considerably in the future. The US Federal Highway Administration projects that vehicle miles traveled will grow at an annual rate of approximately 2.16 % over the next 20 years, resulting in a 24 % growth in 10 years and a 53 % growth in 20 years (FHWA 1996; FHWA 1997; USDOT 1998). Thus traffic planners are concerned that interstate traffic (e.g., I-65) will soon be reaching the point where the vehicle to capacity ratio exceeds 75 %, at which point the interstate will be considered congestion. The objective of this case study is to determine the congestion point as traffic increases on I-65N. This objective is an extension of a previous case study (Harris et al. 2010), and much greater increases in peak hourly traffic will be evaluated. The second objective is to evaluate the feasibility of adding lanes at congestion points.

The north lanes of I-65 from Montgomery to Birmingham were used in this study. Figure 6.16 gives the peak hourly traffic on I-65N from Montgomery to

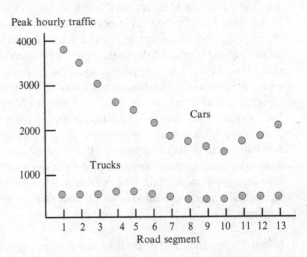

Fig. 6.16 Peak hourly traffic

Birmingham. Note the continual drop in peak traffic as traffic moves north from Montgomery reflecting the commuter traffic going north out of Montgomery. Also peak traffic begins to increase for segments closer to Birmingham. Truck peak hourly traffic remained fairly constant for all road segments.

6.6.2 Process Model

The simulation model was written in ProModel and is a modification to the previous model used to simulate the traffic on Interstate I-65 north (Harris et al. 2010). Since this model is a modification of a previous model, no verification and validation was necessary. The conceptual framework for the ProModel model is documented by Schroer et al. (2009). In Fig. 6.17, the ProModel model for the typical roadway segment is shown. Segment i is a ProModel activity block. The remaining four blocks, $D1i$, $D2i$, $D3i$, and $D4i$, are also ProModel activity blocks and are necessary for routing car and truck entities. Each segment has two label blocks for displaying data in real time during the simulation.

The logic embedded in each segment, or ProModel activity, is as follows: If the entity Vehicle is a car, then the car will wait until the Global Variable Traffic i is less than the Global Variable Capacity $i\text{-}1$. If true, the Global Variable Traffic i will increase by one indicating that a slot on the roadway Segment i is in use. The ProModel logic for calculating segment speed is:

- Speed $Si = 70\,\text{mph}—(20\,\text{mph} * \text{segment density})$, where segment density = number of segment slots in use/total slots in segment.

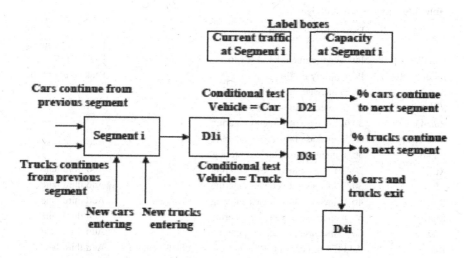

Fig. 6.17 ProModel model for typical roadway segment

The free-flowing speed is 70 mph. If the segment density is one, then the speed is 50 mph. The total slots in a segment are:

- Number of lanes × segment length in miles × 45 slots per lane per mile.

The time the car is in the roadway segment, or the time to travel the segment, is computed by Ti. This time follows the triangular distribution with mean Ti, min value $0.90 \times Ti$, and max value $1.05 \times Ti$. After the car has traveled the roadway segment, the Global Variable Traffic i will decrease by one. The logic is similar if the entity is a truck. The segment travel time in minutes is:

- Segment length Li in miles/segment speed Si in mph × 60 min/h.

Passenger car and truck entities arrive separately at the first roadway segment. Each segment has a roadway capacity based upon the segment length and the number of lanes. The model was constructed such that a truck is equal to 2.5 passenger cars, which is consistent with the *Highway Capacity Manual* for rolling terrain (TRB 2000).

6.6.3 Experimental Design

Table 6.49 gives the experimental design. The Baseline Simulation Run 1 simulates the current peak hourly traffic going north from Montgomery. At peak traffic, 3,770 cars and 419 trucks arrive every hour at the start of the ProModel simulation model.

Table 6.49 Experimental design

Simulation	Interstate segment E173–E176	Interstate segment E181–E186
Baseline Run 1	Peak hourly traffic	Peak hourly traffic
Run 2	10 % increase in vehicles entering	
Run 3	20 % increase in vehicles entering	
Run 4	30 % increase in vehicles entering	
Run 5	40 % increase in vehicles entering	
Run 6	50 % increase in vehicles entering	
Run 7	20 % increase in vehicles entering	Add third lane
Run 8	30 % increase in vehicles entering	Add third lane
Run 9	40 % increase in vehicles entering	Add third lane
Run 10	50 % increase in vehicles entering	Add third lane
Run 11	60 % increase in vehicles entering	Add third lane
Run 12	Add fourth lane and 50 % increase in vehicles entering	Add third lane
Run 13	Add fourth lane and 60 % increase in vehicles entering	Add third lane

Of these numbers, 840 cars and 110 trucks use Exit 173. Therefore, 2,930 cars and 309 trucks actually enter Segment E173–E176.

Simulation Runs 2–6 increase the traffic entering the northbound roadway starting at Segment E173–E176. The traffic increased by 10 %, 20 %, 30 %, 40 %, and 50 % starting with Simulation Run 2.

Simulation Runs 7–11 included the addition of a third lane at the congestion point, Segment E181–E186. This is the segment where the Interstate narrowed from three to two lanes. The traffic increased by 20 %, 30 %, 40 %, 50 %, and 60 % starting with Simulation Run 7.

Simulation Runs 12–13 included the addition of a third lane at Segment E181–E186 and also a fourth lane at a second congestion point, Segment E173–E176. The traffic increased by 50 % and 60 % starting with Simulation Run 12.

6.6.4 Baseline Simulation Run 1 Results

Table 6.50 gives the Baseline Simulation Run 1 results. No congestion occurred at any of the road segments. All volume/capacity ratios were less than 64 % with many of the rural segments below 50 %. Congestion is defined as when the volume/capacity ratio equals or exceeds 75 %.

6.6.4.1 Simulation Runs 2–6: Results

Table 6.51 gives the congestion statistics for Simulation Runs 2–6. Congestion reached 74 % at Segment E181–186 for Simulation Run 3. This is the first segment just after the interstate is reduced from three to two lanes. All other road segments for Simulation Run 3 are well below the congestion levels. Congestion reached 80 % at Segment E181–E186 for Simulation Run 4 and 86 % at the same segment for Simulation Run 5. Congestion reached 100 % at segment E181–E186

Table 6.50 Baseline Run 1 results

Segment	Lanes	Length (miles)	Average speed (mph)	Volume/capacity (%)
E173–E176	3	3	60	52
E176–E179	3	3	62	46
E179–E181	3	2	63	41
E181–E186	2	5	58	64
E186–E200	2	14	62	54
E200–E205	2	5	61	48
E205–E208	2	3	62	44
E208–E212	2	4	63	39
E212–E219	2	7	65	36
E219–E228	2	9	64	41
E228–E231	2	3	53	54
E231–E238	2	7	61	53

Table 6.51 Simulation Runs 2–6 volume/capacity ratio

Road segment	Volume/capacity (%)				
	Simulation Run 2	Simulation Run 3	Simulation Run 4	Simulation Run 5	Simulation Run 6
Increase in traffic (%)	*10*	*20*	*30*	*40*	*50*
E173–E176	55	66	67	72	79
E176–E179	49	54	57	62	68
E179–E181	43	47	50	54	57
E181–E186	68	*74*	*80*	*86*	*100*
E186–E200	56	61	65	69	72
E200–E205	50	54	57	61	62
E205–E208	45	49	52	54	55
E208–E212	40	43	45	47	48
E212–E219	37	40	41	43	44
E219–E228	42	45	47	48	49
E228–E231	54	57	59	60	61
E231–E238	53	57	59	60	61

Table 6.52 Simulation Runs 2–6 average speeds

Road segment	Average speed (mph)				
	Simulation Run 2	Simulation Run 3	Simulation Run 4	Simulation Run 5	Simulation Run 6
Increase in traffic (%)	*10*	*20*	*30*	*40*	*50*
E173–E176	60	58	57	56	54
E176–E179	61	60	59	58	57
E179–E181	62	61	60	59	58
E181–E186	57	55	54	53	47
E186–E200	62	61	59	59	58
E200–E205	61	60	59	58	58
E205–E208	61	61	60	59	59
E208–E212	63	62	62	62	61
E212–E219	64	64	63	62	63
E219–E228	64	63	63	63	62
E228–E231	53	53	53	53	53
E231–E238	61	60	60	59	59

(Simulation Run 6) which indicates that the traffic arrival rates exceed the service rates. Consequently, vehicles had to wait before entering the network.

Table 6.52 gives the average speeds for Simulation Runs 2–6. As congestion increased, the average speed decreased. For example, for Simulation Run 3, the congestion was 74 % at Segment E181–E186. The corresponding average speed was 55 mph.

6.6.4.2 Simulation Runs 7–11 Results

Simulation Runs 7–11 included adding a third lane at congested Segment E181–E186. Table 6.53 gives the congestion for Simulation Runs 7–11. As anticipated, congestion dropped at Segment E181–E186 to 57 % for Simulation Run 9 (+40 % increase in traffic) as compared to 86 % for Simulation Run 5 (+40 % increase in traffic). Adding a third lane at segment E181–E186 allowed a 50 % increase in traffic (Simulation Run 10) before congestion of 79 % occurred at Segment E173–E176, as compared to 74 % congestion for Simulation Run 3 (+20 % increase in traffic). Congestion reached 100 % (Simulation Run 11) for Segment E181–E186. As stated previously, at 100 % congestion, the ProModel simulation model becomes unstable because the traffic arrival rates are greater than the service rates (Table 6.54).

Table 6.53 Runs 7–11 volume/capacity ratios

Volume/capacity (%)					
Segment	R7	R8	R9	R10	R11
Increase in traffic	*20%*	*30%*	*40%*	*50%*	*60%*
E173–E176	62	67	72	79	87
E176–E179	54	58	62	68	74
E179–E181	47	50	54	57	62
E181–E186	50	53	57	72	100
E186–E200	61	66	69	72	73
E200–E205	54	57	61	62	64
E205–E208	49	52	54	55	57
E208–E212	43	45	47	48	50
E212–E219	40	41	43	43	44
E219–E228	45	47	48	49	50
E228–E231	57	59	60	61	62
E231–E238	57	59	60	61	62

Table 6.54 Runs 7–11 average speed

Average speed	R7	R8	R9	R10	R11
Increase in traffic	*Increase in traffic*	*Increase in traffic*	*Increase in traffic*	*Increase in traffic*	*Increase in traffic*
E173–E176	E173–E176	E173–E176	E173–E176	E173–E176	E173–E176
E176–E179	E176–E179	E176–E179	E176–E179	E176–E179	E176–E179
E179–E181	E179–E181	E179–E181	E179–E181	E179–E181	E179–E181
E181–E186	E181–E186	E181–E186	E181–E186	E181–E186	E181–E186
E186–E200	E186–E200	E186–E200	E186–E200	E186–E200	E186–E200
E200–E205	E200–E205	E200–E205	E200–E205	E200–E205	E200–E205
E205–E208	E205–E208	E205–E208	E205–E208	E205–E208	E205–E208
E208–E212	E208–E212	E208–E212	E208–E212	E208–E212	E208–E212
E212–E219	E212–E219	E212–E219	E212–E219	E212–E219	E212–E219
E219–E228	E219–E228	E219–E228	E219–E228	E219–E228	E219–E228
E228–E231	E228–E231	E228–E231	E228–E231	E228–E231	E228–E231
E231–E238	E231–E238	E231–E238	E231–E238	E231–E238	E231–E238

6.6.5   Analysis

Simulation Runs 12–13 included adding a third lane at congested Segment E181–E186 and also a fourth lane at congested Segment E173–E176. Table 6.55 gives the volume/capacity ratios for Simulation Runs 12–13. For Simulation Run 12 (50 % increase in traffic), there was no congestion. However, congestion was 73 % for Segment E181–E186 and 72 % for Segment E186–E200. Note that Segment E186–E200 was the first segment where the number of lanes was reduced from three to two lanes. For Simulation Run 13 (60 % increase in traffic), congestion was 100 % at E181–E186. The system became unstable because traffic arrival rates were greater than the service rates.

Tables 6.56, 6.57, and 6.58 give the four road segments that had congestion greater than or equal to 65 %. All congestion generally occurred at segments E181–E186 and E173–E176. The remaining road segments had considerably lower volume/capacity ratios.

Table 6.55 Volume/capacity ratio

	Volume/capacity (%)	
Road segment	Simulation Run 12	Simulation Run 13
Increase in traffic	*50 %*	*60 %*
E173–E176	59	65
E176–E179	68	74
E179–E181	57	62
E181–E186	73	*100*
E186–E200	72	73
E200–E205	62	64
E205–E208	55	57
E208–E212	48	49
E212–E219	44	44
E219–E228	49	50
E228–E231	61	62
E231–E238	61	62

Table 6.56 Volume/capacity ratios greater than or equal to 65 %

Road segment	Increase in peak traffic (%)				
	Simulation Run 2	Simulation Run 3	Simulation Run 4	Simulation Run 5	Simulation Run 6
Increase in traffic	*20 %*	*30 %*	*40 %*	*50 %*	*60 %*
E173–E176	–	–	67 %	72 %	79 %
E176–E179	–	–	–	–	68 %
E179–E181	68 %	74 %	80 %	86 %	*100 %*
E181–E186	–	–	65 %	69 %	72 %
E186–E200	–	–	67 %	72 %	79 %

Table 6.57 Adding 3rd lane

Road segment	Adding third lane at E181–E186				
	Simulation Run 7	Simulation Run 8	Simulation Run 9	Simulation Run 10	Simulation Run 11
Increase in traffic	*10 %*	*20 %*	*30 %*	*40 %*	*50 %*
E173–E176	–	67 %	72 %	79 %	87 %
E176–E179	–	–	–	68 %	74 %
E179–E181	–	–	–	72 %	*100*
E181–E186	–	–	69 %	72 %	73 %

Table 6.58 Adding 3rd and 4th lane

Road segment	Adding third lane at E181–E186 and fourth lane at E173–E176	
	Simulation Run 12	Simulation Run 13
Increase in traffic	*50 %*	*60 %*
E173–E176	–	65 %
E176–E179	68 %	74 %
E181–E186	73 %	100 %
E186–E200	72 %	73 %

6.6.6 Conclusions

The following conclusions can be drawn:
For Simulation Runs 2–6:

- Peak hourly traffic can increase slightly more than 20 % (Simulation Run 3) before congestion occurs at Segment E181–E186. This segment is the first road segment after the interstate is reduced from three to two lanes (congestion occurs when the volume-to-capacity ratio reaches 75 %).
- The only bottleneck is at the E181–E186 road segment.
- Many of the rural segments have congestion below 50 %.

Simulation Runs 7–11 (adding a third lane at Segment E181–E186):

- Adding a third lane at Segment E181–E186 allowed a 40 % increase in traffic before congestion occurred (Run 10). Congestion occurred at Segment E173–E176, the first northbound segment of the model.
- Congestion at Segment E181–E186 was reduced to 72 % for Run 10 (40 % increase in traffic), down from 74 % for Run 3 (20 % increase in traffic).

Simulation Runs 12–13 (adding a third lane at Segment E181–E186 and a fourth lane at Segment E173–E176):

- There was no congestion for Run 12 (50 % increase in traffic). However, congestion was 73 % for Segment E181–E186 and 72 % for Segment E186–E200. Segment E186–E200 was the first segment where the lanes were reduced from three to two.
- Congestion was 100 % at E181–E186 for Run 13 (60 % increase in traffic). The system became unstable because traffic arrival rates were greater than the service rates. Interestingly, congestion at E176–E179 reached 74 %. This was then the first segment where the number of lanes was reduced from four to three.

In summary, it appears that adding a third lane at Segment E181–E186 would allow for a 50 % increase in peak traffic (Simulation Run 10) for the I-65 north from Montgomery to Birmingham. The USDOT projected a 53 % growth in vehicle miles traveled over the next 20 years. The resulting 79 % congestion at E173–E176 for Simulation Run 10, which was the first model segment, indicated that this may be the maximum traffic increase without increasing entry lanes.

6.7   Interstate Tunnel Traffic Simulation Model

6.7.1   Introduction

The interstate system in the USA is experiencing rapid growth in truck traffic. One reason for this increase is the globalization of international trade. The growth in commuter and truck traffic is significantly increasing the congestion at the I-10 tunnel crossing the Mobile River in Mobile, AL. In addition to the overall effects of growing international trade, area-specific growth of container shipments is occurring at the Mobile Container Terminal at the Port of Mobile, AL. The majority of containers at the recently expanded Port of Mobile are arriving and departing on trucks. Figure 6.18 is a map of the Wallace tunnel on I-10 that crosses the Mobile River in Downtown Mobile, AL.

The Mobile Container Terminal is approximately two miles south of the tunnel and adjacent to I-10. Eastbound truck traffic exiting the tunnel continues on the Jubilee Parkway (I-10) across Mobile Bay. The Jubilee Parkway is a 7.5-mile girder bridge. Depending on the destination, westbound truck traffic exiting the tunnel will stay on I-10 towards Mississippi, take I-10 to I-65 north, or exit at Water Street and travel south to the Mobile Container Terminal or north to I-65 via I-165.

The increase in commuter and truck traffic is significantly increasing the congestion through the Mobile tunnel. A number of alternatives have been suggested to reduce this congestion. Some alternatives that have been proposed are (1) rerouting passenger traffic, (2) encouraging carpooling, and (3) deploying passenger car ferries or transit (bus or rail). Therefore, this case study describes the use of simulation to initiate the evaluation of these alternatives.

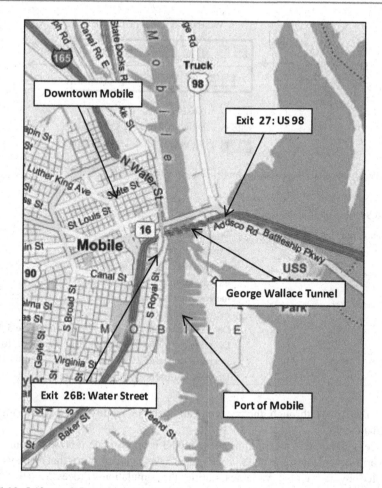

Fig. 6.18 I-10 trough Downtown Mobile, AL

6.7.2   Simulation Model

In the following figure, the conceptual framework for the tunnel simulation model is given. Since traffic moves freely in the off-peak direction, the model only simulates the traffic moving in one direction in the tunnel. The simulation model was implemented in ProModel (Fig. 6.19).

6.7.3   Verification and Validation

Model verification can be defined as determining if the model is correctly represented in the simulation code. Verification is accomplished by eliminating all variation in the model and only using constants for all arrival times and service

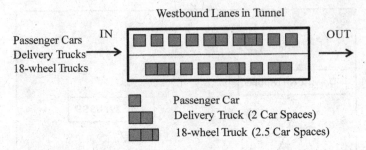

Fig. 6.19 Conceptual framework for tunnel simulation framework

Table 6.59 Experimental design

Simulation run	Description
Baseline	Existing traffic volumes
2	Increase traffic volume to 60 %
3	Increase traffic volume to 65 %
4	Increase truc traffic 5 %
5	Increase truc traffic 10 %
6	Decrease car traffic 5 % and increase truc traffic 5 %
7	Decrease car traffic 5 % and increase truc traffic 10 %
8	Decrease car traffic 10 % and increase truc traffic 15 %
9	Decrease car traffic 10 % and increase truc traffic 20

times. The times through the system can then be readily compared with the input data.

Model validation is determining if the model is an accurate representation of the real-world system (Harris et al. 2008). ProModel has a "label" block that displays data generated by the global variables during the simulation (ProcessModel 1999). By slowing the simulation down, it is possible to observe these values as the entities move through the simulation. A group of transportation experts were placed in front of the computer to observe the model operation. The model only simulated the peak hourly traffic in one direction through the tunnel. A total of 11 % of the daily traffic volume, or 6,560 vehicles, occurred during the peak hour. Fifty-five percent of the peak hourly traffic, or 3,608 vehicles, moved in one direction. The simulation model was run for 1 h to reach steady state and then for another 8 h. The average hourly traffic volume was 3,610 vehicles and compares to the actual volume of 3,608 vehicles.

6.7.4 Experimental Design

Table 6.59 defines the experiment.

Simulation Run 1 was the Baseline Simulation Run, used to simulate existing traffic volumes during peak hour. The second set of simulation runs, Simulation Runs 2 and 3, increased the directional traffic split from 55 % to 60 % and 65 %

while maintaining the 11 % of the total daily traffic. The third set of simulation runs, Simulation Runs 4 and 5, increased truck traffic by 5 % and 10 % while keeping other traffic volumes constant. The fourth set of simulation runs, Simulation Runs 6 and 7, decreased passenger car traffic by 5 % and increased truck traffic by 5 % and 10 %, respectively. Simulation Runs 8 and 9 decreased passenger car traffic by 10 % and increased truck traffic by 15 % and 20 %.

6.7.5   Baseline Simulation Run

The inputted data for the Baseline Run 1 consisted of:

- 1.2-mile tunnel length from US 90/98, Exit 27, through the tunnel to Water Street, Exit 26B. The actual tunnel length is approximately one-half mile. However, once a vehicle passes one of the above exits, the vehicles are committed to proceed through the tunnel. Therefore, for modeling purposes, the length of the tunnel is 1.2 miles.
- Four lanes of traffic in tunnel, two lanes in each direction.
- Assumed speed of 55 mph maximum through the tunnel.
- 59,630 daily volume of vehicles through the tunnel.
- 11 % of daily volume occurs during peak hours.
- 55 %/45 % directional split for peak hourly traffic.
- Daily percentage of truck traffic is 15 % of total traffic volume. During peak hours, the percentage of truck traffic is 11 %.
- 10 % of delivery truck traffic occurs during peak hours.
- 79 % of passenger traffic occurs during peak hours.
- The *Highway Capacity Manual* lists the maximum density for a basic freeway section as 45 passenger cars per mile per lane, which translates to 117 feet per passenger car per mile.
- Vehicle time in tunnel follows a triangular distribution with parameters of 1.243, 1.309, and 1.374 min (based on 55 mph).

It should be noted that the model was run based on the peak hourly traffic and, consequently, models the worst case scenario. Traffic is less during the nonpeak hours and congestion would also be less. The sources of the input data are the Alabama Department of Transportation and the South Alabama Regional Planning Commission.

The traffic volume-to-capacity ratio for the tunnel is defined as the number of vehicles in the tunnel divided by the tunnel capacity for vehicles. The volume-to-capacity ratio is a standard measure used to quantify congestion. A volume-to-capacity ratio of more than 90 % indicates a deficient condition, or congestion, on that segment of highway, according to Alabama Department of Transportation specifications (UAH 2005).

The capacity of vehicles in the tunnel is defined as the number of car spaces, or car slots, in the tunnel. A car space is assumed to be 117 feet. This results in a total of

$$31.2 \text{ miles (length of tunnel)} \times 2 \text{ lanes per tunnel}$$
$$117 \text{ feet per car slot} = 108 \text{ car slots}$$

The size and operating characteristics of trucks cause them to require more space than passenger cars. Since all vehicles must travel on up-/downgrades when using the tunnel, the assumption was made to treat one delivery truck space as equal to 2 passenger cars and one 18-wheel truck space as equal to 2.5 passenger cars (which is consistent with the *Highway Capacity Manual* for rolling terrain). The ProModel simulation tool has a global variable named "Capacity" that is incremented as vehicle spaces are in use (i.e., as vehicles enter the tunnel) and decremented as vehicles exit the tunnel.

6.7.5.1 Simulation Baseline Run 1: Results

Baseline Run 1 had a 2-h warm-up and ran for 8 h. Table 6.60 provides the results of Baseline Run 1. A total of 77 vehicles were in the tunnel at the end of the simulation and occupied 97 car slots. As a result 89 % of the car slots were occupied, or the traffic volume-to-capacity ratio was 89 %. Most transportation planning organizations consider a volume-to-capacity ratio greater than 90 % as congestion. Since the tunnel volume-to-capacity ratio was 89 %, there were no time delays and no queue buildups of vehicles waiting to enter the tunnel. Consequently, there could be a small increase in tunnel traffic volume before experiencing delays.

6.7.6   Increase in Directional Traffic Split

The Baseline Simulation Run 1 indicated a volume-to-capacity ratio of 89 %. Therefore, a small percentage increase in traffic is possible before reaching the congestion level of 90 %. The input data was changed from a 55 % directional traffic split to 60 % and 65 %, respectively.

Table 6.60 Simulation Baseline Run 1: Results

Output	Cars	Delivery trucks	18-wheel trucks	Total
Vehicles in tunnel at the end of simulation	61	8	8	77
Slots in use at the end of simulation	61	16	20	97
Volume-to-capacity ratio				89 %
Average delay entering tunnel (min)	0	0	0	
Average queue entering tunnel	0	0	0	
Hourly traffic	2,857	359	394	3,610

Table 6.61 Simulation results for Simulation Runs 2 and 3

Output	Simulation		
	Run 1	Run 2	Run 3
Directional traffic split	55 %	60 %	65 %
Vehicles in tunnel at the end of simulation			
Passenger cars	61	67	69
Delivery trucks	8	8	8
18-wheel trucks	8	10	9
Slots in use at the end of simulation	97	107	108
Volume-to-capacity ratio	89 %	99 %	100 %
Average delay before entering tunnel (min)			
Passenger cars	0	2	17
Delivery trucks	0	2	17
18-wheel trucks	0	2	17
Queue length before entering tunnel at the end of simulation			
Passenger cars	0	295	1,455
Delivery trucks	0	37	185
18-wheel trucks	0	4-	205
Average hourly traffic			
Passenger cars	2,857	3,158	3,169
Delivery trucks	359	392	404
18-wheel trucks	394	431	445
Hourly traffic	3,610	3,943	4,018

Table 6.61 gives the simulation results for Simulation Runs 2 and 3. The simulation model had a 2-h warm-up and ran for 8 h. The volume-to-capacity ratio for Simulation Run 2 was 99 % indicating congestion. The volume-to-capacity ratio for Simulation Run 3 was 100 %. Both runs had large buildups of traffic waiting to enter the tunnel. The ProModel model for Simulation Run 3 only runs for 4 h because of the queue buildups. A volume-to-capacity ratio of less than 100 % should result in no traffic buildup waiting to enter the tunnel. However, for Simulation Runs 2 and 3, the system can be considered unstable because of the 100 % volume-to-capacity ratio for Simulation Run 3 and the delays beginning to occur for Simulation Run 2. The longer the simulation time for Simulation Runs 2 and 3, the greater these traffic buildups will become.

6.7.7 Increase in Truck Traffic

The input data for Simulation Runs 4 and 5 was modified to include an increase in truck traffic by 5 % and 10 % over the Baseline Simulation Run 1, respectively. The simulation run had a 2-h warm-up and ran for 8 h. Table 6.62 gives the results for Simulation Runs 4 and 5. The volume-to-capacity ratio was 93 % with a 5 % increase in truck traffic and 94 % with a 10 % increase in truck traffic. These ratios suggest that both a 5 % and a 10 % increase in truck traffic result in congestion.

Table 6.62 Simulation results for Simulation Runs 4 and 5

Output	Simulation		
	Run 1	Run 4	Run 5
Directional traffic split	55 %	55 %	55 %
Increase in 18-wheel traffic	0 %	5 %	10 %
Vehicles in tunnel at the end of simulation			
Passenger cars	61	63	64
Delivery trucks	8	8	8
18-wheel trucks	8	9	9
Slots in use at the end of simulation	97	101	102
Volume-to-capacity ratio	89 %	93 %	94 %
Average delays before entering tunnel (min)	0	0	0
Queue length entering tunnel at the end of simulation	0	0	0
Average hourly traffic			
Passenger cars	2,857	2,857	2,856
Delivery trucks	359	359	359
18-wheel trucks	394	413	434
Hourly traffic	3,610	3,629	3,649

6.7.8 Decrease in Passenger Car Traffic and Increase in Truck Traffic

The input data for Simulation Runs 6 and 7 was modified to include a 5 % reduction in passenger car traffic and a continual increase in truck traffic of 5 % and 10 %, respectively. The input data for Simulation Runs 8 and 9 was modified to include a 10 % reduction in passenger car traffic and a continued increase in truck traffic to 15 % and 20 %, respectively. Table 6.63 gives the simulation results for Simulation Runs 6 through 9. The volume-to-capacity ratio for Simulation Run 6 was 89 % and increased to 91 % for Simulation Run 7. The volume-to-capacity ratio for both Simulation Runs 8 and 9 was 89 %.

6.7.9 Conclusions

In Fig. 6.20 a plot of the volume-to-capacity ratios for the simulation runs is shown. In summary the following conclusions are made for peak hourly traffic through the I-10 tunnel:

- The current traffic through the I-10 tunnel during the peak hour is close to being congested with a volume-to-capacity ratio of 89 % (Simulation Run 1).
- An increase in the directional traffic split from 55 % to 60 % (Simulation Run 2) resulted in a volume-to-capacity ratio of 99 % that is above the 90 % congestion.
- An increase in the directional traffic split from 55 % to 65 % (Simulation Run 3) resulted in a volume-to-capacity ratio of 100 % that is above the 90 % congestion. Also, a large number of vehicles were waiting to enter the tunnel.

Table 6.63 Simulation results for Simulation Runs 6 through 9

Output	Simulation			
	Run 6	Run 7	Run 8	Run 9
Directional traffic split	55 %	55 %	55 %	55 %
Decrease in passenger car traffic	−5 %	−5 %	−10 %	−10 %
Increase in 18-wheel traffic	+5 %	+10 %	+15 %	+20 %
Vehicles in tunnel at the end of simulation				
Passenger cars	60	59	56	57
Delivery trucks	7	8	7	7
18-wheel trucks	9	9	11	10
Slots in use at the end of simulation	97	99	97	96
Volume-to-capacity ratio	89 %	91 %	89 %	89 %
Average delays before entering tunnel (min)	0	0	0	0
Queue length entering tunnel at the end of simulation	0	0	0	0
Average hourly traffic				
Passenger cars	2,727	2,727	2,608	2,609
Delivery trucks	359	359	359	359
18-wheel trucks	413	434	454	472
Hourly traffic	3,499	3,520	3,421	3,440

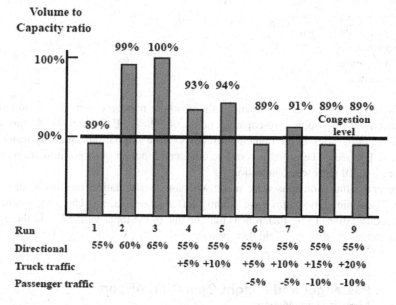

Fig. 6.20 Volume-to-capacity ratios for simulation runs

Simulation Runs 2–3 can be considered unstable where the arrival rate exceeds service rate. Consequently, the queues and delay times will continue to increase.
- A small increase of 5 % in truck traffic (Simulation Run 4) resulted in a volume-to-capacity ratio of 93 %, resulting in congestion.

- A 5 % increase in truck traffic with a 5 % decrease in passenger car traffic (Simulation Run 6) is possible with a volume-to-capacity ratio of 89 %. However, a 10 % increase in truck traffic (Simulation Run 7) resulted in a volume-to-capacity ratio of 91 %.
- A 15 % increase in truck traffic with a 10 % decrease in passenger car traffic (Simulation Run 8) resulted in a volume-to-capacity ratio of 89 %. A further increase in truck traffic to 20 % with a 10 % decrease in passenger car traffic (Simulation Run 9) also resulted in a volume-to-capacity ratio of 89 %.

Once the Baseline Simulation Run 1 is made and the volume-to-capacity ratio determined, it is possible to compute the increase in truck traffic given a reduction in car traffic. For example, Simulation Run 9 had a 10 % decrease in car traffic, or 285 cars. A total of 285 cars/2.5 slots per truck or 114 trucks can occupy these slots. Therefore, a total of 508 trucks (394 from Baseline Simulation Run 1 plus 114) is possible before the volume-to-capacity ratio exceeds 90 %. In summary, the results for Simulation Run 9 are:

Decrease in passenger car traffic per hour:	2,857 (Run 1) to 2,608 (−10 %)
Increase in truck traffic:	394 (Run 1) to 508 (+29 %)
Slots in use at the end of simulation:	98
Volume-to-capacity ratio:	90 %
Average delays and queues:	0
Hourly traffic through tunnel	
Passenger cars	2,609
Delivery trucks	359
18-wheel trucks	508

There is always a danger in using the absolute numbers from the simulation. For example, the volume-to-capacity ratios of 89 %, 90 %, and 91 % may all represent congestion in the tunnel. Variability in the simulation can easily result in these differences. Therefore, it could be concluded that all of these simulation runs (see Fig. 6.20) represent congestion.

Several simulation runs were made with longer simulation run times, such as 40 h. These simulation runs gave slightly different results, especially the volume-to-capacity ratios. These differences may be due to the input variability in the time in the tunnel for different vehicle types.

6.8 Passenger and Freight Operation Airport Simulation Model

6.8.1 Introduction

The behavior of real-world problems as they evolve over time can be studied by developing a simulation model. Therefore, this case study describes the development of a simulation model of the land site processes at airports with regard to dispatch passengers and their luggage. Hence, this model usually takes the form of a set of assumptions concerning the operation of dispatching passengers and luggage. These assumptions can be expressed in mathematical or symbolic relationships between the entities of interest in dispatching passengers and luggage at airports. Once developed and validated, this model can be used to investigate a wide variety of what-if questions about the real-world land-based airport workflows. The achievement of objectives in such a complex system requests for a model-based scenario analysis to investigate potential changes of the airport workflows by simulation in order to predict their impact on the performance of the respective workflow and/or process at the airport. This also allows studying optimization strategies and/or options in an early stage, before anything physically has been done. Thus, modeling and simulation in aviation can be used to predict the effect of changes with regard to the actual existing implementation of workflows and/or layouts of services stations at the airport and as scenario analysis tool to predict the performance of hypothesis under varying sets of circumstances.

The aim of the case study is the development of a simulation model of land site processes at Hamburg Airport with which passengers and their luggage can be dispatched. The numbers of passengers and luggage are generated dynamically depending on the available input data from the different stakeholders involved in the passenger dispatching workflow at the airport. These data are generated dynamically depending on the available input data. For performance testing passenger and luggage flows in the airport terminals are simulated enabling the evaluation on a comprehensive 24-h simulation day to analyze the impact of normal and peak flow situations. With regard to arriving and departing passengers with luggage, two independent simulation models have been developed.

Therefore, at the very first, the case study presents the developed simulation models which are implemented in the commercially available simulation software tool MATLAB Simulink SimEvents. In this simulation analysis Hamburg Airport serves as a template for clustering of terminals and gates. Also travel times and existing handling capacities in dispatching comply with the conditions at Hamburg Airport in this basic version.

In the second step, the case study describes the development of the airport simulation model with regard to the real airport workflows of considered entities, which are introduced in the form of UML activity diagrams as basic modeling building blocks, a technique proper to use for simulating the model in MATLAB Simulink SimEvents. Such a model can also be used to identify possible weaknesses of the passenger and luggage handling workflow at Hamburg Airport by interpreting the obtained simulation results.

In the third step, the paper expands the developed simulation model by embedding the RFID technology for luggage handling. This requires a test procedure which has been developed to test feedback information whether the passenger luggage is allocated to the same gate as the corresponding passenger. Based on the simulation results, assumptions and an assessment of the future use of RFID for luggage identification at airports will be discussed.

As a conclusion of this case study, optimization suggestions to the improvement of the passenger and/or luggage dispatching at Hamburg Airport will be given.

6.8.2 Airport Land Side

The terminal is the part of the airport where passenger traffic is handled. Terminals form the interface between the land and the air side of the airport, whereby the land represents the connection with the ground handling infrastructure, including arriving and parking locations for private vehicles. The air side includes the entire area of air traffic operations including control and maintenance facilities. Allocating the terminal area to the land or air side, the land side covers the publicly accessible area of a terminal, including access roads, ticketing, and check-in. The air side area begins with the security checking of passengers and ends with boarding the aircraft. The airport land side includes the entrance area where the passenger enters the terminal to the point of the boarding gate of his/her flight. The land side can be divided into check-in passenger dispatching and passenger clearance in the security checking area. Thus, the initial point of the simulation model is marked by the passengers entering the terminal. The luggage handling area is also assumed to be located in the terminal. An independent part of passenger check-in, luggage handling plays an important part in the simulation model developed. Therefore, the entire luggage processing sequence of this work is allocated as part of the airport land side.

Based on this workflow, the simulation model developed maps the business processes of passengers and luggage as a chain of stations in the terminal. The aim of the simulation is to obtain valid data on the capacities, process time, and bottlenecks of clearance processes. Based on the results achieved, the potential for optimization and technology recommendations can be identified. This objective meets Level 1 simulation only, and correct input data as well as an adequate process model for formalized model building are imperative. Unified Modeling Language (UML) activity diagrams are suitable for formal mapping of these processes into a model based on semiformal, event-driven process chains. These event-driven process chains are visualized as formal UML activity diagrams which guarantee correct implementation of the real-world processes in the simulation model through a signal transmitter, signal receiver, and time signal receiver. The interaction between multiple processes is characteristic of the formal process representation with UML activity diagrams. The process flow in UML activity diagrams consists of processing series-wise arranged stations. To represent the formal dependencies between multiple processes, the introduced notation elements signal transmitter, signal receiver, and time signal enter the process by a signal transmitter and sent a signal to the corresponding signal receiver of another process. This signal initiates the process flow in the appropriate location of the signal receiver in the process. As long as such a signal fails, processing the process at the signal receiver is interrupted.

Interrupt of processes and its continuation through an external event are typical elements of message queuing systems. The time signal element also is used for the interruption of processes. However, the continuation will be continued not by an external signal, but after a set time delay.

The passenger and luggage of workflows are in the focus of the conceptual model design. This consists of a set of activities that are carried out by different stakeholders in a defined order. Thus, workflow is determined by the order of the individual processing steps and the respective jurisdiction. The workflow controls the unique designation of the competent authority and the provision of all required information. But workflow is more than a process completed in it which means it is the entirety of the processing steps of all interconnected operations.

The passenger departure process model is composed of two substantial components, check-in and security. According to the underlying database, different variants of the check-in process can be introduced: check-in desk, check-in machine, and check-in on the eve. Variations in their implementation are available in the area of processing times of the individual process activities. The workflow security area however is more substantial which covers the entire area of the terminal to which only passengers with a valid ticket have access (requirement is the successful passing through of the security checkpoint). The process model includes several optional branches. Attention should be paid that among other things, an optional manual control of passengers should be implemented in the simulation model (in terms of stochastic considerations).

The luggage handling departure process model is the largest contiguous UML activity diagram. In the context of the fully automated luggage handling system at airports, the entire workflow is considered as a unit. Also the turnoff for bulky luggage is modeled here in terms of completeness. In addition also possible interrupts are modeled, representing detection and sorting of prohibited items and the inadvertent misrouting of luggage.

6.8.3   Model Implementation in SimEvents

SimEvents extends Simulink by a discrete event-driven simulation engine as well as a graphical components library. SimEvents is designed to simulate event-driven communication between system components. The analysis and optimization of latency between components, data throughput, packet loss, and other performance indicators is a major part of SimEvents. Systems can be modeled using the component library correctly and comprehensibly. There are predefined blocks such as queues, servers, and switches, which per drag-and-drop can be assembled to a model file for a simulation model. The block features extensive setting options so that tracing process interruptions, prioritization, and other operations of the simulation system are freely configurable. Thus, SimEvents is suitable for the simulation of event-driven processes with mobile and stationary objects while identifying resource requirements, shortages, and optimization potentials.

The SimEvents block library provides a quantitative manageable, however functionally comprehensive range of blocks. A generator block serves as a starting point of a simulation model in principle. It creates optional entities, events, or signals. Characteristic for SimEvents are the processing entities. Events and signals

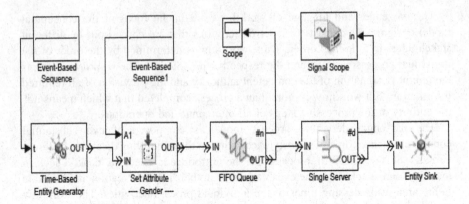

Fig. 6.21 First SimEvents model

are "inherited" Simulink functions and enable the (almost) unlimited combination of components of the various libraries.

In the context of passenger and luggage modeling, the focus is on simulation modeling entities. A first example of such a simulation system comprises, besides the generator block to generate model entities, a final entity sink, in which the entities are captured at the end of the model after they are generated and sent by the model system, as shown in Fig. 6.21.

Between generator and sink, any SimEvents block can be placed to process the entities, to stop, to save, or otherwise to manipulate. The individual blocks are graphically connected by inserting graphical arrows. SimEvents denied automatically the target connection. For example, a block should spend an entity while the receiving block awaits a signal. The simulation flow is formed by the used connections. Another important feature is the granting of individual attributes to an entity. Each entity can have any number of attributes. An attribute can be assigned with positive integer values. This can be overridden by specific blocks or read out to cause, for example, the choice of direction at a point in the simulation model.

Functionally, the implemented simulation model is based on agile object entities, the passengers, which move through a sequence of stations of the simulation model. The used software SimEvents supports precisely these simulation properties by its discrete event-driven core and the production of independent entities. Consequently, the production of passenger entities is the entry point of the model. So far the simulation is started; passengers in the course of time need to enter the airport terminal.

The initialization phase of passenger entities is shown in Fig. 6.22. This model part consists of two subsystems to create entities and set attributes as well as a scope block.

The subsystem that creates entities (Fig. 6.23) consists of a time-based entity-generator block that receives signals from an event-based sequence of blocks. This block sends a signal to the entities generator exactly at the time when a

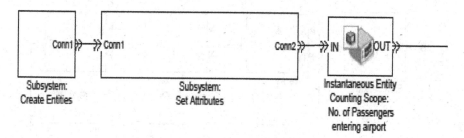

Fig. 6.22 Initialization phase of passenger entities

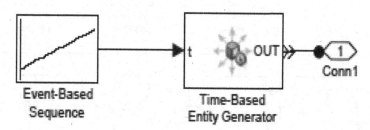

Fig. 6.23 Subsystem that creates entities

passenger entity should enter the airport. The event-based sequence block accesses a matrix with relevant input data. This single-column matrix IntegerTimes contains an entry per generated entity. The comprehended value is the time interval of the arrival entity to its predecessor. For example, the first passenger enters the airport at 2 am, the value of the first IntegerTimes line "2" is accordingly. The second passenger follows at 2:15 am, now the value "0.25" is in row two.

The subsystem consists of eight sequential blocks which each have an attribute to each entity set attributes (Fig. 6.24). The attributes of luggage, flight type, check-in type, terminal, GateCluster, and SchengenArea are from the attribute variables database.

After the passengers have checked in the corresponding terminal, they move to the security control panel. All passengers must pass the security successfully to enter the departure area, where the many departure gates are located. The security control panel is located at the airport of Hamburg in the so-called Airport Plaza, a complex of buildings between Terminal 1 and 2. Passengers of both terminals (i.e., all commercial passengers) complete the security check in the Plaza Complex. Accordingly all passenger entities in the simulation model in subsystem of passenger security area will be merged together via a path combiner block (Fig. 6.25).

To fully map passenger and luggage workflows at Hamburg Airport, the traveling time between the two stations must be modeled in addition to processing and waiting times of the individual stages of the process. The problem exists in the time modeling process which requires the technical processing of entities along a route, with correct averages for individual paths within the airport.

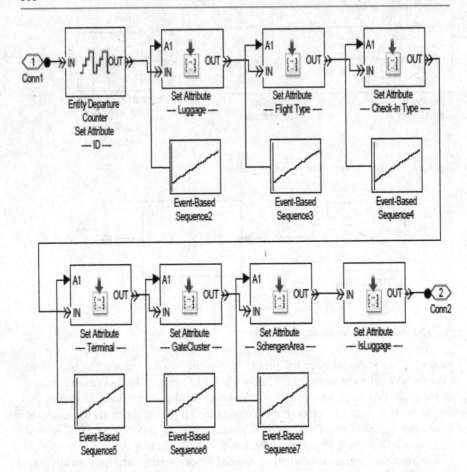

Fig. 6.24 Subsystem set attributes

Fig. 6.25 Routing of the security check subsystem

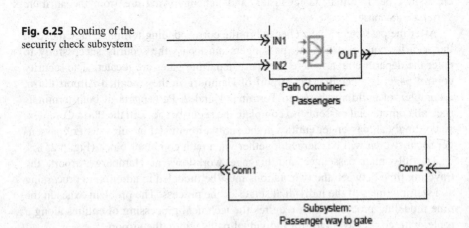

Fig. 6.26 Modeling of passage time with INF Server block

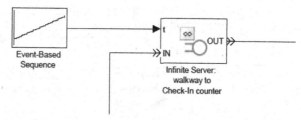

Event-Based
Sequence

Infinite Server:
walkway to
Check-In counter

Modeling of a path has the same characteristic as modeling of a station. In both cases, the entity is set at a fixed value for the workflow period before continuing through the process model (black box). The difference between road and station is based on capacities. While a station can simultaneously edit a certain number of entities, it is assumed that a path at the airport can accommodate as many entities at the same time.

SimEvents possess the INF Server block (Fig. 6.26) modeling stations with unlimited capacity. This processes any entities at the same time and takes, opposed like all other SimEvents Server blocks, the processing time as a parameter. Modeling the passage time is implemented in the entire simulation model with INF Server blocks.

Within the with/without luggage subsystems, a switch point divided the process into the four possible paths of process: check-in counter, check-in machines, check-in other media, and check-in the night before. This routing is based on the entity attribute check-in type. The subsystem passengers with luggage check-in are shown in Fig. 6.27.

A special feature of the check-in with baggage is the premature convergence of paths, check-in machine, and check-in other media. Both workflows include the additional step that placed the luggage at the baggage drop-off counter. Each of the four possible check-ins is an INF Server block on the way to the check-in paths switch, as well as an additional subsystem with the actual process unit (Fig. 6.28). The station consists of the classic components of a queuing system: a FIFO queue block and the block of an N-server with external processing time input signal. The entire model uses basically FIFO queue blocks.

The luggage handling system is the main part of the luggage workflow and consists of multiple subsystems. After the luggage entities have been implemented and initialized into two equal terminal process systems, a path combiner consolidates all luggage entities on a common path of the process. The first subsystem of this path is the luggage passing to the gate system in which the security check of luggage is embedded. Their implementation is analogous to the security control in the area of passenger safety. In the further sequence of luggage handling luggage entities passing the subsystem luggage sorting which models sorting of luggage for departure gates. Here modeling of a possible loss of luggage in the luggage handling system is possible. Using a numeric assortment component, which is identical for the manual control of passengers (Fig. 6.29), each 100th luggage entity "is assumed as lost."

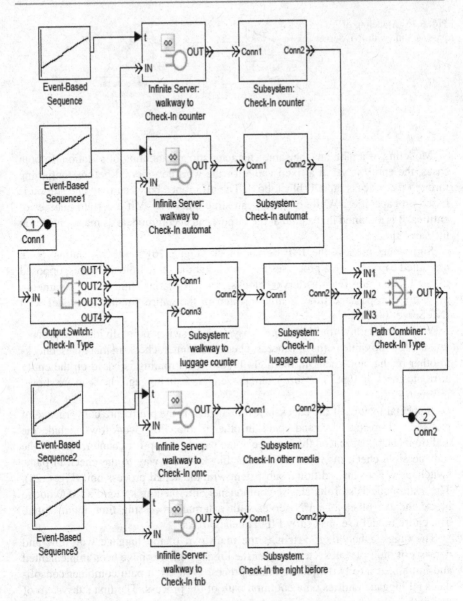

Fig. 6.27 Subsystem check-in for passengers with luggage

Technically, the simulation model has the attribute lost to each affected luggage entity. Therefore, entities not actually lost within the simulation but appear lost in the results listed with special features. Figure 6.30 shows the implementation of lost luggage in the frame of the luggage sorting (gray blocks).

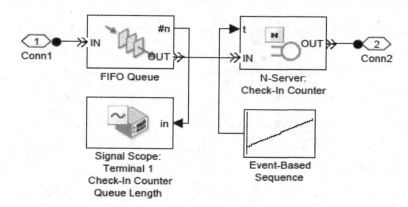

Fig. 6.28 Subsystem check-in counter

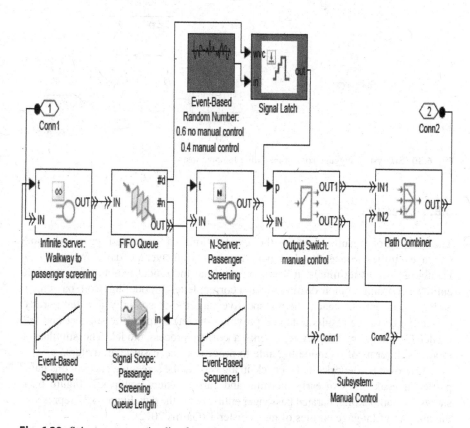

Fig. 6.29 Subsystem screening line for passengers and hand luggage

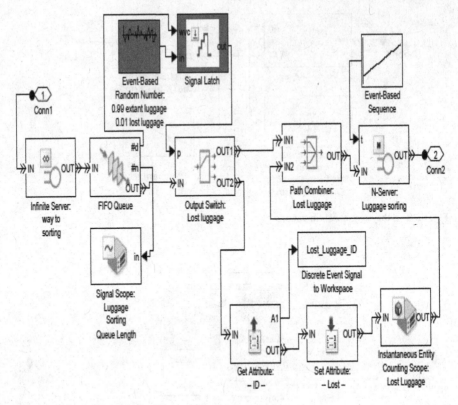

Fig. 6.30 Subsystem luggage sorting including luggage lost

6.8.4   Results

The simulation results confirm the expectations of generated passenger and baggage entities. Since the data base represents an average daily frequency at Hamburg Airport terminals, it was expected that the model is able to process all entities according to their destination at a correct level-1 modeling. Both passenger and luggage entities reach their respective target sink spatially and temporally correctly. Thus, the results obtained provide concrete evidence to suggest that the model for Hamburg Airport represents a correct process model. The simulation model is in terms of its modeling able to process all necessary parameters.

The known bottlenecks in the check-in area and at the screening checkpoints have particular peak times in early morning and early evening confirmed. Figure 6.31 shows the number of generated passenger entities over the day. Figure 6.32 represents the number of luggage entities of gate cluster 1 (Garms 2012).

It should be pointed out that the figures show the result of simulation runs for known input values. The simulation model can be modified with any model values and thus appropriately customized.

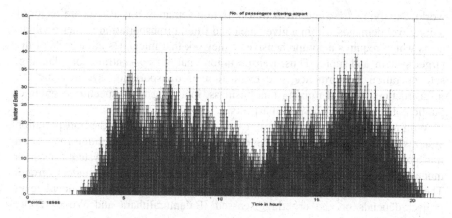

Fig. 6.31 Number of generated passengers over the day

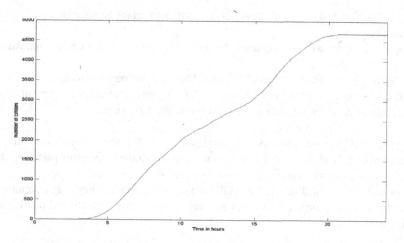

Fig. 6.32 Number of generated baggage over the day: cumulated

6.9 International Student Team Project: Modeling and Simulating an Airport Transportation Operation

6.9.1 Introduction

The transportation systems sector is a vast, open, and interdependent networked system that moves millions of tons of freight and millions of passengers. Every day, cities, hubs, manufacturers, ports, retailers, and so forth are connected through the transportation system network that moves large volumes of freight, goods, and individuals through a complex network of roads and highways, railways and train stations, sea ports and dry ports, and airports and hubs. This is why the transportation systems sector is the most important infrastructure for any economy in the world. Satisfying these enormous transportation demands, it is also a core sector of daily travel demands within a given area and freight transportation in metropolitan areas which requires a comprehensive framework that integrates all aspects of the target system and more. Thus, transportation analysis concentrates on planning, safe operation, performance, and evaluation of transportation systems and the infrastructure required running this business, including the respective economic, public policy, and environmental aspects.

With regard to the importance of a well-skilled workforce in the transportation systems sector, the international student team project in transportation focuses on transportation analysis, modeling, and simulation of a real-world system for which models have been developed and implemented by the international students group. The suitability and efficiency of simulations for the international student team project depends on the criteria discussed (Balamuralithara and Woods 2008; Erugrul 1998). The most important are:

- *Modularity*: test modules developed easily and adapt to specific applications quickly.
- *Executability*: avoid alteration to hide the code or to create standalone applications.
- *Performance*: ensure the modules meet the required performance.
- *Intuitive graphical user interface (GUI)*: enable transatlantic student team members to see a simulation and determine the next steps.

ProModel was selected as the simulation tool for the international student team project. It allows inclusion of comparative studies, scenario planning and analysis, content integration, and modeling and simulation case studies.

In addition to hard skills, soft skills are also important. They require cultural, social, and educational interactions and networking with students and instructors. The goals are to:

- Establish an international program for students to work in interdisciplinary student teams on projects that foster cross-cultural interaction and networking.
- Expose students through this network to cultural, social, and communication issues in intercultural student-based modeling and simulation using a case-based study program.
- Reinforce fundamental scientific concepts, providing opportunities to put them into practice through Computational Modeling and Simulation.
- Provide students with the experience of using the Internet to work as part of an international student project team (Fig. 6.33).

6.9.2 Principles of Operation on the Airport Ramp

This student team transportation project focuses on modeling and simulating the operation on the airport ramp, which include activities at gate areas, ramps, and taxiway and runway systems, strongly influenced by terminal-area operations.

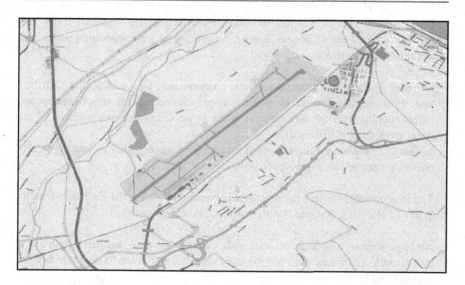

Fig. 6.33 Falconara Airport, Italy (Open Street Map)

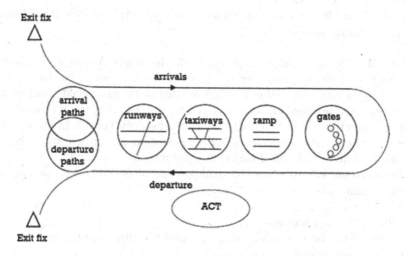

Fig. 6.34 Schematic of the airport system including the terminal area (Balakrishnan 2009)

The different components of the airport system illustrated in Fig. 6.34 have aircraft queues associated with them and interact with each other. The cost per unit of time spent by an aircraft in one of these queues depends on the queue itself. For example, an aircraft waiting in the gate area for pushback clearance predominantly incurs flight crew costs, while an aircraft taxiing to the runway or waiting for departure clearance in a runway queue with its engines running incurs additional fuel costs and increased ramp emissions (Balakrishnan 2009). Arrivals and departures associated with airport operations occur at scheduled times.

Besides real scheduled times of arrivals and departures, random times are defined for an analysis of how to handle the procedure with scheduled times and delays resulting in different scenarios:

- *Scenario 1*: considers random arrivals, characterized by a probability distribution. Poisson distribution is a good approach because it describes events independently with average rates (waiting times between k occurrences of an event are Erlang distributed).
- *Scenario 2*: scheduled arrivals of flights and resources, such as baggage cart, busses, passenger stair, pushback unit, etc.

The international student team project is based on data for the Raffaello Sanzio Airport of Falconara, Ancona (Italy), for the following reasons:

- No large airport: it is easier to start with a simple use case compared with a big hub airport. The simulation can be used as a starting point to simulate bigger airports and hubs.
- International airport: there are more flights to simulate use cases, even if there is not an intercontinental one.
- Used by two low-cost carriers (Ryanair and Volotea) and two Star Alliance carriers (Lufthansa and Alitalia).

From Fig. 6.33, it can be seen that Raffaello Sanzio Airport of Falconara, Ancona (Italy), has only one runway. In the terminal building, two arrival and two departure gates are available. After arriving at the ramp hold position, the plane's engines are turned off and passengers can deplane. They travel to the arrival gate on foot or by bus. The luggage is delivered at the conveyer belt. In between, the plane will be cleaned and refueled. After that procedure, new passengers can board, and the airplane can take off if the taxiway and runway are cleared.

The scenario-based simulation of the airport operation workflow is based on two different simulation tools:

- GPSS for the first scenario
- ProModel for the second scenario, which will be discussed in more detail in the following sections

6.9.3 Data Analysis

The international student team project optimizes the occupancy rate of arrival and departure gates in a cost-efficient way. This requires workflow information on the airport operations available from the airport information office, available at http://www.ancona-airport.com, which shows airport flight plans. Plans used for data analysis were downloaded on June 4 showing the flights for June 3 to 9, 2013. Every week, new flight plans are published; but significant values, such as

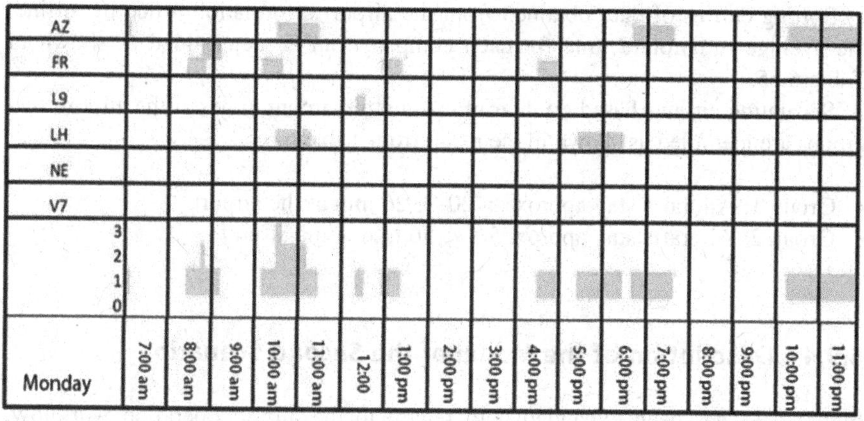

Fig. 6.35 Chart of arrivals and departures of airline flights at Raffaello Sanzio Airport in Falconara, Ancona (Italy)

Table 6.64 Total flights per week

Monday	Tuesday	Wednesday	Thursday	Friday	Saturday	Sunday
11	7	14	8	11	12	9

frequency of flights and workload of airport operations, remain the same for weeks. Therefore, the abovementioned data set can be used as the basis for models developed for the student team project. For this reason, the arrivals are sorted for a weekly schedule and for any further analysis. At first glance, it can be seen that an Alitalia flight departing from Roma Fiumicino and arriving at Ancona Airport stays overnight at the airport. Therefore, to distinguish it from other planes which leave the airport after the time required for the catering, cleaning, deplaning, boarding, refueling, and security check processes, an empty line is used to show the airplane which remains in the airport and leaves on the following day. Figure 6.35 depicts the following for Monday, June 3, 2013:

- Hours of operation for each flight (in other words, how long each remains at the airport)
- Number of flights at the airport at the same time

Times in the chart are restricted to 07:00 am to 11:00 pm because the first arrival is at 08:25 am and the last one is at 10:15 pm, with a departure at 07:05 am.

According to Fig. 6.35, flights are more frequent during the morning, seen in the lower rows of Fig. 6.35, where each block represents 5 min and the total number of planes at the airport is indicated. Furthermore, it can be deduced that between 10:00 am and 11:00 pm, three planes are at the airport, but there are only two gates available, which happens every day of the week due to resource shortages.

The total number of flights per day derived from the airport information office Web site is 72 per week, as shown in Table 6.64.

Sorting entries of data obtained from the airport information office by airline, the average turnaround time for each company can be determined as shown in Table 6.65.

Separating airlines based on their remaining turnaround times at the airport, two groups are identified as shown in the result from Table 6.65:

- Group 1: Aircrafts stay approx. > 20–< 26 min at the airport.
- Group 2: Aircrafts stay approx. 50–< 56 min at the airport.

6.9.4   Description of the Model of the Second Scenario

Figure 6.33 has been expanded with regard to the airport operation workflow, as shown in Fig. 6.36.

Figure 6.36 is the airport operation model. An aircraft arrives at location Airplane_Arrival. Then it is moved to the final parking position, and the engines

Table 6.65 Sorting of entries

Airline					
Agenzia Viaggio	Alitalia	Belleair	Lufthansa	Ryanair	Volotea
50 min	55 min	22.86 min	52.69 min	25.42 min	Remains at airport for the night

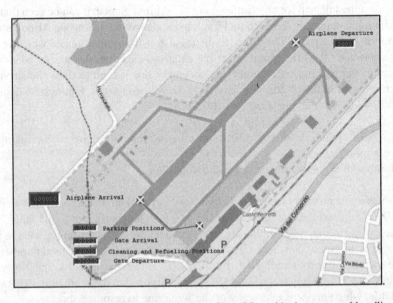

Fig. 6.36 Raffaello Sanzio Airport based on Open Street Map with airport ground handling and terminal positions marked

Table 6.66 Raffaello Sanzio Airport capacity and units

Name	Capacity	Units
Airplane_Arrival	1	1
Parking_Positions	6	1
Gate_Arrival	2	1
Cleaning_and_Refueling Positions	6	1
Gate_Departure	2	1
Airplane_Departure	1	1

are turned off. If all turnaround procedures are complete, the aircraft is ready to be moved to Airplane_Departure and to leave the airport, which is the end of the simulation analysis in ProModel (Processmodel 2011).

6.9.4.1 Locations

To run airport operation simulations, essential locations have to be set keeping the model as realistic as possible, as shown in Table 6.66. The locations are:

- Airplane_Arrival: place where the airplane lands. There is only one runway at Raffaello Sanzio Airport; therefore, the capacity of this location is set to one.
- Parking_Positions: position where aircraft has to wait after landing if both gates are busy. The capacity at this location is six. As the airplane's engines have been turned off, the following three locations coincide with the Parking_Positions:
  - Gate_Arrival: passengers and luggage leave the plane; thus the capacity of this position is set to two, because only two conveyer belts are available in the airport.
  - Cleaning_and_Refueling_Positions: cleaning and refueling are done before passengers enter the plane.
  - Gate_Departure: place from which passengers and their luggage enter the plane.
- Airplane_Departure: beginning of the runway from which the plane takes off if the runway is set to clear. The airport has only one runway; therefore, capacity of this location is set to one.

6.9.4.2 Entities

Airlines using Raffaello Sanzio Airport are classified as:

- *Low-cost carriers*: Turnaround time at the airport is between 20 and 25 min.
- *Regular carriers*: Turnaround time at airport is between 50 and 55 min.

Entities introduced did not distinguish international flights from national flights:

- *Airplane1*: Represents aircraft of regular carriers
- *Airplane2*: Represents aircraft of low-cost carriers

Table 6.67   Raffaello
Sanzio Airport path
network

Name	Paths	Interfaces
Net1	1	2
Net2	1	5

Table 6.68   Model parameters

Entity	Location	Quantity each	First time week 1	Occurrences	Frequency
Airplane 1	Plane_Arrival	1	Mon. 7:00 am	INF	E(2)
Airplane 2	Plane_Arrival	1	Mon. 7:00 am	INF	E(1)

6.9.4.3 Path Network

In order to let planes move through different positions, a path network was created, as shown in Table 6.67:

- *Net1*: path that links Airplane_Arrival with the locations Parking_Positions, Gate_Arrival, Cleaning_and_Refueling_Positions, and Gate_Departure
- *Net2*: path that links Parking_Positions, Gate_Arrival, Cleaning_ and_Refueling_-Positions, and Gate_Departure with Airplane_Departure

6.9.4.4 Shift Assignments

The Raffaello Sanzio Airport is closed during the night. Airplane_Arrival is activated from 7:00 am to 10:00 pm every day, and location Gate_Departure is activated every day from 7:00 am to 11:00 pm.

6.9.4.5 Arrival

Setting arrivals of airplanes requires:

- Using airplane arrival rates in a way which allows two different kinds of airplanes arriving at Raffaello Sanzio Airport to be considered
- Increasing the arrival rate to optimize utilization of the airport

Since the number of arrivals of the two plane categories can be described by a Poisson process, interarrival frequencies are described through exponential distribution, whose parameters are shown in mean time (in hour) between two adjacent arrivals. Parameters for interarrival frequencies are listed in Table 6.65. To analyze a more realistic model of airport operations for scheduling the arrivals, the schedule of the first arrival in Week 1 is repeated, adding a variation of 0.5 for Airplane2. In Table 6.68, the values in the Occurrences column are always INF to show that there is no upper limit on the number of incoming airplanes (if the input was a number, e.g., x, no other plane would arrive after the xth plane landed).

Table 6.69 Process and routing

Process			Routing		
Entity	Location	Operation	Output	Destination	Move logic
ALL	Plane_Arrival		ALL	Parking_Position	Move on Net2
ALL	Parking_Position		ALL	Gate_Arrival	Move on Net2

6.9.4.6 Processing

To run airport operation simulations, some procedures and processes have been defined, such as movements from Airplane_Arrival to Airplane_Departure, for both kinds of planes, whereas some other procedures and processes depend on the type of plane. This distinction is useful to set different mean operation times. After a plane arrives at location Airplane_Arrival, it moves to a Parking_Position to stop if both Gate_Arrival units are busy. The Column Move Logic in Table 6.69 shows on which of the two nets the plane is moving. Actually there is no logic structure behind this movement. As soon as a plane is in Airplane_Arrival, it moves to Parking_Positions.

Once the aircraft is at Gate_Arrival, passengers deplane. Describing deplaning, a normal distribution was chosen, whose parameters depend on the kind of aircraft in the simulation run. Normal distribution is denoted in ProModel in the notation $N(a,b)$, where parameter a is the mean value and parameter b is the standard deviation with parameters:

- Low-cost carrier. The following parameters are set:
  - Mean: 8 min
  - Standard deviation: 2 min
- Regular carrier. The following parameters are set:
  - Mean: 15 min
  - Standard deviation: 3 min

After deplaning, the plane virtually moves to Cleaning_and_Refuling within the simulation model, as shown in Table 6.70.

In the Cleaning_and_Refueling_Positions, the cleaning and fueling processes take place. As before, we used a normal distribution to describe the time needed for the procedures and the parameters depend on the kind of plane being serviced, as shown in Table 6.71:

- Low-cost carrier flights
  - Mean: 7 min
  - Standard deviation: 2 min
- Regular carrier flights
  - Mean: 20 min
  - Standard deviation: 3 min

If plane is ready to move to Gate_Departure, passengers can board the aircraft. Again, a normal deviation is used to describe the time needed and distinguish between the two different kinds of airplanes (Table 6.72).

Table 6.70 Process and routing for Cleaning_and_Refueling_position

Process			Routing		
Entity	Location	Operation	Output	Destination	Move logic
Airplane 1	Gate_Arrival	WAIT N(14,3) max	Airplane 1	Cleaning_and_Refuling_Position	Move on Net2
Airplane 2	Gate_Arrival	WAIT N(8,2) max	Airplane 2	Cleaning_and_Refuling_Position	Move on Net2

Table 6.71 Routing for services

Process			Routing		
Entity	Location	Operation	Output	Destination	Move logic
Airplane 1	Cleaning_and_Refuling_Position	WAIT N(20,3) min	Airplane 1	Gate_Departure	Move on Net2
Airplane 2	Cleaning_and_Refuling_Position	WAIT N(7,2) min	Airplane 2	Gate_Departure	Move on Net2

Table 6.72 Gate and airplane departure

Process			Routing		
Entity	Location	Operation	Output	Destination	Move logic
Airplane 1	Gate_Departure	WAIT N (15,3) min	Airplane 1	Airplane_Departure	Move on Net2
Airplane 2	Gate_Departure	WAIT N (7,2) min	Airplane 2	Airplane_Departure	Move on Net2

Table 6.73 Simulation run for four weeks

Name	Total exits	Average time in system (min)	Average time in operation (min)
Airplane 1	202	54.22	49.22
Airplane 2	429	28.10	23.32

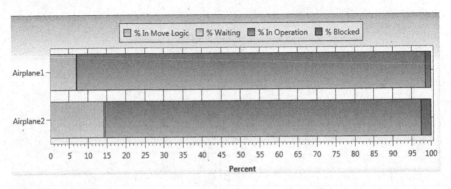

Fig. 6.37 Utilization of locations

Once passengers and baggage are on board, an airplane can move to Airplane_Departure from which it can take off. If a plane has reached Airplane_Departure, it is ready to move to the runway to EXIT the simulation system, meaning takeoff.

6.9.5 Simulation Results

After a simulation run for 4 weeks, we obtained the data shown in Table 6.73 as the most important result.

As assumed, the number of different types of planes relates to the data shown in the analysis of data on the Raffaello Sanzio Airport. Moreover, time that each aircraft spends in the system corresponds with the time obtained from data analysis. Turnaround time for regular carrier planes is approximately 55 min and 28 min for low-cost carriers, which is compatible with the time planes switched their engines off and started them again (turnaround time) which is 20–25 min.

Figure 6.37 shows how often every location is used, apart from Airplane_Arrival and Airplane_Departure, and what percentage of the time it is empty, partly occupied, or fully in use. With regard to the assumed flight schedule, it can be seen that the airport is not used at maximum capacity at this schedule. In fact, Gate_Arrival and Gate_Departure are mostly empty as they are occupied only for a small percentage of time.

6.9.6 Aviation Operation Modeling

In Sect. 6.9.6, we focus on a more realistic model of airport operation modeling and simulation.

6.9.6.1 Locations

The second scenario used different locations in order to represent different steps of the operation flow around the plane, as shown in Fig. 6.38.

This scenario introduces resources which are used during each phase of the work to be done on the plane. This allows locations on the apron where planes park to be used as geographic spots (Fig. 6.39).

Locations Airplane_Arrival and Airplane_Departure are as described in the previous scenario. The other locations are now:

- Three parking positions for regular carrier flights, Parking_PositionBig.1, Parking_PositionBig.2, and Parking_PositionBig.3, which are closer to gates
- Three parking positions for low-cost carrier flights located on outer positions of the apron, Parking_PositionLow.1, Parking_PositionLow.2, and Parking_PositionLow.3

As mentioned before, Parking_PositionBigx is considered to be a unique location that has three different spots; in fact, once an airplane from a regular

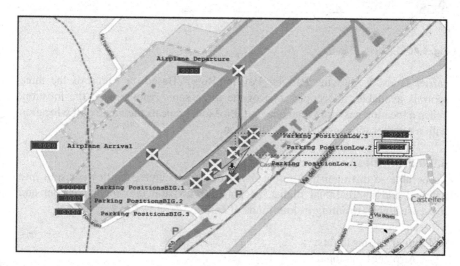

Fig. 6.38 Raffaello Sanzio Airport based on Open Street Map with ground handling and terminal positions marked

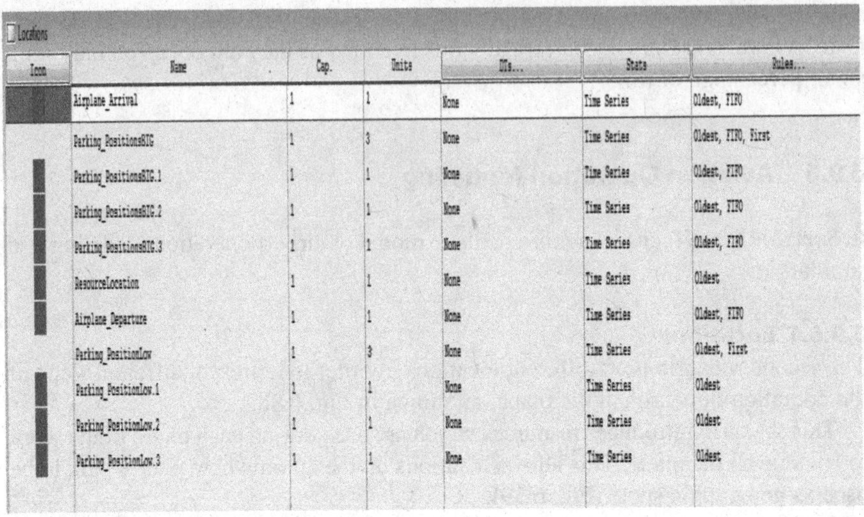

Icon	Name	Cap.	Units	IDs...	Stats	Rules...
	Airplane_Arrival	1	1	None	Time Series	Oldest, FIFO
	Parking_PositionsBIG	1	3	None	Time Series	Oldest, FIFO, First
	Parking_PositionsBIG.1	1	1	None	Time Series	Oldest, FIFO
	Parking_PositionsBIG.2	1	1	None	Time Series	Oldest, FIFO
	Parking_PositionsBIG.3	1	1	None	Time Series	Oldest, FIFO
	ResourceLocation	1	1	None	Time Series	Oldest
	Airplane_Departure	1	1	None	Time Series	Oldest, FIFO
	Parking_PositionLow	1	3	None	Time Series	Oldest, First
	Parking_PositionLow.1	1	1	None	Time Series	Oldest
	Parking_PositionLow.2	1	1	None	Time Series	Oldest
	Parking_PositionLow.3	1	1	None	Time Series	Oldest

Fig. 6.39 Airplane parking positions

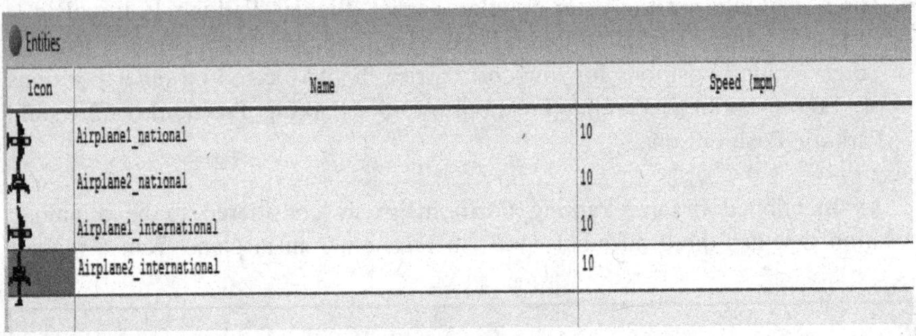

Icon	Name	Speed (mpm)
	Airplane1_national	10
	Airplane2_national	10
	Airplane1_international	10
	Airplane2_international	10

Fig. 6.40 Entities

carrier lands, it goes to "Parking_Position.Bigx." Based on which of the three spots is available, it occupies one of the three spots belonging to the location. When a location has more than one unit, ProModel repeats the name of the location with a number for each unit of the location.

6.9.6.2 Entities
To be able to distinguish different demands for resources, entities are split into national and international flights, keeping the division into regular carrier and low-cost flights, resulting in four kinds of entities, as shown in Fig. 6.40:

- Regular carrier flights:
  - Airplane1_national
  - Airplane1_international

- Low-cost carrier flights:
  - Airplane2_national
  - Airplane2_international

6.9.6.3 Path Networks

In order to move resources and entities, the network has two sets, one for entities and one for resources (Fig. 6.41):

- Net1 is used for movements of planes through the airport.
- Net2 is the path of resources moving from their base to planes in different parking positions.

Planes arrive at Airplane_Arrival; from there they move on to Net1. Based on the kind of airline, they reach one of the free slots in Parking_PositionBig or Parking_PositionLow (Fig. 6.42).

ResourceLocation is where resources are stationed if not in use from which they move to different parking positions serving planes. Planes are located in different spots, according to the air company to which they belong; and ResourceLocation is connected through Net2 to both locations for the plane (Fig. 6.43).

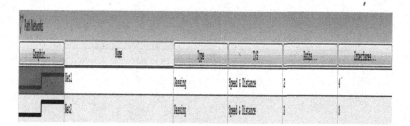

Fig. 6.41 Path networks

Node	Location
N1	Airplane_Arrival
N2	Parking_PositionsBIG
N3	Airplane_Departure
N2	Parking_PositionLow

Fig. 6.42 Interfaces Airplane_Arrival of path networks in Fig. 6.41

Interfaces

Node	Location
N1	ResourceLocation
N2	Parking_PositionsBIG
N3	Parking_PositionLow

Fig. 6.43 Interfaces Parking_Position of path networks in Fig. 6.41

Arrival Cycles

ID	Qty / %	Cumulative	Table...
BigInt	Quantity	No	Defined
BigNat	Quantity	No	Defined
LowInt	Quantity	No	Defined
LowNat	Quantity	No	Defined

Fig. 6.44 Plane arrivals for times from data analysis of flight plans or the schedules

Table for BigNat

Time (Hours)	Qty / %
10.5	0
10.51	1
18.66	0
18.67	1
22.25	0
22.26	1
24	0

Fig. 6.45 Schedule of regular carrier national planes

6.9.6.4 Arrival Cycles

In Scenario 2, plane arrivals are scheduled (Fig. 6.44). Times are from the information derived from data analysis of flight plans. To achieve this, the option of arrival cycles in ProModel is used. There is one schedule for each kind of entity. The flight plan from Wednesday is used as a reference for the schedules.

The schedule of regular carrier national planes is as follows (Fig. 6.45).

The decimal system is used, meaning that 1 h is not divided into 60 min but into 10 tenths, 100 hundredths, 1,000 thousandths, etc. Furthermore, time is cumulative,

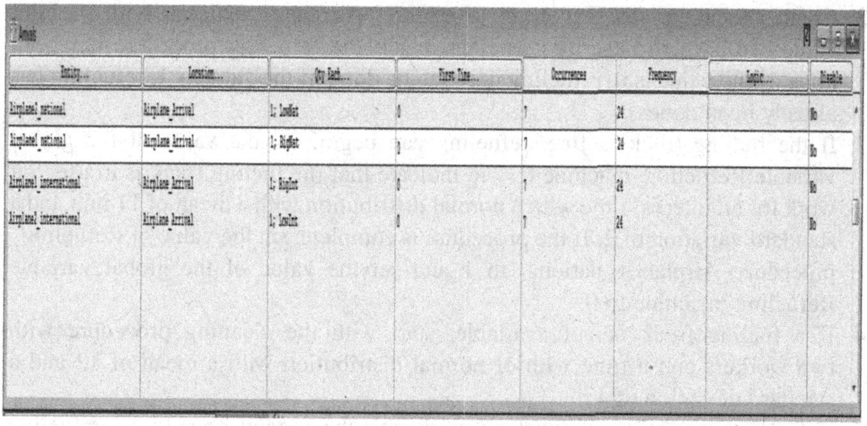
Arrivals

Entity...	Location...	Qty Each...	First Time...	Occurrences	Frequency	Logic...	Disable
Airplane1_national	Airplane_Arrival	1; LowNat	0	7	24		No
Airplane1_national	Airplane_Arrival	1; BigNat	0	7	24		No
Airplane1_international	Airplane_Arrival	1; BigInt	0	7	24		No
Airplane1_international	Airplane_Arrival	1; LowInt	0	7	24		No

Fig. 6.46 Names in Qty. Each column of path networks in Fig. 6.41

meaning that there are 0 arrivals between 0:00 and 10.50 (i.e., 10:30) and 1 between 10.50 and 10.51. Having 0.01 h for each arrival enables an exact arrival schedule. To use the arrival cycles just described, we had to insert the name of the required cycle into the Qty. Each column of the related entity arrival, as shown in Fig. 6.46.

6.9.6.5 Resources

Four different kinds of resources represent the different airport services:

- Fueling represents a fueling truck to refuel planes that have arrived. This quantity was also set to equal 2 to investigate its impact, which was less than 1 % in use.
- PeopleOut are gangways used by passengers to leave a plane via the two arrival gates.
- PeopleIn are gangways used to let passengers enter a plane via the two departure gates.
- Workers represent the staff working with the resources described.

6.9.6.6 Global Variables

Before describing the procedure a plane has to pass through, variables are needed which can be global or local. One global variable is Refueling_machine, with a default value of 0. This variable is used to describe whether the fueling truck is available or in use, which is indicated by the values 0 or 1, respectively.

6.9.6.7 Processes

Scheduling arrivals and defining resources and nets in which airplanes and resources move requires describing the processes a plane is involved in during its stay at the airport. The processes set for Airplane1_national are:

1. Deplane, which is described by a normal (time) distribution with a mean of 13 min and standard deviation of 3 min.

2. Create a local variable Refueling_procedure_ Airplane1_national, with a default value of 0, which describes whether or not the refueling procedure has been done. If the value is 0, refueling needs to be done; if the value is 1, refueling has already been done.
3. If the fueling truck is free, refueling can begin: set the value of the global variable Refueling_machine to 1 to indicate that the fueling truck is in use. Set work for an interval time with a normal distribution with a mean of 11 min and a standard variation of 2. If the procedure is complete, set the value of Refueling_ procedure_Airplane1_national to 1 and set the value of the global variable Refueling_machine to 0.
4. If a fueling truck is not available, start with the cleaning procedure with two workers and a time with of normal distribution with a mean of 12 and a standard deviation of 1.
5. If the first operation is done, proceed with the second one. If the refueling was done first, then clean the plane or vice versa.
6. Board new passengers. This requires a time window with a normal distribution of a mean of 13 and a standard deviation of 3 min.

Numbers in expressions such as "Fueling 150" are priorities for using resources because regular carriers pay more for services the airport offers. Thus priority to services is needed for their flights. A distinction among priorities can also be achieved between international and national flights for the same kind of carriers. International flights are more likely to be connected with intercontinental flights because connected airports, such as Munich, are used by passengers as a hub to reach other destinations/continents.

6.9.6.8 Analysis of the Results
The analysis goal is to identify the required number of ground handling staff for scheduled arrivals of planes to optimize the resources used. The second goal is to identify how airplane delays affect airport efficiency even though regular carriers have to receive good treatment even if there are delays. For this reason, different scenarios of delays are analyzed:

1. The first scenario has only one plane delayed for a few minutes.
2. The next two scenarios assume that planes have a bigger delay and that many planes arrive at the same time or within a short time window.

Scheduled Arrivals
Simulation runs using scheduled arrivals according to the timetable of arrivals at Raffaello Sanzio Airport have been done for a week. First, the simulation was set with ten ground workers, as shown in Fig. 6.47.

The simulation shows that though the refueling machine is one of the most used resources (by percentage), this percentage is not sufficiently large to enable the airport to purchase a new one, even in cases where delays occur. As mentioned before, five workers are used during the simulation when the planes arrive according to the schedule. From the statistic above, we can see that the percentage

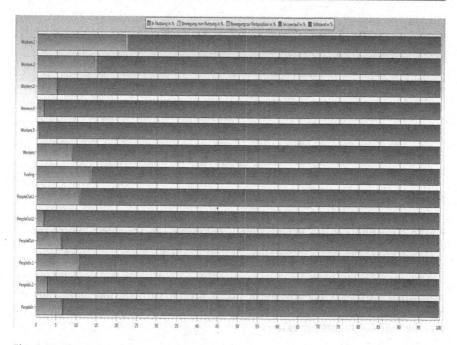

Fig. 6.47 Simulation with ten ramp workers

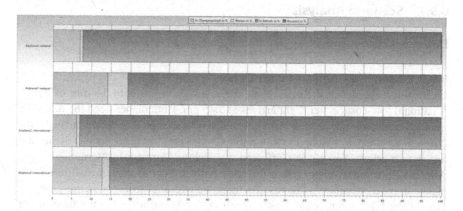

Fig. 6.48 Simulation with five ramp workers

of time when the fifth worker is used is very small (see Fig. 6.48). Therefore, we have tried to implement a scenario with scheduled arrivals of airplanes again but using only four workers instead of five to determine how this affects the results (shown in Fig. 6.49).

The figures show the percentage of time spent by each entity:

- The Time Entity that is in Move is represented by the light-blue-colored bar.
- The Waiting entity is represented by the yellow-colored bar.

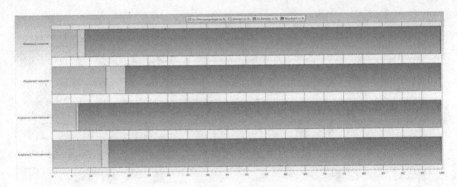

Fig. 6.49 Simulation with four ramp workers

- The Time Entity that is in Operation is represented by the green-colored bar.
- The Time Entity that is Blocked is represented by purple.

Comparing these results with the ones for five workers, it can be deduced that only small variations happen which affect regular carriers: the mean time of international flights decreases, according to their high priority, whereas the mean time for national flights increases.

6.9.7   Scenario Analysis

Different scenarios with delays are investigated.

6.9.7.1 Scenario Analysis I with Delay
The regular international carrier plane arrives at 5:39 pm with 14 min of delay. A national low-cost carrier flight arrives at 5:45 pm. Both planes need the refueling truck. Simulating turnaround times by each plane with scheduled arrivals for different numbers of resources (four and five workers) has been done. Turnaround for five workers for planes affected by delay (Airplane1_international and Airplane2_ national) is not much different with respect to turnaround time if all planes arrive as scheduled. The result is the same with four workers. In this case, the time of the third resource decreases according to the priority; but the time of the second resource increases. This shows that the turnaround process is well established. According to the priorities set, there is no impact on regular carriers. Thus, reducing one unit of workers does not result in changes to the turnaround time of the planes, confirming what was mentioned above. Thus it is reasonable to maintain the turnaround with only four workers (Table 6.74).

6.9.7.2 Scenario Analysis II with Delay
A more critical situation occurs when we suppose that the number of planes arriving at the same time is up to four. Assume two low-cost flights (one national and one international) scheduled to arrive at 8:25 am actually arrive at 10:30 am,

Table 6.74 Impact of different numbers of 4 and 5 workers at ramp position

	Scheduled arrival	Scheduled arrival	Delay one	Delay one
	Five workers	Four workers	Five workers	Four workers
Airplane1_national	60.07	61.05	60.07	61.05
Airplane1_national	29.33	29.33	29.67	29.67
Airplane_1_international	66.29	65.97	66.20	65.88
Airplane_2_international	32.49	32.49	32.49	32.49

Table 6.75 Impact of different numbers of 5, 5, and workers at ramp position

	Scheduled arrival	Scheduled arrival	Delay two	Delay two	Delay two
	Five workers	Four workers	Seven workers	Five workers	Four workers
Airplane1_national	60.07	61.05	59.90	60.33	60.58
Airplane2_national	29.33	29.33	37.90	37.73	37.14
Airplane_1_international	66.29	65.97	66.96	66.84	66.84
Airplane_2_international	32.49	32.49	38.49	39.30	38.94

a 2:05 h delay. At the same time, two regular carrier flights (one national and one international) arrive. In this case, seven workers are needed, although the resources are not used to their maximum level. Thus it can be seen what will happen when the number of workers is reduced. With regard to the resources used, there is no significant difference for regular carrier planes between this scenario and the one without delays. The major difference occurs for low-cost carriers, especially for national flights which have no priority at all. This is compatible with the priorities that were established in this order:

1. Airplane1_international
2. Airplane1_national
3. Airplane2_international
4. Airplane2_national (no priority at all)

Airplanes which are delayed are the ones operated by low-cost carriers, and they have to be rescheduled. Comparing the results, no big changes in turnaround mean time occur if the number of workers has been reduced. Only two kinds of resources affect the turnaround time for only less than a minute. It can be assumed that it is reasonable to use only four workers and to take on one or two more only during peak times when they are really needed (Table 6.75).

6.9.7.3 Scenario Analysis III with Delay
The last scenario involves a case with an increased number of flights in a time window. Suppose that two airplanes belonging to regular carriers are delayed, one national and one international, both scheduled for arrival at 10:30 am and delayed

Table 6.76 Impact of different numbers of 4, 5, and 6 workers at ramp position

	Scheduled arrival	Scheduled arrival	Delay three	Delay three	Delay three
	Five workers	Four workers	Six workers	Five workers	Four workers
Airplane1_national	60.07	61.05	59.68	59.86	59.83
Airplane2_national	29.33	29.33	29.11	29.11	29.11
Airplane_1_international	66.29	65.97	66.83	66.93	66.51
Airplane_2_international	32.49	32.49	38.19	37.16	40.18

until 12:30 pm and 12:50 pm, respectively. Three low-cost carrier international flights coming from different places are affected, arriving at 0:20 pm, 0:45 pm, and 1:05 pm. The simulation run was executed for a week, and the number of resources needed is six. Comparing the simulation results for the required number of workers, it is deduced that, according to the priorities given to the different planes, the only significant impact involves Airplane2_international, whose mean time in the system increases by more than 2 min. The same happens for other entities: Airplane2_international is also affected by changes to the number of workers. Another affected is the result of the three Airplane2_internationals being in the scenario at the same time when the delays take place. Choosing only four workers is a restrictive constraint, but the use staff could be used for cleaning work inside the airport building for a short time, helping to minimize costs (Table 6.76).

6.9.8   Conclusions

The approach of a scientist, especially a specialist in Computational Modeling and Simulation, when dealing with a particular process or situation, is to analyze the reality, try to understand which things influence the environment, and determine whether the process can be optimized with respect to those influences. As international students of Computational Modeling and Simulation, we tried to do the same. By analyzing and optimizing the turnaround process of planes at the airport, we learned how the process works. We implemented the first scenario to see how a simulation of what we had learned could look in ProModel. We assumed the point of view of the airport management with a goal of minimizing the costs while choosing a strategy that would keep customers satisfied. We have decided to give priority to the regular airlines as a marketing strategy; this is similar to what happens in reality. We saw that the factors that influence the costs are resources. For a deeper understanding of how many resources are needed, we implemented a second scenario. At first we let the simulation run with the scheduled time, in order to see how many resources are needed in reality. We saw that up to five workers were needed. However, the statistics showed that the fifth worker was used for a very short percentage of time. Therefore, we tried to implement a scenario with four scheduled workers and saw that the difference when compared to five workers was negligible. Furthermore, since the planes are subject to delays, as every unlucky traveler knows, the administration has to guarantee that the airlines will receive good performance, especially those that pay more for their air tickets, even in the worst case scenarios, when some delays occur. Therefore, we simulated three more scenarios when delays occur. From the analysis of the data resulting from the implementation of the different scenarios, we can conclude:

1. The airport works perfectly with respect to the priorities given to the different kinds of flights.
2. There is no need for the airport to buy a new refueling machine because problems during the simulations affected only the low-cost carriers.
3. The airport can use four workers instead of five and use one or two workers who work inside the building in case there is a delay or on the rare occasions when a fifth worker is needed, even if a delay does not occur.

References and Further Readings

Balakrishnan (2009) www.mit.edu/~hamsa/pubs/SimaiakisBalakrishnanGNC2009.pdf

Balamuralithara B, Woods PC (2008) Virtual laboratories in engineering education: the simulation lab and remote lab. Comput Appl Eng Educ 17:108–118

Berkowitz C, Bragdon C (2006) Advanced simulation technology applied to port safety and security. In: Proceedings of the 9th international conference on applications of advanced technology in transportation, Chicago, IL

Bocca E, Viazzo S, Longo F, Mirabelli G (2005) Developing data fusion systems devoted to security control in port facilities. In: Proceedings of 2005 Winter simulation conference

Chatterjee A (2006) Assessing the impact of port security measures on traffic operations, Final report, University of Tennessee, Knoxville, TN

Crow EL, Shimizu K (eds) (1998) Lognormal distributions: theory and applications. Markel Dekker, Inc., New York

Erugrul N (1998) Towards virtual laboratories: a survey of Lab View-based teaching/learning tools and future trends. Int J Eng Educ 14:1–10

Federal Highway Administration (1996) Highway Statistics, U.S. Department of Transportation, Washington, DC

Federal Highway Administration (1997) Highway Statistics, U.S. Department of Transportation, Washington, DC

Garms CH (2012) Discrete event-based modeling and optimization of the Airport Landsite of Hamburg Airport using MATLAB Simulink SimEvents (in German). Bachelor thesis, University of Hamburg

Harris G, Schroer B, Moeller DPF (2010) Effect of removing passenger car volume on freight movement: a simulation study. In Gauthier JS (ed) Proceedings Huntsville simulation conference. SCS Publication, San Diego, pp 161–166

Harris GA, Jennings L, Schroer B, Moeller DPF (2007) Container terminal simulation. In: Gauthier J (ed) Proceedings Huntsville Simulation Conference. SCS Publ., San Diego, pp. 173–179

Harris GA, Holden A, Schroer B, Moeller DFF (2008) A simulation approach to evaluating productivity improvement at a seaport coal terminal. The Transportation Research Record—Journal of the Transportation Research Board. Paper 1208–1263

Highway Capacity Manual (2000) Transportation Research Board, Washington, DC

Honsi Y, Ramasamy S, Selter J, Anderson P (2005) Vulnerability assessment and emergency evacuation plan simulation training and validation of the port of Jacksonville. Final report, Center for Advanced Transportation Systems Simulation, University of Central Florida, Orlando, FL

Imai M (1986) Kaizen: the key to Japan's competitive success. The KAIZEN Insitute, Ltd. fistributed by Mc Graw Hill Inc.

International Air Transport Association, I.A.T.A. (2012) http://www.iata.org/Pages/default.aspx

Kerr P (2006) Development of an interactive freight mobility and security database structure for research and freight modeling applications. Final report, Center for Advanced Transportation Systems Simulation, University of Central Florida, Orlando, FL

Koch (2007) PortSim-A port security simulation and visualization tool In: Proceedings 41st annual IEEE international Cranahan conference on security technology, pp 109–116

Lewis B, Erera A, White C (2002) Optimization approaches for efficient container security operations at transshipment seaports. Research report, Georgia Institute of Technology

MATLAB (2014) MathWorks: http://www.mathworks.de/products/matlab

Moffatt, Nichol (2002) Development Master Plan (Choctaw Point Terminal), Moffatt Nichol Engineers, Mobile, AL

Möller DPF, Vakilzadian H, Schroer BJ, Crosbie RE (2006) Architectural concepts for integrating simulation into the USE_eNET framework. In: Huntsinger R, Vakilzadian H, Ören T (eds) Proceedings of the 2006 international conference on modeling and simulation-methodology, tools, software applications, SCS Publication, Calgary, pp 111–116

Nakajima S (1988) Introduction to total productive maintenance. Productivity Press, University of Minnesota, Cambridge, MA

National Institute of Standards and Technology (NIST) (1998) Technical advisory committee to develop a federal information processing standard for the federal key management infrastructure

NIST 1998 into Technical Advisory Committee to Develop a Federal Information Processing Standard for the Federal Key Management Infrastructure, National Institute of Standards and Technology (NIST)

Ohno T (1988) Toyota production system: beyond large-scale production. Productivity Inc, Portland

Open Street Map http://www.openstreetmap.org/

ProcessModel (1999) User's manual, ProcessModel Corp., Provo, UT

ProcessModel (2011) ProModel Tutorial 2011. www.promodel.com/solutionscafe/webinars/ProModel%202011%20Tutorial/PM2011Tutorial.html

Sandia (2008) Sandia National Laboratory: Port Simulation Model, Description on the Sandia web site: www.sandia.gov. Albuquerque, NM

Schroer B, Harris G, Anderson M, Spayd M, Moeller DPF (2009) Conceptual framework for discrete event simulation of interstate traffic. In: Gauthier J (ed) Proceedings Huntsville simulation conference, SCS Publication, San Diego

SimEvents (2014) http://www.mathworks.de/products/simevents

Simulink (2014) http://www.mathworks.de/products/simulink

Transportation Research Board (2000) Highway capacity manual. Transportation Research Board, Washington, DC

UAH (2005) Office for Economic Development, 2005, Transportation infrastructure in Alabama—meeting the needs for economic growth. Final report on the requirements for infrastructure and transportation to support the transformation of the Alabama economy. Prepared for the Office of the Secretary, U.S. Department of Transportation, Grant no. DTTS59-03-G-00008

U.S. Department of Transportation (1998) Status of the Nation's surface transportation system: condition & performance report to congress. US DOT, Washington, DC

Womack JP, Jones DT (1996) Lean thinking. Free Press, A Division of Simon & Schuster Inc., New York

Index

A

Acceleration, 5, 14, 181, 184

Access ramp, 126

Accuracy, 7, 20, 182, 201, 236

ACSL, 202, 203

Adaptability, 6, 187

Airport
    capacity, 178, 317
    hub, 1, 311, 314
    landside, 302–303
    operation simulation, 165–183, 317, 319
    ramp, 7, 167, 168, 175, 312–314
    ramp control, 178, 179
    surveillance radar (ASR), 178
    technology, 180, 181
    topography, 180, 182

Air traffic
    control (ATC), 85, 167, 180, 181, 302
    shortage, 167

Air transportation sector, 3

Alabama State Docks, 229, 238, 247,
    274, 275

Algorithm
    numerical, 141, 195, 209
    sorting, 201

Ambassador
    federate, 148
    RTI, 148

Animation, 202–205, 207, 208, 210, 216,
    218, 234

Arbitrary distance, 13

Arena, 204, 207, 208, 210

Arrival
    customer, 26, 27, 61, 71, 73
    event, 28, 63, 65
    inter, 12, 28, 30, 36, 37, 40, 64, 65, 68–70,
        73, 158, 206, 208, 318
    pattern, 57

rate, 58, 59, 61, 63, 85, 131, 268, 288–290,
    292, 299, 318
successive, 57
time, 63, 64, 92, 104, 155, 185, 268, 293

Arterial street, 121, 126

Artificial swarm system, 186

A-SMGCS, 178–180

Assessment, 32, 46, 48, 140, 301

Assignment
    all-or-nothing, 111, 112
    dynamic, 111
    multiclass, 109, 117–119
    path, 112, 122–124
    problem, 109, 110, 117
    stochastic travel, 111

Attribute, 5, 30, 36, 37, 61–65, 78, 94, 95, 112,
    117, 118, 120, 144–146, 151, 152,
    158, 171, 175, 210–212, 219, 220,
    223, 224, 232, 240, 249, 255, 259,
    260, 268, 275, 276, 304–308

AutoMod, 204, 207, 208

B

Backcasting, 2

Backtracking, 2

Barge, 40, 157, 239, 240, 242–247

Baseline
    condition, 47, 76
    run, 128–131, 136, 237, 244–246, 254, 255,
        260, 262–266, 268, 270, 272, 273,
        276, 277, 280–283, 286–289,
        294–297, 300

Bellman principle, 9

Bertalanffy's General System Theory, 12, 14

Bijection, 81

Boarding, 57, 168, 302, 315

Bond graph theory, 9

© Springer-Verlag London 2014
D.P.F. Möller, *Introduction to Transportation Analysis, Modeling and Simulation*,
Simulation Foundations, Methods and Applications,
DOI 10.1007/978-1-4471-5637-6

Printed in the United States
By Bookmasters